To my mother
Professor Regina Tyshkevich
on the occasion of her 80th birthday

MEMOIRS
of the
American Mathematical Society

Number 939

The Minimal Polynomials of Unipotent Elements in Irreducible Representations of the Classical Groups in Odd Characteristic

I. D. Suprunenko

July 2009 • Volume 200 • Number 939 (fourth of 6 numbers) • ISSN 0065-9266

American Mathematical Society
Providence, Rhode Island

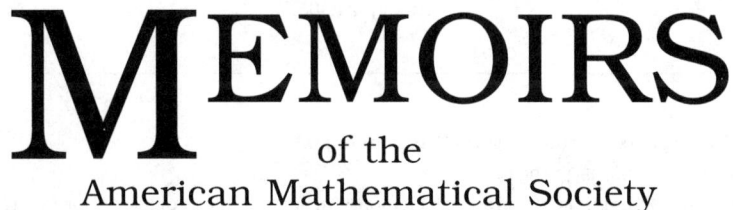

2000 *Mathematics Subject Classification.*
Primary 20G05.

Library of Congress Cataloging-in-Publication Data

Suprunenko, I. D. (Irina D.), 1954–
The minimal polynomials of unipotent elements in irreducible representations of the classical groups in odd characteristic / I. D. Suprunenko.
p. cm. — (Memoirs of the American Mathematical Society, ISSN 0065-9266 ; no. 939)
"Volume 200, number 939 (fourth of 6 numbers)."
Includes bibliographical references and index.
ISBN 978-0-8218-4369-7 (alk. paper)
1. Linear algebraic groups. 2. Irreducible polynomials. 3. Representations of groups. I. Title.

QA179.S87 2009
512′.55—dc22 2009008895

Memoirs of the American Mathematical Society

This journal is devoted entirely to research in pure and applied mathematics.

Subscription information. The 2009 subscription begins with volume 197 and consists of six mailings, each containing one or more numbers. Subscription prices for 2009 are US$709 list, US$567 institutional member. A late charge of 10% of the subscription price will be imposed on orders received from nonmembers after January 1 of the subscription year. Subscribers outside the United States and India must pay a postage surcharge of US$65; subscribers in India must pay a postage surcharge of US$95. Expedited delivery to destinations in North America US$57; elsewhere US$160. Each number may be ordered separately; *please specify number* when ordering an individual number. For prices and titles of recently released numbers, see the New Publications sections of the *Notices of the American Mathematical Society.*

Back number information. For back issues see the *AMS Catalog of Publications.*

Subscriptions and orders should be addressed to the American Mathematical Society, P. O. Box 845904, Boston, MA 02284-5904, USA. *All orders must be accompanied by payment.* Other correspondence should be addressed to 201 Charles Street, Providence, RI 02904-2294, USA.

Memoirs of the American Mathematical Society (ISSN 0065-9266) is published bimonthly (each volume consisting usually of more than one number) by the American Mathematical Society at 201 Charles Street, Providence, RI 02904-2294, USA. Periodicals postage paid at Providence, RI. Postmaster: Send address changes to Memoirs, American Mathematical Society, 201 Charles Street, Providence, RI 02904-2294, USA.

This publication is indexed in *Science Citation Index®*, *SciSearch®*, *Research Alert®*, *CompuMath Citation Index®*, *Current Contents®/Physical, Chemical & Earth Sciences.*
Printed in the United States of America.

♾ The paper used in this book is acid-free and falls within the guidelines established to ensure permanence and durability.
Visit the AMS home page at http://www.ams.org/

10 9 8 7 6 5 4 3 2 1 14 13 12 11 10 09

Contents

Abstract

The minimal polynomials of the images of unipotent elements in irreducible rational representations of the classical algebraic groups over fields of odd characteristic are found. These polynomials have the form $(t-1)^d$ and hence are completely determined by their degrees. In positive characteristic the degree of such polynomial cannot exceed the order of a relevant element. It occurs that for each unipotent element the degree of its minimal polynomial in an irreducible representation is equal to the order of this element provided the highest weight of the representation is large enough with respect to the ground field characteristic. On the other hand, classes of unipotent elements for which in every nontrivial representation the degree of the minimal polynomial is equal to the order of the element are indicated. In the general case the problem of computing the minimal polynomial of the image of a given element of order p^s in a fixed irreducible representation of a classical group over a field of characteristic $p > 2$ can be reduced to a similar problem for certain s unipotent elements and a certain irreducible representation of some semisimple group over the field of complex numbers. For the latter problem an explicit algorithm is given. Results of explicit computations for groups of small ranks are contained in Tables I–XII. The article may be regarded as a contribution to the programme of extending the fundamental results of Hall and Higman (1956) on the minimal polynomials from p-solvable linear groups to semisimple groups.

Received by the editor August 4, 2004; and in revised form February 12, 2007.

2000 *Mathematics Subject Classification.* 20G05.

Key words and phrases. Classical algebraic groups, modular representations, unipotent elements, minimal polynomials.

This research has been supported by the Institute of Mathematics of the National Academy of Sciences of Belarus in the framework of the State Basic Research Programme "Mathematical Structures" (2001–2005) and partially supported by the Belarus Basic Research Foundation, Project F98-180.

1. Introduction

The minimal polynomials of the images of unipotent elements in irreducible rational representations of the classical algebraic groups over fields of odd characteristic are found. It is well known that for rational representations of arbitrary algebraic groups such polynomials have the form $(t-1)^d$ and hence are completely determined by their degrees. If the ground field characteristic is positive, the order of a unipotent element is equal to some power of this characteristic and it is clear that the degree of its minimal polynomial in any representation over the same field is at most the order of this element. It occurs that for each unipotent element the degree of its minimal polynomial in an irreducible representation is equal to the order of this element provided the highest weight of the representation is large enough with respect to the ground field characteristic. On the other hand, classes of unipotent elements for which in every nontrivial representation the degree of the minimal polynomial is equal to the order of the element are indicated. In the general case the problem of computing the minimal polynomial of the image of a given element of order p^s in a fixed irreducible representation of a classical group over a field of characteristic $p > 2$ can be reduced to a similar problem for certain s unipotent elements and a certain irreducible representation of some semisimple group over the field of complex numbers.

There exists an explicit algorithm for computing the minimal polynomials of the images of unipotent elements in irreducible representations of simple algebraic groups in characteristic 0 (see Algorithm 1.6 below) which for the classical groups requires at most $O(n^2)$ operations where n is the dimension of the standard realization of the relevant group. Results of explicit computations in positive characteristic for groups of ranks at most 8 are contained in Tables I–XII.

We want to give some brief comments on the current state of the minimal polynomial problem before stating our main results. Let K be an algebraically closed field of positive characteristic p. Due to famous Steinberg's theorem [**Ste63**, Theorem 1.1], any irreducible K-representation of a finite classical group defined over a field of characteristic p is the restriction of some representation of the relevant classical algebraic group. This enables us to transfer our results to K-representations of finite classical groups in describing characteristic. So the article may be regarded as a contribution to the programme of extending the fundamental results of Hall and Higman [**HH56**] on the minimal polynomials of p-elements in finite irreducible p-solvable groups in characteristic p to groups which are not p-solvable. According to [**HH56**, Theorem B], the degree of the minimal polynomial of an element of order p^s in such p-solvable group is at least $(p-1)p^{s-1}$ for $p > 2$ and $3p^{s-2}$ for $p = 2$, $s > 1$. If p is odd and is not a Fermat prime, then this degree is always p^s. Shult [**Shu65**] has extended these results to the complex case. They found numerous applications. Probably, similar results for groups close to simple will occur useful as well. For a progress in this direction researchers have to consider different classes of groups separately using specific machinery in each case. Actually, the research in this direction goes back to Blichfeldt (1917) who studied complex finite linear groups containing matrices with the minimal polynomial of degree 2. Nowadays Robinson [**Rob95**] has shown that the degree of the minimal polynomial of a noncentral element of prime order p in a complex finite primitive linear group is at least $(p-1)/2$; notice also that the estimate $(p+3)/4$ was obtained in [**Rob83**] without using the classification of finite simple groups. The study of the minimal polynomials of unipotent elements in linear groups in positive characteristic p that are not p-solvable, probably, began with the famous Thompson's note [**Tho71**] where the classification of linear groups over finite fields of characteristic $p > 3$ generated by p-elements with the quadratic minimal polynomial was announced. The case $p = 3$ was settled by Ho [**Ho76**] and Chermak [**Che04**]. Zalesski has obtained a series of results on minimal polynomials that led to the complete solution of the problem in some important cases. Recently he has proved that for a finite irreducible complex linear group and its element g that does not lie in a proper normal subgroup and has prime order p modulo the group centre, the degree of the minimal polynomial of g can be equal only to p, $p-1$, $p-2$, $(p-1)/2$, or $(p+1)/2$ [**Zal06**, Theorem 1]; in [**Zal06**, Theorem 2] a classification of such groups and elements with the minimal polynomials of degree less than $p-1$ is given. Below we give references concerning the situations where, according to the author's knowledge, at present the minimal polynomial problem is solved for representations of finite quasisimple and related groups. Notice that we mean absolutely irreducible representations. The case of elements of prime order modulo the centre in characteristic 0 representations of quasisimple finite groups is settled by Zalesski [**Zal**, Theorems 1.1 and 1.2]. For elements of order p in cross characteristic representations of quasisimple finite groups of Lie type in characteristic p Zalesski [**Zal88**] reduced the problem to certain representations of symplectic groups, then Guralnick, Magaard, Saxl, and Tiep [**GMST02**, Theorem 3.1] obtained the final result. Di Martino and Zalesski [**DMZ08**, Theorem 1.1]have solved the problem for arbitrary p-elements in such representations. In [**DMZ08**] asymptotic estimates for eigenvalue multiplicities of relevant elements are given as well. For some small groups the results of [**Zal**] and [**DMZ08**] indicate the dimensions of representations where some element in question has the minimal polynomial of degree less than its order rather than representations themselves, the relevant representations can be read off from ordinary and modular character tables of these groups. For projective representations of symmetric and alternating groups the assertions of [**Zal**, Theorems 1.1 and 1.2]) were proved in [**Zal96**] and for groups with cyclic p-Sylow subgroups in [**Zal99**]. Tiep and Zalesski have obtained lower estimates for the degrees of the minimal polynomials for semisimple elements of prime power order in cross characteristic irreducible representations of finite classical groups and their extensions [**TZ**, Theorem 1.3], for such elements of prime order and quasisimple classical groups these degrees are indicated explicitly, except some small groups [**TZ**, Theorem 1.2]. For cross characteristic representations, the problem is also solved for those semisimple elements in finite symplectic and general unitary groups and central extensions of subgroups of finite orthogonal groups containing the derived subgroups that have prime power order modulo the centre and stabilize a nonzero totally isotropic (totally singular) subspace of the standard module (Di Martino and Zalesski [**DMZ01**, Theorem 1.1]: identification of elements that can have minimal polynomials of degrees less than their order, Guralnick, Magaard, Saxl, and Tiep [**GMST02**, Theorem 3.2]: the final result). For p-elements in p-modular representations of finite quasisimple groups with cyclic p-Sylow subgroups the problem is settled by Zalesski [**Zal99**] and for elements of order p in irreducible projective representations of symmetric and alternating groups in odd characteristic p and elements of order 4 in such representations in characteristic 2 by Kleshchev and Zalesski [**KZ04**]. Relevant results for elements of order p in absolutely irreducible representations of groups of Lie type in defining characteristic p follow from [**Sup96**] and for arbitrary unipotent elements in such representations of finite classical groups for odd p from this article (see the comments above on connections of representations of algebraic groups with those of finite groups of Lie type). A detailed analysis of available results on the minimal polynomials and related problems can be found in a survey of Tiep and Zalesski [**TZ00**, Section 9]; notice also [**GMST02**, Section 3] and comments in the Introduction of [**TZ**].

Now we need some notation to state our main results. In what follows $\mathbb{C}$ is the field of complex numbers, K is an algebraically closed field of characteristic $p > 2$, G is a simply connected simple algebraic group of a classical type (A, B, C or D) over K. If $\mathcal{G}$ is a semisimple algebraic group over K, then $\mathcal{G}_{\mathbb{C}}$ is a simply connected semisimple algebraic group over $\mathbb{C}$ with the same root system as $\mathcal{G}$. Throughout the article $r(\mathcal{G})$, $\operatorname{Irr}\mathcal{G}$ and $\mathbf{X}(\mathcal{G})$ are the rank, the set of irreducible rational representations (considered up to equivalence), and the weight system of an algebraic group $\mathcal{G}$; $\omega(\varphi)$ is the highest weight of a representation φ, ω_i and α_i, $1 \leq i \leq r(\mathcal{G})$, are the fundamental weights and the simple roots of $\mathcal{G}$, for simple groups they are labelled in the standard way as in [**Bou68**]. We set $r = r(G)$, $\operatorname{Irr} = \operatorname{Irr} G$, $\operatorname{Irr}_{\mathbb{C}} = \operatorname{Irr} G_{\mathbb{C}}$, and $\mathbf{X} = \mathbf{X}(G)$. One can naturally identify the systems $\mathbf{X}$ and $\mathbf{X}(G_{\mathbb{C}})$. For an element $x \in \mathcal{G}$ and a representation φ of $\mathcal{G}$ denote by $d_\varphi(x)$ the degree of the minimal polynomial of the matrix $\varphi(x)$. The symbol $d_\varphi(C)$ is used similarly for a conjugacy class $C \subset \mathcal{G}$ and $d_M(x)$ denotes such degree for an element $x \in \mathcal{G}$ acting on an $\mathcal{G}$-module M. As usually, $|x|$ denotes the order of an element x. We assume that $r > 2$ for $G = B_r(K)$ and $r > 3$ for $G = D_r(K)$ (recall that $B_1(K) \cong D_1(K) \cong A_1(K)$, $B_2(K) \cong C_2(K)$, $D_2(K) \cong A_1(K) \times A_1(K)$, and $D_3(K) \cong A_3(K)$).

Let n be the dimension of the standard realization of G (the standard G-module). It is well known that $n = r + 1$ for $G = A_r(K)$, $2r + 1$ for $G = B_r(K)$ and $2r$ for $G = C_r(K)$ or $D_r(K)$.

Denote by $\langle \omega, \alpha \rangle$ the value of a weight $\omega \in \mathbf{X}(\mathcal{G})$ at a root α of $\mathcal{G}$ (the canonical pairing in the sense of [**Ste68**, §3]). If $\mathcal{G}$ is defined over K, a dominant weight $\omega \in \mathbf{X}(\mathcal{G})$ is called p-restricted if $\omega = \sum_{i=1}^{r(\mathcal{G})} a_i\omega_i$ and all $a_i < p$. Every dominant weight ω can be written in the form $\sum_{j=0}^{t} p^j\lambda_j$ with p-restricted λ_j. Set $\overline{\omega} = \sum_{j=0}^{t} \lambda_j$. For $\varphi \in \operatorname{Irr}\mathcal{G}$ denote by $\varphi_{\mathbb{C}}$ the representation in $\operatorname{Irr}\mathcal{G}_{\mathbb{C}}$ with highest weight $\overline{\omega(\varphi)}$.

DEFINITION 1.1. A dominant weight $\omega \in \mathbf{X}(\mathcal{G})$ is called p-large if $\langle \overline{\omega}, \alpha \rangle \geq p$ for the maximal root α of $\mathcal{G}$. A representation $\varphi \in \operatorname{Irr}\mathcal{G}$ is called p-large if $\omega(\varphi)$ is p-large.

Now we can start stating our results on the minimal polynomials of unipotent elements in irreducible representations of G. In certain cases one can see that $d_\varphi(x) = |x|$ for a unipotent element x without special computations.

THEOREM 1.1. *Let $\varphi \in \operatorname{Irr}$ be p-large. Then $d_\varphi(x) = |x|$ for each unipotent element $x \in G$.*

The following lemma shows that the assumptions of Theorem 1.1 are threshold with respect to the degree of the minimal polynomial of a unipotent element.

LEMMA 1.2. *Assume that $n = p^s + 1$ for $G = A_r(K)$ and $C_r(K)$, $n = p^s + 2$ for $G = B_r(K)$, and $n = p^s + 3$ for $G = D_r(K)$; $s > 0$ in all cases. Let $\varphi \in \operatorname{Irr}$*

with $\omega(\varphi) = (p-1)\omega_1$ *for* $G = A_r(K)$ *and* $C_r(K)$ *and* $\omega(\varphi) = \frac{1}{2}(p-1)\omega_2$ *for* $G = B_r(K)$ *and* $D_r(K)$. *Then* $\langle \omega, \alpha \rangle = p-1$ *for the maximal root* α *of* G *and for a regular unipotent element* $x \in G$ *the degree* $d_\varphi(x) = (p-1)p^s + 1 < |x| = p^{s+1}$.

Set $F = K$ or $\mathbb{C}$, $\widehat{G} = G$ or $G_{\mathbb{C}}$ and assume that $u \in \widehat{G}$ is unipotent. Let

$$k_1 \geq k_2 \geq \ldots \geq k_t \tag{1.1}$$

denote the sizes of all Jordan blocks of u in the standard realization of G. Here $k_1 + k_2 + \ldots + k_t = n$. Set $J(u) = (k_1, k_2, \ldots, k_t)$. Recall that unipotent elements in G are exactly p-elements. Assume that $x \in G$ and $|x| = p^{s+1}$, $s \geq 0$. (We use $s+1$ rather than s here and throughout the text to simplify many formulae.)

THEOREM 1.3. *Let* $k_1 = p^{s+1}$. *Then* $d_\varphi(x) = p^{s+1}$ *for every nontrivial representation* $\varphi \in \mathrm{Irr}$, *except the cases where* $k_1 = 3$ *or* 5, $k_2 = 1$, $G = B_r(K)$ *or* $D_r(K)$ *and* $\omega(\varphi) = p^j\omega_r$ *for* $G = B_r(K)$ *and* $p^j\omega_{r-1}$ *or* $p^j\omega_r$ *for* $G = D_r(K)$.

In the general case we reduce the minimal polynomial problem for an element x of order p^{s+1} to computing the minimal polynomials of certain $s+1$ unipotent elements in some representations of a semisimple group in characteristic 0 constructed for x in a certain special way.

For our constructions, we need to recall some information on the unipotent conjugacy classes in $\widehat{G}$. Denote by $\mathcal{U}$ the set of such classes. In this paragraph the terminology from [**Car85**, ch.5] is used. As usually, if $L \subset \widehat{G}$ is a Levy subgroup and $P \subset L$ is a distinguished parabolic subgroup in the commutator subgroup of L, the pair (L, P) is called a distinguished pair. Let $\mathcal{DC}$ be the set of $\widehat{G}$-conjugacy classes of distinguished pairs and $\mathcal{F} : \mathcal{DC} \to \mathcal{U}$ be the map that sends the class containing a pair (L, P) to the class containing the dense orbit of P on its unipotent radical. By [**BC76**] and [**Pom80**], $\mathcal{F}$ is a bijection since p is good for G. Hence we have the same parametrization of the unipotent conjugacy classes both for G and $G_{\mathbb{C}}$ (see comments in [**Sei00**, Section 2]) and hence define the labelled Dynkin diagram of a unipotent conjugacy class $C \subset G$ as such diagram for the class in $G_{\mathbb{C}}$ associated with the "same" distinguished pair (strictly speaking, with the pair determined by the same roots). One can consult [**Car85**, Section 5.6] on labelled Dynkin diagrams in characteristic 0. It is well known that if $C \subset G$ and $C' \subset G_{\mathbb{C}}$ are classes with the same labelled Dynkin diagram, then $J(x) = J(x')$ for $x \in C$ and $x' \in C'$.

In what follows we say that a positive integer k is proper for G if $G = A_r(K)$, if k is odd for $G = B_r(K)$ or $D_r(K)$, and if k is even for $G = C_r(K)$; otherwise we call k improper.

LEMMA 1.4. *Let* $u \in \widehat{G}$ *be a unipotent element.*

i) [**TZ02**, Lemma 2.31] *If* $\widehat{G} \neq D_r(F)$ *or some integer* k_j *in* $J(u)$ *is odd, the conjugacy class of* u *is uniquely determined by* $J(u)$.

ii) (follows from [**Hes76**, Propositions 3.3 and 3.5], [**SS70**, Ch.IV, Exercise 2.15 and Item 2.27(ii)]*) Let* $\widehat{G} = D_r(F)$ *and all the integers in the sequence* (1.1) *be even. Then there are just two unipotent conjugacy classes* C_1 *and* $C_2 \subset \widehat{G}$ *with the same sequence* (1.1); *if* δ_i^j *is the label on the labelled Dynkin diagram of* C_j *corresponding to the root* α_i $(1 \leq i \leq r,\ j = 1, 2)$, *then* $\delta_i^1 = \delta_i^2$ *for* $1 \leq i \leq r-2$, $\delta_{r-1}^1 = \delta_r^2 = 0$, *and* $\delta_{r-1}^2 = \delta_r^1 = 2$.

iii) ([**SS70**, Ch. IV, Item 2.19] *and* [**Sei00**, Proposition 2.6]*) Integers* $k_1, k_2, \ldots$, $k_t \in \mathbb{Z}^+$ *with* $k_1 \geq k_2 \geq \ldots \geq k_t$ *and* $k_1 + k_2 + \ldots + k_t = n$ *yield the sequence* (1.1) *for a unipotent element* $u \in \widehat{G}$ *if and only if each improper* k_j *appears in this sequence an even number of times.*

Naturally, in [**TZ02**], [**Sei00**], and [**Sup96**] the authors do not regard relevant items of Lemma 1.4 as new results, but give concise statements of well-known facts they need. A more detailed comments on unipotent conjugacy classes in the classical groups will be given later in this Introduction and in Section 2.

Now we describe the situation with the minimal polynomials of unipotent elements in characteristic 0. The following proposition enables one to reduce the problem to fundamental representations. In Items 1.5 and 1.6 ρ_i are the fundamental representations of a semisimple group $\mathcal{G}$ over $\mathbb{C}$ labelled in the standard way for classical $\mathcal{G}$, $m_i(C) = d_{\rho_i}(C) - 1$.

PROPOSITION 1.5. ([**Sup96**, Propositions 1.3 and 2.6]) *Let* $\mathcal{G}$ *be a semisimple group of rank* l *over* $\mathbb{C}$, C *be a unipotent conjugacy class in* $\mathcal{G}$, $\pi, \psi_1, \psi_2 \in \operatorname{Irr}\mathcal{G}$, $\omega(\pi) = \omega(\psi_1) + \omega(\psi_2)$. *Then* $d_\pi(C) = d_{\psi_1}(C) + d_{\psi_2}(C) - 1$. *In particular, if* $\omega(\pi) = \sum_{i=1}^{l} a_i\omega_i$, *we have* $d_\pi(C) = 1 + \sum_{i=1}^{l} a_i m_i(C)$.

ALGORITHM 1.6. (Computing the indices $d_{\rho_i}(C)$ for the classical groups [**Sup96**, Algorithms 1.4]) *In the assumptions of Proposition* 1.5 *let* $\mathcal{G}$ *be a classical group and* $C \subset \mathcal{G}$ *be a unipotent conjugacy class. Suppose that* $u \in C$ *and* $J(u) = (k_1, \ldots, k_t)$. *Let*

$$N(C) = (k_1 - 1, k_1 - 3, \ldots, 1 - k_1, k_2 - 1, \ldots, 1 - k_2, \ldots, k_t - 1, \ldots, 1 - k_t) \quad (1.2)$$

(a sequence).

1. *Assume that* $i \leq l$ *for* $\mathcal{G} = A_l(\mathbb{C})$ *and* $C_l(\mathbb{C})$, $i < l$ *for* $\mathcal{G} = B_l(\mathbb{C})$, *and* $i < l - 1$ *for* $\mathcal{G} = D_l(\mathbb{C})$. *Then* $m_i(C)$ *is equal to the sum of* i *largest members of the sequence* $N(C)$.

2. *Let* $\mathcal{G} = B_l(\mathbb{C})$ *and* $i = l$. *Then* $m_i(C)$ *is the half of the sum in Item* 1.

3. *Let* $\mathcal{G} = D_l(\mathbb{C})$ *and* $i = l-1$ *or* l. *If there is an odd integer among* $k_1, \ldots, k_t$, *then* $m_i(C)$ *is such as in Item* 2. *However, if* $k_1, \ldots, k_t$ *are all even, there exist two distinct unipotent classes* $C_1, C_2 \subset \mathcal{G}$ *with* $N(C_1) = N(C_2)$. *One can assume that the following holds for the labels* δ_i^j *on the labelled Dynkin diagrams of the classes* C_j *corresponding to the roots* α_i $(1 \leq i \leq l,\ j = 1, 2)$*:* $\delta_i^1 = \delta_i^2$ *for* $1 \leq i \leq l - 2$, $\delta_{l-1}^1 = \delta_l^2 = 0$, $\delta_{l-1}^2 = \delta_l^1 = 2$. *Then* $m_l(C_1)$ *is equal to the halfsum of the* l *largest members of* $N(C_1)$, $m_l(C_2) = m_l(C_1) - 1$, *and* $m_{l-1}(C_i) = m_l(C_j)$ *for* $\{i, j\} = \{1, 2\}$.

4. *If* $t = 1$, *we have* $m_i(C) = i(n - i)$ *if* $\mathcal{G} \neq B_l(\mathbb{C})$ *or* $i < l$ *and* $m_l(C) = l(l+1)/2$ *for* $\mathcal{G} = B_l(\mathbb{C})$.

These facts will be heavily used for solving the problem in positive characteristic. Now we pass to the modular case. Let $x \in G$ be an element of order p^{s+1} with $s \geq 0$. Assume that $z_i \in G_{\mathbb{C}}$ are elements with the same labelled Dynkin diagram as x^{p^i}, $0 \leq i \leq s$. By Lemma 1.4 i), if $G \neq D_r(K)$ or some of the integers k_i in $J(x)$ is odd, it suffices to suppose that $J(x^{p^i}) = J(z_i)$. Analyzing the connection of an irreducible G-module with the Weyl module and the irreducible $G_{\mathbb{C}}$-module with the same highest weight, we shall show that $d_\varphi(x^{p^i}) \leq d_{\varphi_{\mathbb{C}}}(z_i)$ (Corollary 2.40). On the other hand, one easily deduces that $d_\varphi(x) \leq p^i d_\varphi(x^{p^i})$ (Proposition 2.5 a)). Naturally, $d_\varphi(x) \leq |x| = p^{s+1}$. This yields that

$$d_\varphi(x) \leq \min\{p^{s+1}, p^i d_{\varphi_{\mathbb{C}}}(z_i), \mid 0 \leq i \leq s\}. \quad (1.3)$$

It occurs that for regular unipotent elements the equality holds in (1.3), except some special cases.

THEOREM 1.7. *Let* $x \in G$ *be a regular unipotent element,* $|x| = p^{s+1}$*, and let* $z_i \in G_{\mathbb{C}}$*,* $0 \leq i \leq s$*, be an element with the same canonical Jordan form as* x^{p^i}*. Let* $\varphi \in \mathrm{Irr}$*. Then*

$$d_\varphi(x) = \min\{p^{s+1}, p^i d_{\varphi_{\mathbb{C}}}(z_i) \mid 0 \leq i \leq s\}, \tag{1.4}$$

except the cases where $G = A_r(K)$ *or* $C_r(K)$*,* $n = p^s + p$ *with* $s > 0$*,* $\overline{\omega(\varphi)} = \omega_p$ *or* ω_{n-p} *for* $G = A_r(K)$ *and* $\overline{\omega(\varphi)} = \omega_p$ *for* $G = C_r(K)$*. In the exceptional cases* $d_\varphi(x) = p^{s+1} - p + 2$*.*

It is well known that for $G \neq D_r(K)$ a regular unipotent element has a single Jordan block in the standard realization of G , i.e. the sequence (1.1) has one member (see, for instance, comments in [**SS97**]), and for $G = D_r(K)$ there are two such blocks, $k_1 = 2r - 1$, and $k_2 = 1$. Emphasize that Theorem 1.7 holds for p-large representations and for $u_1 = |x|$. However, Theorem 1.7 cannot be extended to arbitrary unipotent elements even for $G = A_r(K)$. Lemma 1.8 below yields a relevant example for elements that have two or three Jordan blocks in the standard realization of G.

LEMMA 1.8. *Let* $n = 2p + l$ *for* $G = A_r(K)$ *or* $D_r(K)$ *and* $n = 3p + l$ *for* $G = B_r(K)$ *or* $C_r(K)$ *with* $1 < l < p - 1$*,* l *even for* $G = B_r(K)$ *or* $D_r(K)$ *and* l *odd for* $G = C_r(K)$*. Assume that* $J(x) = (p + l, p)$ *for* $G = A_r(K)$ *or* $D_r(K)$ *and* $J(x) = (p + l, p, p)$ *for* $G = B_r(K)$ *or* $C_r(K)$*. Let* $\varphi \in \mathrm{Irr}$ *and* $\omega(\varphi) = \omega_l$*. Then* $d_\varphi(x) = lp + 1$ *and is less than the value given by Formula* (1.4)*.*

Now we pass to the general case. If $s > 0$ and $k_1 \geq p^s + p$, there is a rather large class of cases where the form of the highest weight guarantees that $d_\varphi(x) = |x|$ and no computations are needed.

THEOREM 1.9. *Let* $\varphi \in \mathrm{Irr}$*,* $\omega(\varphi) = \sum_{i=1}^r a_i \omega_i$*, and let* $a_j \neq 0$ *for some* j *with* $p \leq j \leq r + 1 - p$ *for* $G = A_r(K)$ *and* $p \leq j$ *for other groups. Assume that* $s > 0$*,* $p^s + p \leq k_1 \leq p^{s+1}$ *and, moreover, for* $G = B_r(K)$ *either* $k_1 \geq p^s + 2p$ *or* $j < r$ *and for* $G = D_r(K)$ *either* $k_1 \geq p^s + 2p$ *or* $j < r - 1$*. Then either* $d_\varphi(x) = |x| = p^{s+1}$ *or the following holds:* $G = A_r(K)$ *or* $C_r(K)$*,* $s > 0$*,* $k_1 = p^s + p$*,* x *is a regular unipotent element or* $s > 1$ *and* $k_2 \leq p^s - p$*,* $\omega(\varphi) = p^j \omega_p$ *or* $p^j \omega_{r+1-p}$ *for* $G = A_r(K)$ *and* $p^j \omega_p$ *for* $G = C_r(K)$*,* $d_\varphi(x) = p^{s+1} - p + 2$*.*

By Theorems 1.1 and 1.3, one can assume that φ is not p-large and $k_1 < p^{s+1}$. It is well known that each unipotent element $x \in G$ is a regular unipotent element of a semisimple subgroup $S(x) \subset G$ such that the restriction to $S(x)$ of the standard G-module is a direct sum of standard modules for simple components of $S(x)$ and, may be, several copies of the trivial $S(x)$-module. The type of $S(x)$ depends upon non-unit integers k_l in $J(x)$. Some unipotent elements are regular unipotent elements in semisimple subgroups of different types with such restrictions of the standard modules. Here we choose a type suitable for our arguments. We prove in Section 11 that if the pair (x, φ) does not satisfy the assumptions of Theorem 1.9, than usually $d_\varphi(x) = d_\chi(x)$ for some explicitly indicated composition factor χ of the restriction $\varphi|S(x)$ (since the restriction is not necessarily completely reducible, it is not clear a priori that there exists a factor χ with this property). In fact, some upper bound for $d_\varphi(x)$ is found and we prove that $d_\chi(x)$ equals this bound for certain χ. In exceptional cases some special arguments are used, see Results 11.33–11.35.

To compute $d_\chi(x)$, one can use induction on r applying Theorem 1.7 to groups of smaller ranks, and the formulae for the Jordan block structure of tensor products of unipotent Jordan blocks (Theorem 2.9 and Results 2.10–2.18).

The construction of $S(x)$ is described in details in Proposition 2.27. It is clear that $d_\varphi(x)$ is the same for all representatives of a fixed conjugacy class, so in fact for each unipotent conjugacy class we construct a semisimple subgroup with regular unipotent elements lying in this class. It is crucial that the construction in Proposition 2.27 allows one to get a desired factor χ immediately. Recall that in positive characteristic branching rules describing the composition factors of the restriction $\varphi|S(x)$ are unavailable in general even if $S(x)$ is a simple group of rank $r-1$. The group $S(x)$ constructed in Proposition 2.27 has the following properties: for some maximal torus $T \subset G$ the intersection $T \cap S(x)$ is a maximal torus in $S(x)$ and the intersections of $S(x)$ with the subgroups U^+ and $U^- \subset G$ generated by the positive and the negative root subgroups with respect to T are maximal unipotent subgroups in $S(x)$. Assume that M is an irreducible G-module affording a representation φ. Then a nonzero highest weight vector of M generates an indecomposable $S(x)$-module with an irreducible head. It occurs that the required representation χ is realized in this head. Let $\theta : \mathbf{X} \to \mathbf{X}(S(x))$ be the restriction of weights from T to $T \cap S(x)$. One can observe that $\omega(\chi) = \theta(\omega(\varphi))$. For $\varphi \in \mathrm{Irr}_p$ the representation χ is usually p-restricted as well, exceptions occur only for $G = C_r(K)$ (Lemma 11.4) and are handled in Proposition 11.32. The Steinberg tensor product theorem and the formulae for tensor products of Jordan blocks mentioned above permit one to transfer to not p-restricted representations.

Naturally, one would like to have an explicit algorithm for computing $d_\varphi(x)$. A procedure reducing the problem to a similar one for the characteristic 0 case is presented below. Let $H^p = H^p(x)$ be the simply connected semisimple algebraic group over K that is a central extension of $S(x)$ and $H = H(x)$ be the simply connected group over $\mathbb{C}$ with the same root system as H^p. Then one can regard θ as a homomorphism from $\mathbf{X}$ to $\mathbf{X}(H^p) = \mathbf{X}(H)$. We describe the groups H^p and H and the homomorphism θ and construct unipotent elements $h_0, \ldots, h_s \in H$ such that in the general case for a not p-large representation $\varphi \in \mathrm{Irr}$ and the irreducible representation ψ of H with $\omega(\psi) = \theta(\overline{\omega(\varphi)})$

$$d_\varphi(x) = \min\{p^{s+1}, p^i d_\psi(h_i) \mid 0 \le i \le s\}.$$

All exceptions will be indicated explicitly.

First, basing on (1.1), construct a collection $u_1, \ldots, u_l$, $u_1 \ge \ldots \ge u_l$, $l \le t$, as follows. Find all the maximal subsequences in (1.1) consisting of equal improper integers and replace each such subsequence of $2m$ integers a by a subsequence of m integers a (by Lemma 1.4 iii), all maximal subsequences in question have even lengths). Denote the ith member of the collection obtained by u_i. For instance, for $G = B_{37}(K)$ the sequence (1.1) of the form $(13, 12, 12, 11, 11, 4, 4, 4, 4)$ yields $(13, 12, 11, 11, 4, 4)$ and for $G = C_{37}(K)$ the sequence (1.1) of the form $(12, 11, 11, 10, 10, 5, 5, 5, 5)$ yields $(12, 11, 10, 10, 5, 5)$. Naturally, if all k_j are proper, we have $l = t$ and $k_j = u_j$, $1 \le j \le t$. Fix maximal c with $u_c > 1$. Set $c(x) = c$ and $Se(x) = \{u_1, u_2, \ldots, u_c\}$.

If $1 \le j \le c$, put

$$H_j^p = \begin{cases} SL_{u_j}(K) & \text{if } G = A_r(K) \text{ or } u_j \text{ is improper for } G, \\ Spin_{u_j}(K) & \text{if } G = B_r(K) \text{ or } D_r(K) \text{ and } u_j \text{ is odd}, \\ Sp_{u_j}(K) & \text{if } G = C_r(K) \text{ and } u_j \text{ is even}. \end{cases}$$

In all cases set

$$H_j = (H_j^p)_{\mathbb{C}},\ H^p = \prod_{j=1}^{c} H_j^p,\ H = \prod_{j=1}^{c} H_j.$$

Let $x_{0j} \in H_j^p$ be a regular unipotent element, $x_{fj} = x_{0j}^{p^f}$ for $1 \le f \le s$, $h_{fj} \in H_j$ be a unipotent element with the same canonical Jordan form as x_{fj}, and $h_f = \prod_{j=1}^{c} h_{fj}$ (in the latter two cases $0 \le f \le s$). Here we mean the canonical Jordan form of x_{fj} and h_{fj} acting on the standard H_j^p and H_j-modules, respectively.

Now we start a formal description of θ. In what follows ε_i, $1 \le i \le r+1$ for $G = A_r(K)$ and $1 \le i \le r$ otherwise, are weights of the standard realization of G, their labelling is standard and corresponds to [**Bou75**, §13]; $\mathcal{E}(G) = \{\varepsilon_1, \dots, \varepsilon_w\}$ with $w = r+1$ for $G = A_r(K)$ and $w = r$ otherwise. We describe the set $\mathcal{E}(G)$ in more details in Section 2. Denote by the symbols ε_i^j similar weights of H_j. Naturally, the action of the Weyl group of $S(x)$ changes θ, but we shall prove in Proposition 2.27 that one can assume that the values $\theta(\varepsilon)$ with $\varepsilon \in \mathcal{E}(G)$ are determined by the procedure described below. Observe that these values completely determine θ as each weight of G is a linear combination of elements of $\mathcal{E}(G)$ with integer or rational coefficients.

First suppose that either $G \ne D_r(K)$, or some k_j is odd. Then by Lemma 1.4 i), the conjugacy class of x is uniquely determined by the sequence (1.1). A weight $\lambda \in \mathbf{X}(H)$ can be written in the form $(\lambda_1, \dots, \lambda_c)$ where $\lambda_j \in \mathbf{X}(H_j)$ is the restriction of λ to a relevant maximal torus of H_j; here the weights $\lambda_1, \dots, \lambda_j$ completely determine λ. Set $pr_j(\lambda) = \lambda_j$, $\theta_j = pr_j\theta$. Put $\nu(\varepsilon_i^j) = u_j - 2i + 1$, $\nu(-\varepsilon_i^j) = -\nu(\varepsilon_i^j)$, $\nu(0) = 0$. For $1 \le j \le c$ write down the p-adic expansions $u_j = b_{s,j}p^s + \dots + b_{0,j}$ with $0 \le b_{f,j} < p$. Naturally, $b_{s,j}$ can be zero for some $j > 1$. Let

$$m = \max\{j \mid u_j \ge b_{s,1}p^s + \dots + b_{1,1}p\},$$
$$m_1 = \max\{j \mid u_j \ge b_{s,1}p^s + \dots + b_{2,1}p^2 + (b_{1,1} - 1)p\},$$

$b'_j = b_{0,j}$ if u_j is proper for G and $2b_{0,j}$ otherwise, $b = \sum_{j=1}^{m} b'_j$. The values $\theta(\varepsilon_i)$ are determined by the Rules 1–9 below.

1). If $\theta(\varepsilon_i) \ne 0$, there exists unique j, $1 \le j \le c$, with $\theta_j(\varepsilon_i) \ne 0$; then $\theta_j(\varepsilon_i) = \pm\varepsilon_q^j$ and the "minus" sign can occur only for $G \ne A_r(K)$ and improper u_j. In this case set $\mathbf{j}(i) = j$ and $\mathbf{q}(i) = \pm q$ (the same sign as in the formula for θ_j).

2). If $i < r$, $\mathbf{j}(i) = j$, $\mathbf{q}(i) = q > 0$, u_j is improper for G, and $\nu(\varepsilon_q^j) > 0$, then $\mathbf{j}(i+1) = j$ and $\mathbf{q}(i+1) = -(u_j + 1 - q)$.

3). If $k > i$ and $\mathbf{j}(k) = \mathbf{j}(i)$, one of the following holds:

a) $\mathbf{q}(k) > \mathbf{q}(i) > 0$;

b) $\mathbf{q}(i) > 0$, $0 > \mathbf{q}(k) \ge -(u_{\mathbf{j}(i)} + 1 - \mathbf{q}(i))$;

c) $\mathbf{q}(i) < 0$, $\mathbf{q}(k) > u_{\mathbf{j}(i)} + 1 + \mathbf{q}(i)$;

d) $\mathbf{q}(i) < \mathbf{q}(k) < 0$.

4). For $G = A_r(K)$ and $i < (r+2)/2$, if $\theta(\varepsilon_i) \neq 0$, we put $\mathbf{j}(r+2-i) = \mathbf{j}(i)$ and $\mathbf{q}(r+2-i) = u_{\mathbf{j}(i)} + 1 - \mathbf{q}(i)$, except the case where $\mathbf{q}(i) = (u_{\mathbf{j}(i)}+1)/2$. In the exceptional case $\theta(\varepsilon_{r+2-i}) = 0$ or $\mathbf{q}(r+2-i) = (u_{\mathbf{j}(r+2-i)}+1)/2$.

5). If $\theta(\varepsilon_i) = 0$ and $i \leq (r+2)/2$ for $G = A_r(K)$, then for $G = A_r(K)$ we have $\theta(\varepsilon_k) = 0$ for $i \leq k \leq r+2-i$; for other groups in this case $\theta(\varepsilon_k) = 0$ for $i \leq k \leq r$.

6). If $s > 0$ and $i \leq b$, then $\mathbf{j}(i) \leq m$ and $\mathbf{q}(i) \leq b_{0,\mathbf{j}(i)}$.

7). If $s > 0$, $u_1 \geq p^s + p$ and $i \leq p$, then $\mathbf{j}(i) \leq m_1$; if in this case $\mathbf{j}(i) > m$, we have $\mathbf{q}(i) \leq b_{0,\mathbf{j}(i)}$.

8). $\mathbf{j}(1) = \mathbf{q}(1) = 1$.

9). Let $i > 1$. Suppose that the values $\theta(\varepsilon_g)$ with $g < i$ are already determined. Assume that $i \leq (r+2)/2$ for $G = A_r(K)$, that $\theta(\varepsilon_{i-1}) \neq 0$ and that $\mathbf{q}(i-1) < 0$ if $u_{\mathbf{j}(i-1)}$ is improper for G. Denote by A the set of all pairs (j, ε_k^j) with $\varepsilon_k^j \in \mathbf{X}(H_j)$ and $\nu(\varepsilon_k^j) \geq 0$ that can be taken for $(\mathbf{j}(i), \theta_j(\varepsilon_i))$ under assumptions 1–8. If A is empty, put $\theta(\varepsilon_i) = 0$. If A is nonempty, denote by A_1 the subset in A consisting of all the pairs with the maximal value of $\nu(\varepsilon_k^j)$; to determine $\mathbf{j}(i)$ and $\theta_j(\varepsilon_i)$, choose (a unique) pair in A_1 with minimal j.

Now let $G = A_r(K)$ and $(r+2)/2 < i \leq r+1$. Set $i' = r+2-i$. If Rules 4 and 5 determine $\theta(\varepsilon_i)$, take the value of $\theta(\varepsilon_i)$ prescribed by these rules. Otherwise $\mathbf{q}(i') = (u_{\mathbf{j}(i')}+1)/2$. If there exist j and k such that $k = (u_j+1)/2$ and $\varepsilon_k^j \neq \theta_j(\varepsilon_g)$ for some $g < i$, fix such pair with minimal j and set $\mathbf{j}(i) = j$ and $\theta_j(\varepsilon_i) = \varepsilon_k^j$. Otherwise put $\theta(\varepsilon_i) = 0$.

One can deduce that Rules 1–9 completely determine the values $\theta(\varepsilon_i)$.

Now let $G = D_r(K)$ and all the integers in the sequence (1.1) be even. Then there are two unipotent conjugacy classes C_1 and $C_2 \subset G$ with the same sequence (1.1) mentioned in Item ii) of Lemma 1.4. If $x \in C_1$, proceed as above. For $x \in C_2$ define a map $\theta' : \mathcal{E}(G) \to \mathbf{X}(H)$ using Rules 1–9 and set $\theta(\varepsilon_i) = \theta'(\varepsilon_i)$ for $1 \leq i \leq r-1$ and $\theta(\varepsilon_r) = -\theta'(\varepsilon_r)$.

Now we can state other main results of the article.

THEOREM 1.10. *Let $\varphi \in \mathrm{Irr}$ and let $\omega(\varphi)$ be not p-large. Assume that $u_1 < p^{s+1}$. Denote by ψ an irreducible representation of H with highest weight $\theta(\overline{\omega(\varphi)})$. Then*

$$d_\varphi(x) = \min\{p^{s+1}, p^f d_\psi(h_f) \mid 0 \leq f \leq s\}, \tag{1.5}$$

except the following cases:

1) G, x, and φ are such as in the exceptional case of Theorem 1.9;

2) $G = D_{2p}(K)$, $u_1 = 2p$, $x \in C_1$ and $\omega(\varphi) = p^j\omega_r$ or $x \in C_2$ and $\omega(\varphi) = p^j\omega_{r-1}$.

In the exceptional cases $d_\varphi(x) = p^{s+1} - p + 2$.

Unfortunately, due to an inaccuracy Case 2 of Theorem 1.10 was omitted in the announcement of these results in [**Sup01**].

In Section 12 examples proving that the minimum in Formula (1.5) can be equal to each value in the braces are given, see Lemma 12.1. However, the following proposition shows that it is worth to start computing $d_\varphi(x)$ from finding $d_{\varphi_{\mathbb{C}}}(z_s)$.

PROPOSITION 1.11. *In the assumptions of Theorem 1.10 let $d_{\varphi_{\mathbb{C}}}(z_s) > p$. Then $p^f d_\psi(h_f) > p^{s+1}$ for all f with $0 \leq f \leq s$ and $d_\varphi(x) = p^{s+1}$ unless one of the exceptional cases in Theorem 1.10 occurs and $d_\varphi(x) = p^{s+1} - p + 2$.*

Quite often in Formula (1.5) $d_\psi(h_f) = d_{\varphi_{\mathbb{C}}}(z_f)$.

PROPOSITION 1.12. *Let φ and x be as in Theorem 1.10. Suppose that the pair $(u_1, \omega(\varphi))$ does not satisfy the assumptions of Theorem 1.9. Then either $d_\psi(h_s) = d_{\varphi_{\mathbb{C}}}(z_s)$ or $G = D_r(K)$, all blocks in $J(x)$ are even, $2p \leq u_1 < 3p$, $c > 1$, $u_c = 2p$, and both $d_\psi(h_s)$ and $d_{\varphi_{\mathbb{C}}}(z_s) > p$. Furthermore, if $u_1 \geq p^s + p$ and $\omega(\varphi) = \sum_{i=1}^r a_i\omega_i$ with $a_i = 0$ for $p \leq i \leq r+1-p$ for $G = A_r(K)$ and $a_i = 0$ for all $i \geq p$ otherwise, then $d_\psi(h_f) = d_{\varphi_{\mathbb{C}}}(z_f)$ for $0 < f \leq s$.*

So in many cases a complicated construction of θ is used to find $d_\psi(h_0)$ only. If $u_1 < p^s + p$, one has to compute only 3 parameters in order to find the value of $d_\varphi(x)$ for $\varphi \in \mathrm{Irr}$.

PROPOSITION 1.13. *In the assumptions of Theorem 1.10 let $p^s < u_1 < p^s + p$. Then*

$$d_\varphi(x) = \min\{p^{s+1}, d_\psi(h_0), pd_\psi(h_1), p^s d_{\varphi_{\mathbb{C}}}(z_s)\}.$$

We emphasize that Theorem 1.9 holds for p-large representations and for $u_1 = |x|$. Formally Theorems 1.7 and 1.9 are corollaries of Theorems 1.1, 1.3, and 1.10, but actually in the general case they will be proved simultaneously with Theorem 1.10. One could make the construction of τ and the assertion of Theorem 1.10 still more involved to include the cases covered by Theorems 1.1 and 1.3, but this would harden the understanding of the results.

The following theorem yields an induction base for the proofs of Theorems 1.1, 1.3, 1.7, 1.9, and 1.10.

THEOREM 1.14. ([**Sup96**, Theorem 1.1]) *Let $\mathcal{G}$ be a semisimple algebraic group over an algebraically closed field $\mathcal{K}$ of characteristic $q > 0$. Assume that $q > 2$ if not all simple components of $\mathcal{G}$ are groups of types A_l or E_6 and $q > 3$ if there are groups of types G_2 or E_8 among these components. Then there exists a bijection f from the set of unipotent conjugacy classes of $\mathcal{G}$ to the similar set for $\mathcal{G}_{\mathbb{C}}$ such that*

$$d_\varphi(C) = \min\{p,\ d_{\varphi_{\mathbb{C}}}(f(C))\}$$

for all classes C which consist of elements of order q and representations $\varphi \in \mathrm{Irr}\,\mathcal{G}$. The bijection f preserves the labelled Dynkin diagram of C. If $\mathcal{G}$ is classical, then for every $x \in C$ and $y \in f(C)$ the canonical Jordan forms of x and y in the standard realizations of $\mathcal{G}$ and $\mathcal{G}_{\mathbb{C}}$, respectively, coincide.

For $p = 3$ results of Premet and the author [**PS83**] on irreducible representations of simple algebraic groups containing elements with the minimal polynomials of degree 2 enable one to find the minimal polynomials of elements of order 3 in all irreducible representations of such groups.

Naturally, one would like to extend the results of the paper to the classical groups in characteristic 2 and to exceptional ones. At present this problem does not seem untractable. A part of the machinery elaborated in Section 2 for proving the main results of the paper can be used in these situations as well, we shall indicate this where appropriate. However, a substantial work still has to be done.

The case where $p = 2$ will be considered in a consequent article. There usually $d_\varphi(x) = |x|$ for all unipotent elements x if φ is nontrivial and $\varphi(G)$ does not coincide with the standard realization of G. The author expects to indicate explicitly all exceptions. Though, probably, the statements of the expected results will sound much easier for $p = 2$ than in the general case, this special situation requires some specific machinery. Recall that for $p = 2$ and $G \neq A_r(K)$ there is no bijection between the set of unipotent conjugacy classes for G and such set for $G_{\mathbf{C}}$. So the labelled Dynkin diagram of a unipotent element x is not defined and one cannot determine the elements z_j as in odd characteristic. On the other hand, for $p = 2$ there is no need to deal with a complicated construction of the groups $S(x)$ and $H(x)$ and the homomorphism θ as we expect to find an explicit answer in all cases rather than to reduce the problem to a similar one in characteristic 0. Here much attention will be put on analyzing the action of some special elements of a fixed order and some special embeddings of classical groups that occur only in characteristic 2.

Exceptional groups have unipotent elements of order greater than p only for some not very large p ($p \leq 29$ for $G = E_8(K)$ and is smaller for other exceptional groups). Since for elements of order p the problem was solved in [**Sup96**], the Steinberg tensor product theorem and the formulae for the Jordan block structure of a tensor product of unipotent Jordan blocks (Theorem 2.9) yield that it suffices to consider a finite number of representations. So here one should search for explicit values of the degree of the minimal polynomial for given representations and fixed conjugacy class. For the majority of these classes their representatives can be embedded into proper semisimple subgroups, then one can try to find a suitable composition factor ρ of the restriction of a representation φ to the relevant subgroup, to compute $d_\rho(x)$ for a chosen element x using results of the current article and induction by rank and to prove that $d_\varphi(x) = d_\rho(x)$. For regular unipotent elements and some other elements that do not lie in proper semisimple subgroups some machinery and approaches elaborated in Sections 2 and 3 can be applied, but a detailed information on the action of the centralizers of p-powers of x in relevant modules will be required. A lot of case-by-case analysis is inevitable.

There is a striking difference between the behaviour of unipotent elements in irreducible representations of finite groups of Lie type in describing characteristic and other situations mentioned earlier in this Introduction. In the former case Theorem 1.10, Theorem 1.14, Proposition 1.5, Algorithm 1.6, and Lemma 12.1 imply that there are many representations where some elements have the minimal polynomials of degrees much less than their orders. In all other situations discussed before usually the degree of the minimal polynomial is equal to the order of an element and the exceptional cases are more or less explicitly described.

The author is grateful to the referee whose report inspired some informal comments on the main results and helped to improve the exposition and thanks A.E. Zalesski for stimulating discussions and sending his unpublished articles.

2. Notation and preliminary facts

We keep all the notation introduced in the Introduction. Throughout the text $\mathbb{Z}$ and $\mathbb{Z}^+$ are the sets of integers and nonnegative integers, E_l is the identity matrix of degree l, $\binom{a}{b}$ is the relevant binomial coefficient, we set $\binom{a}{0} = 1$ for all $a \in \mathbb{Z}^+$; J_l is the upper Jordan block of degree l with the identity on the diagonal. Many formulae occurring in the text include sums of the form $\sum_{i=a}^{b} f_i$ where $a, b \in \mathbb{Z}^+$ are some parameters. These sums are assumed to be zero in all cases where $b < a$.

For a semisimple algebraic group $\mathcal{G}$ denote by $L(\mathcal{G})$, $W(\mathcal{G})$, $\mathbf{X}(\mathcal{G})$, and $R(\mathcal{G})$ the Lie algebra, the Weyl group, the weight and root systems of $\mathcal{G}$, respectively. $\mathbf{X}^+(\mathcal{G})$ and $R^+(\mathcal{G})$ are the sets of dominant weights and positive roots (with respect to some fixed maximal torus of $\mathcal{G}$), $R^-(\mathcal{G})$ is the set of negative roots, and $\Pi(\mathcal{G})$ is a basis in $R(\mathcal{G})$. For $x \in \mathcal{G}$ the symbol $\mathrm{cl}(x)$ denotes the conjugacy class containing x. Let $\bar{C}$ be the Zariski closure of a conjugacy class C. If $\alpha \in R(\mathcal{G})$ and t is an element of the ground field, then $\mathcal{X}_\alpha$, X_α, and $x_\alpha(t)$ are the root subgroup and the root elements in $L(\mathcal{G})$ and $\mathcal{G}$ associated with α and t, $X_{\alpha,d}$ are the elements in the hyperalgebra of $L(\mathcal{G})$ associated with α and $d \in \mathbb{Z}^+$. Recall that $X_{\alpha,0} = 1$ and that $X_{\alpha,d} = X_\alpha^d/d!$ for groups in characteristic 0, in characteristic p the latter holds for $d < p$. For subgroups $\mathcal{G}_1, \dots, \mathcal{G}_t$ of $\mathcal{G}$ and vectors $v_1, \dots, v_t$ of some vector space we denote by $\langle \mathcal{G}_1, \dots, \mathcal{G}_t \rangle$ and $\langle v_1, \dots, v_t \rangle$ the subgroup generated by $\mathcal{G}_1, \dots, \mathcal{G}_t$ and the linear span of $v_1, \dots, v_t$. If $\beta_1, \dots, \beta_l \in R^+(\mathcal{G})$, the subgroup $\langle \mathcal{X}_{\beta_i}, \mathcal{X}_{-\beta_i} \mid 1 \leq i \leq l \rangle$ is denoted by $\mathcal{G}(\beta_1, \dots, \beta_l)$. Set $\mathcal{G}(i_1, \dots, i_s) = \mathcal{G}(\alpha_{i_1}, \dots, \alpha_{i_s})$, $X_{\pm i} = X_{\pm\alpha_i}$, $\mathcal{X}_{\pm i} = \mathcal{X}_{\pm\alpha_i}$, $x_{\pm i}(t) = x_{\pm\alpha_i}(t)$, $X_{\pm i,d} = X_{\pm\alpha_i,d}$. Put $U^+(\mathcal{G}) = \langle \mathcal{X}_\alpha \mid \alpha \in R^+(\mathcal{G}) \rangle$ and $U^-(\mathcal{G}) = \langle \mathcal{X}_\alpha \mid \alpha \in R^-(\mathcal{G}) \rangle$. If $\mathcal{G}$ is a group of type A_1, we identify $\mathbf{X}(\mathcal{G})$ with $\mathbb{Z}$ mapping a weight $a\omega_1 \in \mathbf{X}(\mathcal{G})$ onto $a \in \mathbb{Z}$.

Only finite dimensional rational representations and modules are considered. In what follows $\mathrm{Irr}_p(\mathcal{G}) \subset \mathrm{Irr}\,\mathcal{G}$ is the set of p-restricted representations; $\varphi|\Gamma$ is the restriction of a representation φ to a subgroup Γ; φ^* is the representation dual to φ, $\dim \varphi$ ($\dim M$) is the dimension of a representation φ (a module M); $\mathbf{X}(\varphi)$ ($\mathbf{X}(M)$) is the set of weights of a representation φ (a module M). The notation $\langle \mu, \alpha \rangle$ is used for a weight $\mu \in \mathbf{X}(\mathcal{G})$ and a root $\alpha \in R(\mathcal{G})$ in the same way as for G. If $\omega \in \mathbf{X}^+(\mathcal{G})$, then $\varphi(\omega)$, $M(\omega)$, and $V(\omega)$ are the representation in $\mathrm{Irr}\,\mathcal{G}$, the irreducible module, and the Weyl module with highest weight ω. For a $\mathcal{G}$-module M and a weight vector $v \in M$ the symbol $\omega(v)$ denotes the weight of v, $M_\mu \subset M$ is the weight subspace of weight μ. If $\Gamma = \mathcal{G}(\beta_1, \dots, \beta_k)$, then $\omega_\Gamma(v)$ is the weight of v regarded as an element of the Γ-module V. If $\beta_1, \dots, \beta_k \in R^+(\mathcal{G})$ and form a basis of $R(\Gamma)$, we assume that $\langle \omega_i, \beta_j \rangle = \delta_{ij}$ (the Kronecker symbol) for the fundamental weight $\omega_i \in \mathbf{X}(\Gamma)$. By [**Bor70**, Proposition 5.13], for $m \in M$, $\alpha \in R(\mathcal{G})$, and an element t of the ground field one has

$$x_\alpha(t)m = \sum_{d=0}^{\infty} t^i X_{\alpha,d} m \tag{2.1}$$

with $X_{\alpha,d} M_\mu \subset M_{\mu+d\alpha}$.

If $\mathcal{G} = \mathcal{G}_1 \dots \mathcal{G}_m$ where $\mathcal{G}_i$ are the simple components of $\mathcal{G}$, the weight system $\mathbf{X}(\mathcal{G})$ can be canonically identified with the set of all m-tuples $(\mu_1, \dots, \mu_m)$ with $\mu_i \in \mathbf{X}(\mathcal{G}_i)$ (here we fix a maximal torus T in $\mathcal{G}$ and for a weight $\mu \in \mathbf{X}(\mathcal{G})$ set μ_i equal to the restriction of μ to $T \cap \mathcal{G}_i$). Throughout the text we often write $\mu = (\mu_1, \dots, \mu_m)$ in this case. It follows from [**Ste68**, Corollary a) of Lemma 68] that each representation $\rho \in \mathrm{Irr}\,\mathcal{G}$ can be realized as a tensor product $\rho_1 \otimes \dots \otimes \rho_m$ where $\rho_i \in \mathrm{Irr}\,\mathcal{G}_i$. Extend the notation $J(x)$ to unipotent elements x of an arbitrary simple algebraic group of a classical type. A similar notation is used for unipotent conjugacy classes.

If $\mathcal{G} = G$, we often omit the indication of a group in the notation above and write L, $\mathbf{X}$, R, W, etc. instead of $L(G)$, $\mathbf{X}(G)$, $R(G)$ and $W(G)$. In this case set $\Pi = \{\alpha_1, \dots, \alpha_r\}$. Denote by w_i ($1 \leq i \leq r$) the reflection in W associated with the root α_i. For a vector m in a G-module M put $m(i_1 \cdot d_1, \dots, i_t \cdot d_t) = X_{-i_1,d_1} \dots X_{-i_t,d_t} m$. If m is a weight vector, set $\omega_i(m) = \langle \omega(m), \alpha_i \rangle$. A unipotent element of G is regular if it is conjugate to $\prod_{i=1}^r x_i(1)$ (and to $\prod_{i=1}^r x_{-i}(1)$), see, for instance, [**Car85**, Proposition 5.13]. Denote by π_i, $1 \leq i \leq r$, the fundamental representation of $G_{\mathbb{C}}$ with highest weight ω_i.

Some preliminary results of the article are valid for characteristic 2 as well. For these results we fix the following notation: $\tilde{p}$ is a fixed prime, $\tilde{K}$ is an algebraically closed field of characteristic $\tilde{p}$, and $\tilde{G}$ is a semisimple simply connected algebraic group of rank $\tilde{r}$ over $\tilde{K}$. If $\tilde{G}$ is a classical group, observe that for unipotent $x \in \tilde{G}$ with $J(x) = (k_1, \dots, k_l)$ one has $|x| = \tilde{p}^a$ if and only if $\tilde{p}^{a-1} < k_1 \leq \tilde{p}^a$. Denote by Fr the Frobenius morphism of $\tilde{G}$ associated with raising the elements of $\tilde{K}$ to the $\tilde{p}$th power. It will always be clear from the context which group is considered.

We need some facts on the standard realizations (standard modules) of the classical groups. Recall that n denotes the dimension of the standard G-module. It is well known that $G \cong SL_n(K)$ for $G = A_r(K)$, $G \cong Sp_n(K)$ for $G = C_r(K)$; $G \cong Spin_n(K)$ and $SO_n(K)$ is a quotient of G by a central subgroup of order 2 for $G = B_r(K)$ and $D_r(K)$. Let V be the standard G-module. There exists a base

$$v_1, v_2, \dots, v_n \tag{2.2}$$

of V with the following properties.

A. The elements of some maximal torus $T \subset G$ have diagonal matrices in the base (2.2).

B. Denote by ε_i the weight of v_i with respect to T with $1 \leq i \leq n$ for $G = A_r(K)$ and $1 \leq i \leq r$ otherwise. Then the weights ε_i, $1 \leq i \leq r$, are independent elements of $\mathbf{X}$; $\sum_{i=1}^n \varepsilon_i = 0$ for $G = A_r(K)$; $\omega(v_{r+1}) = 0$ for $G = B_r(K)$; in all other cases where $G \neq A_r(K)$ one has $\omega(v_i) = -\varepsilon_{n+1-i}$ for $r < i \leq n$. The weights ε_i were mentioned in the Introduction.

C. One can assume that

$$R^+ = \begin{cases} \{\varepsilon_i - \varepsilon_j \mid 1 \leq i < j \leq n\} & \text{for } G = A_r(K), \\ \{\varepsilon_i \pm \varepsilon_j,\ \varepsilon_i \mid 1 \leq i < j \leq r\} & \text{for } G = B_r(K), \\ \{\varepsilon_i \pm \varepsilon_j,\ 2\varepsilon_i \mid 1 \leq i < j \leq r\} & \text{for } G = C_r(K), \\ \{\varepsilon_i \pm \varepsilon_j \mid 1 \leq i < j \leq r\} & \text{for } G = D_r(K); \end{cases}$$

$\alpha_i = \varepsilon_i - \varepsilon_{i+1}$ for $G = A_r(K)$ or $i < r$; and

$$\alpha_r = \begin{cases} \varepsilon_r & \text{for } G = B_r(K), \\ 2\varepsilon_r & \text{for } G = C_r(K), \\ \varepsilon_{r-1} + \varepsilon_r & \text{for } G = D_r(K). \end{cases}$$

This choice of Π gives the standard labelling of the simple roots (as in [**Bou68**]). Then $\omega_i = \sum_{j=1}^i \varepsilon_j$ for $G = A_r(K)$ or $C_r(K)$, for $G = B_r(K)$ with $i < r$, and for $G = D_r(K)$ with $i < r-1$; $\omega_r = (\sum_{j=1}^r \varepsilon_j)/2$ for $G = B_r(K)$ or $D_r(K)$; and $\omega_{r-1} = ((\sum_{j=1}^{r-1} \varepsilon_j) - \varepsilon_r)/2$ for $G = D_r(K)$.

D. The action of the elements $X_\alpha \in L$, $\alpha \in R^+$, on the vectors v_j is determined by the following equalities:

$$
\begin{aligned}
X_{\varepsilon_i-\varepsilon_j}v_j &= v_i \quad \text{for } G = A_r(K) \text{ or } i < j \leq r,\\
X_{\varepsilon_i-\varepsilon_j}v_{n+1-i} &= -v_{n+1-j} \quad \text{for } G \neq A_r(K),\ i < j \leq r,\\
X_{\varepsilon_i}v_{r+1} &= 2v_i \quad \text{for } G = B_r(K),\ i \leq r,\\
X_{\varepsilon_i}v_{n+1-i} &= v_{r+1} \quad \text{for } G = B_r(K),\ i \leq r,\\
X_{2\varepsilon_i}v_{n+1-i} &= v_i \quad \text{for } G = C_r(K),\ i \leq r,\\
X_{\varepsilon_i+\varepsilon_j}v_{n+1-i} &= v_j \quad \text{for } G \neq A_r(K),\ i \neq j,\ i,j \leq r,\ i > j \text{ or } G = C_r(K),\\
X_{\varepsilon_i+\varepsilon_j}v_{n+1-i} &= -v_j \quad \text{for } G = B_r(K) \text{ or } D_r(K),\ i < j \leq r,\\
X_\alpha v_t &= 0 \quad \text{in all other cases.}
\end{aligned}
\tag{2.3}
$$

For zero characteristic such base is described in [**Bou75**, §13]. Using the construction of irreducible G-modules described in [**Ste68**, §3 and §12] and discussed in Lemma 2.38 below, one easily concludes (and it is well known) that a desired base exists for a standard KG-module as well (here for $G = B_r(K)$ it is essential that $p \neq 2$). In what follows we call the base (2.2) the standard base of V and suppose that the weight system and the root system of G are determined with respect to the maximal torus T mentioned above. For an arbitrary classical algebraic group $\mathcal{G}$ define the weights $\varepsilon_{i,\mathcal{G}}$ and the set $\mathcal{E}(\mathcal{G})$ in the same manner as the weights ε_i and the set $\mathcal{E}$ were defined for G. Throughout the text we assume that the ordering of elements in $\mathcal{E}(\mathcal{G})$ is similar to that in $\mathcal{E}$. If we need to emphasize the type of $\mathcal{G}$, the notation $\varepsilon_{i,A}$, $\varepsilon_{i,B}$, etc., is used for a group $\mathcal{G}$ of type A_k, B_k, etc., it will be clear from the context what group is considered. For $I \subset \{1, 2, \ldots, |\mathcal{E}(\mathcal{G})|$ denote by $\mathcal{G}_I$ the subgroup in $\mathcal{G}$ generated by all root subgroups associated with the roots that are linear combinations of $\varepsilon_{i,\mathcal{G}}$ with $i \in I$. The symbols $V(\mathcal{G})$ and $n(\mathcal{G})$ are used to denote the standard $\mathcal{G}$-module and its dimension.

If $G \neq A_r(K)$, let Φ be a nondegenerate bilinear form on V preserved by G and set $v_{-i} = v_{n+1-i}$ for $i \leq r$. If $G = B_r(K)$, put $v_0 = v_{r+1}$. One can suppose that $\Phi(v_i, v_j) = 0$ if $i + j \neq n + 1$.

The Jordan block structure of the regular unipotent elements of G can be easily deduced from Formulae (2.3).

LEMMA 2.1. ([**Bor70**, Lemma 5.14], [**Sei87**, 1.5], and [**Sup97**, 2.1]) *(i) For the operators $X_{\alpha,d}$ in (2.1) the following equalities hold:*

$$X_{-\alpha}X_{\alpha,d} = X_{\alpha,d}X_{-\alpha} - H_\alpha X_{\alpha,d-1} + (d-1)X_{\alpha,d-1},$$

$$X_{\alpha,d}X_\beta = X_\beta X_{\alpha,d} + \sum_{t=1}^{d} c_t X_{t\alpha+\beta}X_{\alpha,d-t},\ c_t \in \mathbb{Z}$$

(here $H_\alpha = [X_\alpha, X_{-\alpha}]$). If $\mathcal{G} = G$, then $X_{i,k}X_{-j,d} = X_{-j,d}X_{i,k}$ for $i \neq j$.

(ii) Let V be a $\mathcal{G}$-module, $\mu \in \mathbf{X}(\mathcal{G})$, $v \in V_\mu \setminus \{0\}$, $\alpha \in R(\mathcal{G})$, $X_{\alpha,b}v = 0$ for $b > 0$, and $\langle \mu, \alpha \rangle = c \geq 0$. Then $X_\alpha X_{-\alpha,b}v = (c-b+1)X_{-\alpha,b-1}v$ and $X_{-\alpha,c}v \neq 0$. In particular, if $0 < c < p$, one has $X_{-\alpha}v \neq 0$.

THEOREM 2.2. (Steinberg Tensor Product Theorem [**Ste63**, Theorem 1.1]) *Let $\varphi \in \operatorname{Irr} \tilde{G}$. Assume that $\omega(\varphi) = \sum_{j=0}^{t} \tilde{p}^j \lambda_j$ with $\tilde{p}$-restricted $\lambda_j \in \mathbf{X}^+(\tilde{G})$. Then $\varphi \cong \bigotimes_{j=0}^{t} \varphi(\lambda_j)\operatorname{Fr}^j$.*

LEMMA 2.3. *Let $\mu = \sum_{i=1}^{r} a_i\omega_i \in \mathbf{X}$ and let α be the maximal root of G. Then*

$$(2.4)\qquad \langle \mu, \alpha\rangle = \begin{cases} \sum_{i=1}^{r} a_i & \text{for } G = A_r(K) \text{ or } C_r(K),\\ a_1 + a_r + 2\sum_{i=1}^{r-1} a_i & \text{for } G = B_r(K),\\ a_1 + a_{r-1} + a_r + 2\sum_{i=1}^{r-2} a_i & \text{for } G = D_r(K).\end{cases}$$

PROOF. This follows immediately from the formulae in [**Bou68**, Tables I–IV] for the maximal roots of the classical groups (Lie algebras). □

LEMMA 2.4. *Let $x_{\mathbb{Z}} \in GL_m(\mathbb{Z})$ be a unipotent matrix and $x \in GL_m(K)$ be the matrix obtained from x by reduction of its elements modulo p. Denote by d and $d_{\mathbb{Z}}$ the degrees of the minimal polynomials of x and $x_{\mathbb{Z}}$, respectively, and by d' and $d'_{\mathbb{Z}}$ the maximal block sizes in $J(x)$ and $J(x_{\mathbb{Z}})$ that are smaller than d and $d_{\mathbb{Z}}$ (we assume that d or $d' = 0$ if all blocks in $J(x)$ or $J(x_{\mathbb{Z}})$ are of the same size). Then $d \le d_{\mathbb{Z}}$. If $d = d_{\mathbb{Z}}$, the number k of blocks of size d in $J(x)$ is not bigger than such number $k_{\mathbb{Z}}$ for $x_{\mathbb{Z}}$. If $d = d_{\mathbb{Z}}$ and $k = k_{\mathbb{Z}}$, we have $d' \le d'_{\mathbb{Z}}$.*

PROOF. Denote by r_l and $r_{l,\mathbb{Z}}$ the ranks of the matrices $(x-1)^l$ and $(x_{\mathbb{Z}}-1)^l$, respectively. Obviously, for each l the matrix $(x-1)^l$ can be obtained from $(x_{\mathbb{Z}}-1)^l$ by reduction modulo p. The same holds for submatrices of the relevant matrices that consist of the elements occupying the same positions. This forces $r_l \le r_{l,\mathbb{Z}}$. Since d is equal to minimal l for which $r_l = 0$ and $d_{\mathbb{Z}}$ is equal to minimal t with $r_{t,\mathbb{Z}} = 0$, we get $d \le d_{\mathbb{Z}}$. Now assume that $d = d_{\mathbb{Z}}$. Then $r_{d-1} = k$ and $r_{d-1,\mathbb{Z}} = k_{\mathbb{Z}}$. This yields $k \le k_{\mathbb{Z}}$. Finally, let $d = d_{\mathbb{Z}}$ and $k = k_{\mathbb{Z}}$. If $m = kd$, there is nothing to prove. So assume that $m > kd$. One easily observes that $r_a > k(d-a)$ for $a < d'$ and that $r_{d'_{\mathbb{Z}},\mathbb{Z}} = k(d - d'_{\mathbb{Z}})$. Since $r_{d'_{\mathbb{Z}}} \le r_{d'_{\mathbb{Z}},\mathbb{Z}}$, this implies that $d' \le d'_{\mathbb{Z}}$ and completes the proof. □

We want to emphasize that Results 2.5–2.18 stated below are valid in characteristic 2 as well.

PROPOSITION 2.5. *Let M be a $\tilde{G}$-module, $x \in \tilde{G}$ be unipotent, and $|x| = \tilde{p}^{s+1} > \tilde{p}$.*

a) Assume that $l \le s$ and $z = x^{\tilde{p}^l}$. Then

$$\tilde{p}^l(d_M(z) - 1) < d_M(z) \le \tilde{p}^l d_M(z).$$

b) Set $y = x^{\tilde{p}^s}$, $d_M(y) = a+1$, $M_y = (y-1)^a M$, and $d_{M_y}(x) = b$. Then $b \le \tilde{p}^s$, $d_M(x) = a\tilde{p}^s + b$ and

$$\dim(x-1)^{a\tilde{p}^s+b-1}M = \dim(x-1)^{b-1}M_y.$$

If x acts as a single Jordan block on M, then y has b Jordan blocks of size $a+1$ and $(\tilde{p}^s - b)$ Jordan blocks of size a. In the latter case $d_{M_1}(x) = b$.

PROOF. a) One easily observes that $(x-1)^{\tilde{p}^f g}M = (x^{\tilde{p}^f}-1)^g M$ (as elements of $\operatorname{End} M$) for every $f, g \in \mathbb{Z}^+$. Hence $(x-1)^{\tilde{p}^f g}M = 0$ if and only if $(x^{\tilde{p}^f}-1)^g M = 0$. This implies Item a) of the proposition.

b) Write $M = \oplus_{j=1}^{l} L_j$ where L_j are indecomposable x-invariant subspaces. Label the subspaces L_j such that $\dim L_j \ge \dim L_{j+1}$ for all $j < l$. Choose maximal i with $\dim L_i = \dim L_1$ and denote $d_M(x)$ by d. Set $d_j = \dim L_j$ and write $d_j = a_j\tilde{p}^s + b_j$ with $a_j, b_j \in \mathbb{Z}^+$ and $0 < b_j \le \tilde{p}^s$. Then it is clear that $d = d_1$ and $\dim(x-1)^{d-1}M = i$. Set $N_j = M_y \cap L_j$, $1 \le j \le l$, and $N = \oplus_{j=1}^{l} N_j$. Since x commutes with y, the subspaces M_y, N_j, and N are x-invariant. Fix a base $e_1^j, \dots, e_{d_j}^j$ in L_j where x acts as the upper Jordan block J_{d_j}. Observe that $(x-1)^{\tilde{p}^s} e_i^j = 0$ for $i \le \tilde{p}^s$ and $(x-1)^{\tilde{p}^s} e_i^j = e_{i-\tilde{p}^s}^j$ for $i > \tilde{p}^s$ and that $(x-1)^{\tilde{p}^s} = y-1$. We have $s > 0$. Obviously, y acts trivially on L_j if $a_j = 0$. Next, assume that $a_j > 0$. Now it is clear that the subspaces

$$\langle e_1^j, e_{\tilde{p}^s+1}^j, \dots, e_{a_j\tilde{p}^s+1}^j \rangle, \dots, \quad \langle e_{b_j}^j, e_{\tilde{p}^s+b_j}^j, \dots, e_{d_j}^j \rangle,$$

$$\langle e_{b_j+1}^j, e_{\tilde{p}^s+b_j+1}^j, \dots, e_{(a_j-1)\tilde{p}^s+b_j+1}^j \rangle, \dots, \quad \langle e_{\tilde{p}^s}^j, \dots, e_{a_j\tilde{p}^s}^j \rangle$$

are invariant with respect to y and the element y has a single Jordan block on each of them (if $b_j = p^s$, here we have no subspaces of dimension a_j). So $a = a_1$. Hence $d_{L_j}(y) = a_j + 1$ and $d_M(y) = a_1 + 1$. One also observes that $(y-1)^{a_j} L_j = \langle e_1^j, \dots, e_{b_j}^j \rangle$. This forces that $N_j = 0$ if $a_j < a_1$ and that $d_{N_j}(x) = b_j$ if $a_j = a_1$. So $d_{M_y}(x) = d_N(x) = b_1 = b$, $(x-1)^{b-1} M_y = (x-1)^{b-1} N$, and $\dim(x-1)^{b-1} N = i$. This completes the proof. □

REMARK 2.6. This proposition is a key tool that enables us to apply induction for proving the main results (see Section 3). In fact, the last equality in the assertion of Proposition 2.5 is not required in this article, but we expect to apply it in a subsequent paper for obtaining asymptotic estimates for the number of Jordan blocks of the maximal size in the image of a unipotent element in a p-large representation valid for groups of rank large enough with respect to the order of this element.

For some computations we need the following property of binomial coefficients.

LEMMA 2.7. ([**Jam78**, Lemma 22.4]) *Let $a = \sum_{j=0}^{s} a_j \tilde{p}^j$ and $b = \sum_{j=0}^{s} b_j \tilde{p}^j$ with $a_j, b_j \in \mathbb{Z}^+$. Assume that $a_j + b_j < \tilde{p}$ for $0 \le j \le s$. Then $\binom{a+b}{a} \not\equiv 0 (\mod \tilde{p})$.*

Now we give a number of facts on the minimal polynomials of tensor products of unipotent Jordan blocks. In Results 2.8–2.18 stated below $d(u)$ denotes the degree of the minimal polynomial of a unipotent element u, u is regarded as an element of $GL_t(\tilde{K})$ with fixed t which is clear from the context. In Lemma 2.8 and Theorem 2.9 it is inconvenient to write down the canonical Jordan forms of tensor products considered as sequences of blocks of nonincreasing sizes. So we do not use our notation $J(x)$ here, but write $J_a \otimes J_b \cong J_{i_1} \oplus \dots \oplus J_{i_t}$ if $J_{i_1}, \dots, J_{i_t}$ constitute the complete collection of the Jordan blocks of the canonical Jordan form of the matrix $J_a \otimes J_b$ (multiplicities taken into account); in this case we also write kJ_a instead of a sum $J_a \oplus \dots \oplus J_a$ (k times).

LEMMA 2.8. ([**Fei82**, ch. VIII, Theorem 2.7]) *Let $1 \le f \le g \le \tilde{p}$. Then*

$$J_f \otimes J_g \cong \oplus_{i=0}^{h-1} J_{g-f+2i+1} \oplus N J_{\tilde{p}}$$

where $h = \min\{f, \tilde{p} - g\}$, $N = 0$ if $f + g \le \tilde{p}$, and $N = f + g - \tilde{p}$ if $f + g > \tilde{p}$. In particular, $d(J_f \otimes J_g) = f + g - 1$ if $f + g \le \tilde{p}$ and $d(J_f \otimes J_g) = \tilde{p}$ for $f + g > \tilde{p}$.

THEOREM 2.9. ([**GR85**, Lemma 6.14 and Theorem 6.4]) *Set $q = \tilde{p}^s$, $s \ge 1$. Assume that $0 < g, h \le q$ and*

$$J_g \otimes J_h \cong \oplus_{i=1}^{l} J_{n_i} \oplus N J_q$$

with all $n_i < q$. Then $l = \min\{g, h, q-g, q-h\}$.

Let $a = uq + g$ and $b = vq + h$ with $0 \leq u \leq v \leq \tilde{p} - 1$. For $0 \leq j \leq u$ set $f_j = v - u + 2j$. If $a + b \leq \tilde{p}q$, one has

$$J_a \otimes J_b \cong \bigoplus_{i=1}^{l} \bigoplus_{j=0}^{u} J_{f_j q + n_i} \bigoplus \bigoplus_{i=1}^{l} \bigoplus_{j=0}^{u-1} J_{(f_j+2)q - n_i} \bigoplus \bigoplus_{j=0}^{u-1} |g - h| J_{(f_j+2)q}$$
$$\bigoplus \bigoplus_{j=0}^{u-1} |q - g - h| J_{(f_j+1)q} \bigoplus P,$$

where

$$P = \begin{cases} 0 & \text{if } l = g, \\ (g-h)J_{(v-u)q} & \text{if } l = h, \\ (g+h-q)J_{(u+v+1)q} & \text{if } l = q - h, \\ (g-h)J_{(v-u)q} \oplus (g+h-q)J_{(u+v+1)q} & \text{if } l = q - g. \end{cases}$$

Hence $d(J_a \otimes J_b) < \tilde{p}q$.

Next, let $a + b > \tilde{p}q$. Set $a_1 = \tilde{p}q - a$ and $b_1 = \tilde{p}q - b$. Then $a_1 + b_1 < \tilde{p}q$ and

$$J_a \otimes J_b \cong J_{a_1} \otimes J_{b_1} \oplus (a + b - \tilde{p}q) J_{\tilde{p}q}.$$

So $d(J_a \otimes J_b) = \tilde{p}q$.

The following two lemmas are, probably, well known, but we cannot find an explicit reference.

LEMMA 2.10. *Let $f \geq a$ and $g \geq b$. Then $d(J_f \otimes J_g) \geq d(J_a \otimes J_b)$.*

PROOF. Let V_1 and V_2 be $\tilde{K}$-modules, $\dim V_1 = f$, and $\dim V_2 = g$. Set $x = J_f \otimes J_g$ and consider the natural action of x on $V_3 = V_1 \otimes V_2$. There exists a $\tilde{K}\langle x \rangle$-invariant submodule in V_3 where x acts as $J_a \otimes J_b$. □

LEMMA 2.11. *Let $F = \tilde{K}$ or $\mathbb{C}$. In this lemma we expand the notation J_a to denote a relevant block over F. Let $a, b > 0$ and let $x = J_a \otimes J_b \in GL_{ab}(F)$. If $F = \tilde{K}$, assume also that $\binom{a+b-2}{a-1} \not\equiv 0(\mod \tilde{p})$. Then $d(x) = a + b - 1$.*

PROOF. Let M be the standard $GL_{ab}(F)$-module. We can identify M with $M_a \otimes M_b$ where M_a and M_b are vector spaces over F of dimensions a and b, respectively. Choose a base f_i, $1 \leq i \leq a$, in M_a such that $f_i = (x-1)^{a-i} f_a$, and a base g_j, $1 \leq j \leq b$, in M_b such that $g_j = (x-1)^{b-j} g_b$. Then $\{f_i \otimes g_j,\ 1 \leq i \leq a,\ 1 \leq j \leq b\}$ yields a base in M. Using induction on l, one can deduce that for $u \in M_a$ and $v \in M_b$

$$(x-1)^l (u \otimes v) = (\sum_{i=0}^{l} \binom{l}{i} (x-1)^i u \otimes (x-1)^{l-i} v) + A \tag{2.5}$$

where A is a sum of vectors of the form $c_{ij}(x-1)^i u \otimes (x-1)^j v$ with $c_{ij} \in F$ and $i + j > l$. (Recall that $(x-1)^0 w = w$ for each vector w and that $\binom{l}{i} = \binom{l-1}{i-1} + \binom{l-1}{i}$ for $l > 1$, $i > 0$.) Hence $(x-1)^{a+b-1} M = 0$ as $(x-1)^a M_a$ and $(x-1)^b M_b = 0$. Formula (2.5) implies that $(x-1)^{a+b-2}(f_a \otimes g_b) = \binom{a+b-2}{b-1} f_1 \otimes g_1 \neq 0$. This completes the proof. □

REMARK 2.12. For $F = \mathbb{C}$ Lemma 2.11 follows from the Clebsch-Gordan formula for the group $A_1(\mathbb{C})$ (see [**Bou75**, Ch. VIII, §9, Example in Item 4]) as the matrices J_a and J_b are conjugate to the images of a unipotent element of this group in the irreducible representations of dimensions a and b. For $a+b \leq \tilde{p}$ the lemma follows from Lemma 2.8. In the general case with $F = \tilde{K}$ it, probably, could be deduced from Theorem 2.9, but this would require some extra arguments.

COROLLARY 2.13. *We have* $d(J_a \otimes J_b) < a+b$.

PROOF. In the notation of Lemma 2.4 this degree is the minimal integer k such that $r_k = 0$. Now apply Lemmas 2.11 and 2.4. □

COROLLARY 2.14. *For* $1 \leq i \leq g$ *let* $a_i = \sum_{j=0}^{s} a_{ij}\tilde{p}^j$ *with* $a_{ij} \in \mathbb{Z}^+$. *Assume that* $\sum_{i=1}^{g} a_{ij} \leq \tilde{p}-1$ *for all* j. *Then* $d(\bigotimes_{i=1}^{g} J_{a_i+1}) = 1 + \sum_{i=1}^{g} a_i$. *If* $n_i \geq a_i + 1$, *then* $d(\bigotimes_{i=1}^{g} J_{n_i}) \geq 1 + \sum_{i=1}^{g} a_i$.

PROOF. Apply induction on g. For $g = 1$ there is nothing to prove. Let $g = 2$. By Lemma 2.7, $\binom{a_1+a_2}{a_1} \not\equiv 0(\mod \tilde{p})$. Then the first assertion follows from Lemma 2.11 and the second one from Lemma 2.10. Now assume that $g > 2$ and the assertion of the lemma holds for tensor products of less than g Jordan blocks. Set $a = \sum_{i=1}^{g-1} a_i$ and $d = d(\bigotimes_{i=1}^{g-1} J_{n_i})$. By the induction conjecture, $d(\bigotimes_{i=1}^{g-1} J_{a_i+1}) = a+1$ and $d \geq a+1$. To complete the proof, apply the fact just proved for $g = 2$ to $J_{a+1} \otimes J_{a_g+1}$ and $J_d \otimes J_{n_g}$. □

COROLLARY 2.15. *Let* $0 < a < \tilde{p}$ *and* $b = \tilde{p}^s - a$ *with* $s > 0$. *Then* $d(J_a \otimes J_b) = \tilde{p}^s - 1$.

PROOF. This follows directly from Corollary 2.14. □

Proposition 2.16 below is an important technical tool for proving Theorem 1.10 for representations that are not p-restricted and for nonregular unipotent elements.

PROPOSITION 2.16. *For* $1 \leq i \leq a$ *and* $0 \leq f \leq s$ *let* $D_{fi} \in \mathbb{Z}^+$ *and* $D_{0i} < \tilde{p}^{s+1}$. *Assume that* $\tilde{p}^f D_{fi} \geq \tilde{p}^g D_{gi}$ *if* $f < g$. *Set* $n_i = \min\{\tilde{p}^{s+1}, \tilde{p}^f(D_{fi}+1) \mid 0 \leq f \leq s\}$ *and* $D_f = \sum_{i=1}^{a} D_{fi}$. *Then*

$$d(\bigotimes_{i=1}^{a} J_{n_i}) = \min\{\tilde{p}^{s+1}, \tilde{p}^f(D_f+1) \mid 0 \leq f \leq s\}. \tag{2.6}$$

PROOF. For $a = 1$ this is trivial. Apply induction on a. Denote by d the left side of Formula (2.6) and by d_m the minimal value in its right side. First suppose that $a = 2$. If n_1 or $n_2 = \tilde{p}^{s+1}$, we get $d = \tilde{p}^{s+1}$ by Theorem 2.9. Since $D_f + 1 \geq D_{fi} + 1$, one has $\tilde{p}^f(D_f+1) \geq \tilde{p}^{s+1}$ in this case for $0 \leq f \leq s$. Now assume that $n_1, n_2 < \tilde{p}^{s+1}$. Fix minimal j and k such that $n_1 = \tilde{p}^j(D_{j1}+1)$ and $n_2 = \tilde{p}^k(D_{k2}+1)$. We can suppose that $j \leq k$. If $d_m = \tilde{p}^{s+1}$, then $n_1 + n_2 \geq \tilde{p}^k(D_{k2} + 1) + \tilde{p}^k D_{k1} + \tilde{p}^j > \tilde{p}^k(D_k+1) \geq \tilde{p}^{s+1}$. Hence $d = \tilde{p}^{s+1} = d_m$ by Theorem 2.9. Let $d_m < \tilde{p}^{s+1}$. Fix minimal l such that $d_m = \tilde{p}^l(D_l+1)$. We claim that $l \geq k$. Indeed, suppose that $l < k$. Then $\tilde{p}^l(D_l+1) > \tilde{p}^k D_{k1} + \tilde{p}^k(D_{k2}+1) = \tilde{p}^k(D_k+1)$ which yields a contradiction. Hence $l \geq k$. Let $h \geq k$. Since $\tilde{p}^j D_{j1} \geq \tilde{p}^h D_{h1}$, $\tilde{p}^j(D_{j1}+1) \leq \tilde{p}^h(D_{h1}+1)$ and similar inequalities hold for $\tilde{p}^k D_{k2}$ and $\tilde{p}^h D_{h2}$, one can deduce that $n_i = \tilde{p}^h D_{hi} + t_{ih}$ with $0 < t_{ih} \leq \tilde{p}^h$ for $i = 1, 2$. Set $t_i = t_{il}$. Now we shall show that

$$t_1 + t_2 > \tilde{p}^l. \tag{2.7}$$

If $l = k$, we have $t_2 = \tilde{p}^l$ and hence (2.7) holds. Let $l > k$. Then $n_1 + n_2 > \tilde{p}^k(D_k + 1) > \tilde{p}^l(D_l+1)$ which yields (2.7). We have $n_i - 1 = \sum_{f=0}^{s} b_{fi}\tilde{p}^f$ with $0 \leq b_{fi} < \tilde{p}$

and $\sum_{f=h}^{s} b_{fi}\tilde{p}^f = \tilde{p}^h D_{hi}$ for $h \geq k$. Next, we claim that $b_{f1} + b_{f2} < \tilde{p}$ for $f \geq l$. Indeed, if $b_{s1} + b_{s2} \geq \tilde{p}$, one gets $\tilde{p}^l(D_l + 1) > \tilde{p}^l(D_{l1} + D_{l2}) \geq \tilde{p}^s(b_{s1} + b_{s2}) \geq \tilde{p}^{s+1}$; if $b_{h1} + b_{h2} \geq \tilde{p}$ for $l \leq h < s$, we have

$$\tilde{p}^l(D_l+1)>\tilde{p}^h(D_{h1}+D_{h2})=\tilde{p}^{h+1}(D_{h+1,1}+D_{h+1,2})+\tilde{p}^h(b_{h1}+b_{h2}) \geq \tilde{p}^{h+1}(D_{h+1}+1).$$

In both cases this yields a contradiction.

Now our aim is to prove that $d \geq d_m$. Set $n' = \tilde{p}^l D_{l2} + \tilde{p}^l + 1 - t_1$. Then $n' \leq n_2$ by (2.7). By Corollary 2.14, $d(J_{n_1} \otimes J_{n'}) = d_m$. Hence $d \geq d_m$ by Lemma 2.10.

It remains to show that $d \leq d_m$. Let $x_i = J_{n_i}^{\tilde{p}^l}$. Proposition 2.5 b) yields that $d(x_i) = D_{li} + 1$. Then $d(x_1 \otimes x_2) \leq D_l + 1$ by Corollary 2.13. Hence $d \leq d_m$ by Proposition 2.5 a). This settles the case $a = 2$.

Now assume that $a > 2$ and that the assertion of the lemma holds for a tensor product of less than a elements. Put $y = \bigotimes_{i=1}^{a-1} J_{n_i}$ and $D'_f = \sum_{i=1}^{a-1} D_{fi}$, $0 \leq f \leq s$. Then $d(y) = \min\{\tilde{p}^{s+1}, \tilde{p}^f(D'_f + 1) \mid 0 \leq f \leq s\}$ by the induction conjecture. To complete the proof, apply the induction conjecture to $y \otimes J_{n_a}$. □

LEMMA 2.17. *Let a_1, $a_2 \in \mathbb{Z}^+$ and $a_i < \tilde{p}$. Assume that $a_i\tilde{p}^s < n_i \leq (a_i + 1)\tilde{p}^s$ for $i = 1, 2$ and that $n_1 + n_2 > (a_1 + a_2 + 1)\tilde{p}^s$. Then $d(J_{n_1} \otimes J_{n_2}) = \min\{(a_1 + a_2 + 1)\tilde{p}^s, \tilde{p}^{s+1}\}$.*

PROOF. Set $d(J_{n_1} \otimes J_{n_2}) = d$. Arguing as in the proof of Proposition 2.16 for $m = 2$, one can show that $d \leq (a_1 + a_2 + 1)\tilde{p}^s$. Since $n_i \leq \tilde{p}^{s+1}$, Theorem 2.9 forces $d \leq \tilde{p}^{s+1}$. If $a_1 + a_2 \geq \tilde{p} - 1$, we get $d = \tilde{p}^{s+1}$ by Theorem 2.9. Let $a_1 + a_2 < \tilde{p} - 1$. Set $n' = (a_1 + a_2 + 1)\tilde{p}^s + 1 - n_1$. Then $n' \leq n_2$. By Lemma 2.10 and Corollary 2.14, $d \geq d(J_{n_1} \otimes J_{n'}) = (a_1 + a_2 + 1)\tilde{p}^s$. This yields the lemma. □

LEMMA 2.18. *For $1 \leq i \leq f$ let $a_i \in \mathbb{Z}^+$ and $a_i < \tilde{p}$. Assume that $a_i\tilde{p}^s < n_i \leq (a_i + 1)\tilde{p}^s$ and $\sum_{i=1}^{f} a_i \geq \tilde{p}$. Then $d(\bigotimes_{i=1}^{f} J_{n_i}) = \tilde{p}^{s+1}$.*

PROOF. Fix maximal j with $\sum_{i=1}^{j} a_i < \tilde{p}$. If $j = 1$, the claim of the lemma follows immediately from Theorem 2.9. If $j > 1$, set $x = \bigotimes_{i=1}^{j} J_{n_i}$. By Corollary 2.14, $d(x) \geq 1 + (\sum_{i=1}^{j} a_i)\tilde{p}^s$. To complete the proof, apply Theorem 2.9 to $x \otimes J_{n_{j+1}}$. □

The following inequality is well known:

$$a(c - a) \leq b(c - b) \quad \text{if} \quad 0 \leq \mathrm{a} \leq \mathrm{b} \leq \mathrm{c}/2. \tag{2.8}$$

LEMMA 2.19. ([**Sup96**, Lemma 2.20]) *Let $\mathcal{G}$ be a semisimple algebraic group, $x, y \in \mathcal{G}$ be unipotent elements and $y \in \overline{\mathrm{cl}(x)}$. Then $d_\varphi(y) \leq d_\varphi(x)$ for each rational representation φ of $\mathcal{G}$.*

In Definition 2.1 and Items 2.20–2.22 below the group $\mathcal{G}$ is such as in Lemma 2.19, $r' = r(\mathcal{G})$.

DEFINITION 2.1. For an irreducible $\mathcal{G}$-module M set $d(M) = l$ if the difference between $\omega(M)$ and the lowest weight ν of M is equal to $\sum_{i=1}^{r'} l_i\alpha_i$ with $\sum_{i=1}^{r'} l_i = l$. Similarly define $d(\varphi)$ for $\varphi \in \operatorname{Irr} \mathcal{G}$.

LEMMA 2.20. *Let M be an irreducible $\mathcal{G}$-module with $d(M) = l$. Assume that $x \in U^-(\mathcal{G})$ and $d_M(x) = l + 1$. Then $(x - 1)^l v \neq 0$ for a nonzero highest weight vector $v \in M$.*

PROOF. For an integer $a \in \mathbb{Z}^+$ set

$$\mathbf{X}_a = \{\mu \in \mathbf{X}(\mathcal{G}) \mid \omega(M) - \mu = \sum_{i=1}^{r'} c_i\alpha_i, \ \sum_{i=1}^{r'} c_i = a\}, \quad M_a = \langle M_\mu \mid \mu \in \mathbf{X}_a \rangle.$$

Then $M_a = 0$ for $a > l$. By [**Ste68**, Lemma 72], $(x-1)^f M_a \subset \langle M_e \mid e \geq a+f \rangle$. Hence $(x-1)^l M_a = 0$ for $a > 0$. Since $d_M(x) = l+1$ and $(x-1)^l M \neq 0$, this yields the lemma. □

Recall that $\mathbf{X}(A_1(\mathbb{C}))$ has been canonically identified with $\mathbb{Z}$ via the mapping $a\omega_1 \to a$. It is well known (see, for instance, [**Car85**, Ch.5]) that for a semisimple group Δ over $\mathbb{C}$ and a unipotent conjugacy class $C \subset \Delta$ there exist a subgroup $A \subset \Delta$ of type A_1 and a homomorphism $\mathcal{D}_C : \mathbf{X} \to \mathbb{Z}$ such that A contains a representative of C, $\mathcal{D}_C$ can be obtained by restricting the weights from some maximal torus of Δ to a maximal torus of A, $\mathcal{D}_C(\alpha) \in \{0,1,2\}$ for all $\alpha \in \Pi(\Delta)$ and coincides with the label on the labelled Dynkin diagram of C that corresponds to α; if C consists of regular unipotent elements, then $\mathcal{D}_C(\alpha) = 2$ for all $\alpha \in \Pi(\Delta)$. We shall call $\mathcal{D}_C$ the Dynkin homomorphism for C. For $z \in C$ set $\mathcal{D}_z = \mathcal{D}_C$. By [**Sup96**, Lemma 2.12],

$$d_\rho(C) = \mathcal{D}_C(\omega(\rho)) + 1 \tag{2.9}$$

for each $\rho \in \operatorname{Irr}\Delta$.

LEMMA 2.21. *Assume that $\mathcal{G}$ is a group over K and $\varphi \in \operatorname{Irr}_p(\mathcal{G})$. Then $d(\varphi) = d_{\varphi_\mathbb{C}}(z) - 1$ for a regular unipotent element $z \in \mathcal{G}_\mathbb{C}$.*

PROOF. Set $\mathcal{D} = \mathcal{D}_z$ and $\omega = \omega(\varphi)$. Recall that $\omega(\varphi_\mathbb{C}) = \omega$. By (2.9), $d_{\varphi_\mathbb{C}}(z) - 1 = \mathcal{D}(\omega)$. Put $m_+ = \max\{\mathcal{D}(\mu) \mid \mu \in \mathbf{X}(\varphi_\mathbb{C})\}$ and $m_- = \min\{\mathcal{D}(\mu) \mid \mu \in \mathbf{X}(\varphi_\mathbb{C})\}$. It follows from well-known facts of the representation theory of $A_1(\mathbb{C})$ that $m_+ = -m_-$. As $\mathcal{D}(\alpha_i) = 2$ for all i, we get $m_+ = \mathcal{D}(\omega)$, $m_- = \mathcal{D}(\nu)$ (ν is such as in Definition 2.1) and hence $d(M) = \mathcal{D}(\omega)$. This yields the lemma. □

COROLLARY 2.22. *Let M be a $\mathcal{G}$-module affording a representation $\varphi \in \operatorname{Irr}_p(\mathcal{G})$. Suppose that x, z, and v satisfy the assumptions of Lemmas 2.20 and 2.21 and x is a regular unipotent element. Set $d = d_{\varphi_\mathbb{C}}(z) - 1$. Then $(x-1)^d v \neq 0$.*

PROOF. This follows immediately from Lemmas 2.20 and 2.21. □

We need a number of facts on the embeddings of classical groups and relevant restrictions of weights. A series of results below enables us for each nontrivial unipotent $x \in G$ to construct a subgroup $S(x) \subset G$ described in the Introduction. Here $S(x)$ is a quotient of $H^p(x)$ by a central subgroup and the restriction of weights from T to $T \cap S(x)$ yields the homomorphism $\theta(x)$. In Lemmas 2.23 and 2.25 by the abuse of notation, the symbols V, v_i, and T mean for the group Δ considered there the same as for G throughout the text. So T is a maximal torus of Δ. Lemma 2.23 is well known, but we failed to find a reference with all details required for further arguments.

LEMMA 2.23. *Let $\Delta = D_e(K)$ with $e \geq 2$. Assume that $f + g = e - 1$. Let $I_1 = \{i_1, i_2, \ldots, i_f\}$ and $I_2 = \{j_1, j_2, \ldots, j_g\} \subset \{1, 2, \ldots, e\}$ (set $I_1 = \varnothing$ if $f = 0$ and $I_2 = \varnothing$ if $g = 0$). Suppose that $|I_1 \cup I_2| = e - 1$. There exists a subgroup $B \subset \Delta$ such that $B \cong B_f(K)B_g(K)$ (a central product) (here we set $B_0(K) = 1$), $T \cap B$ is a maximal torus in B, and the restriction of weights from T*

to $T \cap B$ *induces a homomorphism* $\rho_B : \mathbf{X}(\Delta) \to \mathbf{X}(B)$ *with* $\rho_B(\varepsilon_{i_h,D}) = (\varepsilon_{h,B}, 0)$, $\rho_B(\varepsilon_{j_h,D}) = (0, \varepsilon_{h,B})$, *and* $\rho_B(\varepsilon_{k,D}) = 0$ *for* $k \notin I_1 \cup I_2$.

PROOF. Put $V' = \langle v_k, v_{-k}\rangle$. Fix a nonisotropic vector $w \in V'$. There exists a nonzero $w' \in V'$ such that $\Phi(w, w') = 0$ and $V' = \langle w, w'\rangle$. Set $V_1 = \langle v_{i_1}, v_{i_2}, \dots, v_{i_f}, w, v_{-i_f}, \dots, v_{-i_2}, v_{-i_1}\rangle$ and $V_2 = \langle v_{j_1}, v_{j_2}, \dots, v_{j_g}, w', v_{-j_g}, \dots, v_{-j_2}, v_{-j_1}\rangle$. Let $B_1 \subset SO_{2r}(K)$ be the subgroup consisting of all elements that fix V_1 and V_2 and whose restrictions to V_1 and V_2 have the determinant 1. Then $B_1 \cong SO_{2f+1}(K) \times SO_{2g+1}(K)$. Denote by B the commutator subgroup of the inverse image of B_1 in Δ. Observe that V_1 and V_2 are isomorphic to the standard modules for the groups $B_f(K)$ and $B_g(K)$, respectively, and the sequences of vectors used to determine these modules can be regarded as standard bases for them. Analyzing the restriction of weights from T to $T \cap B$, one can conclude that $B \cong B_f(K)B_g(K)$, $T \cap B$ is a maximal torus in B, and this restriction induces the homomorphism ρ_B with the required properties. Notice that $B \cong B_{c-1}(K)$ if f or $g = 0$. □

In the assumptions of Lemma 2.23 denote by B^1 and B^2, respectively, the subgroups of B consisting of all elements that act trivially on V_2 (respectively, on V_1).

LEMMA 2.24. *In the assumptions of Lemma 2.23 let* $\Delta = G$, $f \geq g$, $I_1 = \{1, 2, \dots, f\}$, $I_2 = \{f+1, \dots, r-1\}$ *if* $f < r-1$ *and* $I_2 = \varnothing$ *otherwise. Set* $x_1 = x_1(1)x_2(1)\dots x_{f-1}(1)x_{\varepsilon_f - \varepsilon_r}(1)x_{\varepsilon_f+\varepsilon_r}(-1)$, $x_2 = \prod_{i=f+1}^{r} x_i(1)$ *if* $f < r-1$ *and* $x_2 = 1$ *for* $f = r-1$*; put* $x = x_1x_2$*. Then one can assume that* x_j *is a regular unipotent element of* B^j *for* $j = 1, 2$ *and* x *is a regular unipotent element of* B.

PROOF. In Lemma 2.23 take $w = v_r - v_{-r}$. Then one can put $w' = v_r + v_{-r}$. Analyzing the action of x_j on V for $j = 1, 2$, now we can conclude that $x_j \in B^j$, $d_{V_1}(x_1) = 2f + 1$, and $d_{V_2}(x_2) = 2g + 1$. This implies the lemma. □

We also need to deal with the standard embeddings of $A_{t-1}(K)$ to $C_t(K)$ or $D_t(K)$ twisted by the action of the Weyl group of the bigger group.

LEMMA 2.25. *1) Let* $\Delta = C_t(K)$ *with* $t \geq 2$*. There exists a subgroup* $\Sigma \subset \Delta$ *such that* $\Sigma \cong A_{t-1}(K)$, $T \cap \Sigma$ *is a maximal torus in* Σ, *and the restriction of weights from* T *to* $T \cap \Sigma$ *induces a homomorphism* $\rho_\Sigma : \mathbf{X}(\Delta) \to \mathbf{X}(\Sigma)$ *with* $\rho_\Sigma(\varepsilon_{2i-1,C}) = \varepsilon_{i,A}$ *and* $\rho_\Sigma(\varepsilon_{2i,C}) = -\varepsilon_{t-i+1,A}$.

2) Let $\Delta = D_t(K)$ *with* $t \geq 2$*. There exist subgroups* Σ_1 *and* $\Sigma_2 \subset \Delta$ *such that* $\Sigma_j \cong A_{t-1}(K)$, $T \cap \Sigma_j$ *is a maximal torus in* Σ_j *for* $j = 1$ *and* 2, *and the restriction of weights from* T *to* $T \cap \Sigma_j$ *induces a homomorphism* $\rho_j : \mathbf{X}(\Delta) \to \mathbf{X}(\Sigma_j)$ *with* $\rho_1(\varepsilon_{2i-1,D}) = \varepsilon_{i,A}$, $\rho_1(\varepsilon_{2i,D}) = -\varepsilon_{t-i+1,A}$, $\rho_2(\varepsilon_{j,D}) = \rho_1(\varepsilon_{j,D})$ *for* $j < t$, *and* $\rho_2(\varepsilon_{t,D}) = -\rho_1(\varepsilon_{t,D})$.

PROOF. Fix the integers a and d such that $2a - 1 \leq t < 2a + 1$ and $2d \leq t < 2d + 2$. Observe that $a + d = t$. Set $e_i = v_{2i-1}$ for $1 \leq i \leq a$, $e_{a+j} = v_{-2(d-j+1)}$ for $1 \leq j \leq d$, $e_{-k} = v_{-h}$ if $e_k = v_h$, and $e_{-k} = v_h$ if $e_k = v_{-h}$ for $1 \leq k \leq t$. Put $V_1 = \langle e_k \mid 1 \leq k \leq t\rangle$ and $V_2 = \langle e_{-k} \mid 1 \leq k \leq t\rangle$. Denote by Σ the commutator subgroup of the stabilizer of V_1 and V_2 in Δ. Then $\Sigma \cong A_{t-1}(K)$. Observe that V_1 is isomorphic to the standard Σ-module and the sequence $(e_1, \dots, e_t)$ plays the role of the standard base of this module. Next, consider the action of T and $T \cap \Sigma$ in the base $e_1, \dots, e_t, e_{-t}, \dots, e_{-1}$. One can conclude that $U^+(\Delta) \cap \Sigma = U^+(\Sigma)$

and identify $\omega_\Sigma(e_i)$ with $\varepsilon_{i,A}$ for $1 \le i \le t$. Analyzing the restriction of weights from T to $T \cap \Sigma$, we can deduce that the assertion 1) holds for $\Delta = C_t(K)$ and that Σ can be taken for Σ_1 if $\Delta = D_t(K)$. To construct Σ_2, consider the base $e'_1, \dots, e'_t, e'_{-t}, \dots, e'_{-1}$ obtained from $e_1, \dots, e_t, e_{-t}, \dots, e_{-1}$ by interchanging v_t and v_{-t} and the subspaces $V'_1 = \langle e'_1, \dots, e'_t \rangle$ and $V'_2 = \langle e'_{-t}, \dots, e'_{-1} \rangle$ and apply the arguments above to these base and subspaces. □

COROLLARY 2.26. *In the assumptions of Lemma 2.25 for a group $\Sigma^+ = \Sigma$, Σ_1 or Σ_2 there exist roots $\beta_1, \dots, \beta_{t-1} \in R^+(\Delta)$ such that $\Sigma^+ = \Delta(\beta_1, \dots, \beta_{t-1})$ and one can set $\Pi(\Sigma^+) = \{\beta_1, \dots, \beta_{t-1}\}$.*

PROOF. This follows directly from the proof of Lemma 2.25. Indeed, let ρ be the homomorphism of weight systems in Lemma 2.25 associated with Σ^+. The definition of ρ enables to find roots $\beta_1, \dots, \beta_{t-1} \in R^+(\Delta)$ such that $\rho(\beta_i)$, $1 \le i \le t-1$, constitute a base of $R(\Sigma^+)$. To complete the proof, consider the action of the root subgroups $\mathcal{X}_{\pm\beta_i}$ on V_1 and V_2 or V'_1 and V'_2 if relevant. □

In Proposition 2.27 and Corollary 2.28 below the sequence $u_1, \dots, u_l$ with $u_1 \ge u_2 \ge \dots \ge u_l$, the integer $c = c(x)$, the groups H_j, $H = H(x)$, and H^p, and the homomorphism $\theta = \theta(x)$ are constructed for a unipotent element $x \in G$ such as in the Introduction. If $\mathrm{cl}(x)$ is not determined by $J(x)$, the classes C_1 and C_2 are also defined as in the Introduction. Set $r_j = r(H_j)$ for $1 \le j \le c$.

PROPOSITION 2.27. *Let $x \neq 1$ be a unipotent element of G. There exists a subgroup $S = S(x) \subset G$ isomorphic to a quotient of H^p by its central subgroup and such that $\mathrm{cl}(x)$ contains a regular unipotent element of S, $T \cap S$ is a maximal torus in S, and the homomorphism $\mathbf{X} \to \mathbf{X}(H(x))$ induced by the restriction of weights from T to $T \cap S$ coincides with $\theta(x)$.*

PROOF. Set $r' = r + 1$ for $G = A_r(K)$ and $r' = r$ otherwise. Throughout this proof for a set $I \subset \{1, 2 \dots, r'\}$ the subgroup G_I is determined such as in the notation in the beginning of this section. We assume that $|I| > 1$ for $G = A_r(K)$ or $D_r(K)$. If $G = B_r(K)$ and $|I| > 1$, denote by G^1_I the subgroup generated by all long root subgroups contained in G_I. Observe that $G_I \cong A_{|I|-1}(K)$ for $G = A_r(K)$, for $\mathcal{G} \in \{B, C, D\}$ and $G = \mathcal{G}_r(K)$ one has $G_I \cong \mathcal{G}_{|I|}(K)$; $G^1_I \cong D_{|I|}(K)$. Set $e_j = |\mathcal{E}(H_j)|$ for $1 \le j \le c$. The construction of θ implies that for each $k \le e_j$ the weight $\varepsilon^j_k = \pm\theta_j(\varepsilon_i)$ for some i. For every $j \le c$ we construct a subgroup $S_j \cong H^p_j$ such that S_j and S_k commute for $j \neq k$. Then put $S = \prod_{j=1}^c S_j$.

For $1 \le j \le c$ set $I_j = \{i \mid \mathbf{j}(\varepsilon_i) = j\}$.

First let $G = A_r(K)$ or $C_r(K)$. Put $G_j = G_{I_j}$. If u_j is proper for G, put $S_j = G_j$. Assume that u_j is improper for G. Then $G = C_r(K)$, $e_j > 1$, and $G_j \cong C_{e_j}(K)$. Now identify G_j with $C_{e_j}(K)$ and construct a subgroup $A_j \subset G_j$ as the subgroup Σ in Lemma 2.25 1). Recall that $A_j \cong A_{e_j-1}(K)$. Set $S_j = A_j$.

For $G = B_r(K)$ or $D_r(K)$ set $I_0 = \{i \mid \theta(\varepsilon_i) = 0\}$.

Now suppose that $G = B_r(K)$. Then the number of proper u_j is odd. For improper u_j set $G_j = G^1_{I_j}$. In this case $G_j \cong D_{e_j}(K)$. Now construct a subgroup $A_j \subset G_j$ as the subgroup Σ_1 in Item 2) of Lemma 2.25 and set $S_j = A_j$. If all u_j with $j \le c$ are improper, we are done. If there is just one proper $u_j > 1$, set $S_j = G_{I_j}$ for this j. Next, assume that there are q proper $u_j > 1$ with $q > 1$. Set $q' = [q/2]$. Then Rules 1—9 imply that $|I_0| \ge q'$. Let $j_1 < j_2 < \dots < j_q$ exhaust the set of $j \le c$ for which u_j is proper. Assume that $i_1 < i_2 < \dots < i_{q'}$ are the q'

minimal members of $|I_0|$. If $q' > 1$, for $1 \leq k < q'$ set $F_k = I_{j_{2k-1}} \cup I_{j_{2k}} \cup \{i_k\}$ and $G_k = G^1_{F_k}$. Put $r_{k,1} = (u_{j_{2k-1}} - 1)/2$ and $r_{k,2} = (u_{j_{2k}} - 1)/2$. Construct subgroups $B^1 \cong B_{r_{k,1}}(K)$, $B^2 \cong B_{r_{k,2}}(K)$, and $B = B^1 B^2 \subset G_k$ as in Lemma 2.23 taking $I_{j_{2k-1}}$ for the set I_1 of that lemma, $I_{j_{2k}}$ for I_2, and i_k for the exceptional index k. Set $S_j = B^1$ for $j = j_{2k-1}$ and $S_j = B^2$ for $j = j_{2k}$. If q is odd, define the set $F_{q'}$ and construct the subgroups S_j for $j = j_{q-2}$ and $j = j_{q-1}$ in the similar way. For even q set $F_{q'} = I_{j_{q-1}} \cup \{i_{q'}\}$, $G_{q'} = G^1_{F_{q'}}$, $r_{q'} = (u_{j_{q-1}} - 1)/2$ and construct a subgroup $B \cong B_{r_{q'}}(K) \subset G_{q'}$ as in Lemma 2.23 taking $I_{j_{q-1}}$ for I_1, $i_{q'}$ for the exceptional index and putting $I_2 = \varnothing$. In all cases set $S_{j_q} = G_{I_{j_q}}$.

Finally, let $G = D_r(K)$. Now the number of proper u_j is even. First assume that some u_j is proper for G. For improper j (if any) set $G_j = G_{I_j}$ and construct the subgroup $S_j \subset G_j$ using Lemma 2.25 as for $G = B_r(K)$. We get $S_j \cong A_{u_j-1}(K)$. If there exist proper $u_j > 1$, define q as for $G = B_r(K)$, but now take $q' = [q+1]/2$. Argue as for $G = B_r(K)$ with the following changes: now $G_k = G_{F_k}$ for $1 \leq k \leq q'$; if q is even, all the subgroups S_j are constructed in pairs as described in the previous paragraph, and if q is odd, we construct $q'-1$ such pairs (0 for $q = 1$) and at the last step take $F_{q'} = I_{j_q} \cup \{i_{q'}\}$, construct a subgroup $B \cong B_{r_{q'}}(K)$ as in the previous paragraph and put $B = S_{j_q}$.

Now let all u_j be improper. Then $r \in I_j$ for some $j \leq a$. If $x \in C_1$ or $r \notin I_j$, construct S_j as before for improper u_j. Next, assume that $x \in C_2$ and $r \in I_j$. Then construct a subgroup $A_j \subset G_j$ as the subgroup Σ_2 in Lemma 2.25 2) and set $S_j = A_j$.

In all cases the arguments above yield that S_j and S_t commute for $j \neq t$. Hence one can define $S = S_1 S_2 \ldots S_c$.

To complete the proof, we need some identifications of weights. If $j \leq c$ and u_j is proper for G, set $d = r_j + 1$ for $G = A_r(K)$ and $d = r_j$ otherwise. Then there exist just d indices $l_1, \ldots, l_d$ for which the restrictions of the weights ε_{l_i} with $1 \leq i \leq d$ from T to $T \cap S_j$ are nonzero. Assume that $l_1 < \ldots < l_d$ and $S_j = \mathcal{G}_d(K)$ with $\mathcal{G} \in \{A, B, C\}$. Identify the restriction of ε_{l_i} to $T \cap S_j$ with $\varepsilon_{i,\mathcal{G}} \in \mathcal{E}(S_j)$. For improper u_j the subgroup $S_j \cong A_{u_j-1}(K)$ and lies in the subgroup $G_S = G_j$ or G^1_j. Here $G_S \cong C_{u_j}(K)$ or $D_{u_j}(K)$ and there exist just u_j indices $l_1, \ldots, l_{u_j}$ for which the restrictions of the weights ε_{l_i} with $1 \leq i \leq u_j$ from T to $T \cap G_S$ are nonzero. As before, assume that $l_1 < \ldots < l_{u_j}$. First identify the restrictions of ε_{l_i} to $T \cap G_S$ with $\varepsilon_{i,C}$ or $\varepsilon_{i,D} \in \mathcal{E}(G_S)$ and then identify the restrictions of the latter weights in $\mathcal{E}(G_S)$ to $T \cap S_j$ with some weights of S_j as in the course of constructing the subgroup Σ in Lemma 2.25. Here for $G = D_r(K)$ one must distinguish whether Σ is constructed as the subgroup Σ_1 in Lemma 2.25 2) or the subgroup Σ_2. Well-known facts on the structure of the classical groups imply that all these identifications are correctly determined.

Now the subgroup S is constructed. It is clear that S is isomorphic to a central quotient of H^p and $T \cap S$ is a maximal torus in S. For $G = A_r(K)$ or $C_r(K)$ we have $S \cong H^p$. The construction of θ and the identifications above imply the assertion of the proposition on the restriction of weights from T to $T \cap S$.

It remains to prove that $\mathrm{cl}(x)$ contains a regular unipotent element of S. It can be deduced from the construction of S that the latter element has the same canonical Jordan form as x. Then Lemma 1.4 implies our claim if $G \neq D_r(K)$ or some u_j is odd. Let $G = D_r(K)$ and all u_j be even. Then there exists a subset $R_S \subset R$ such that $S = \langle \mathcal{X}_\alpha \mid \alpha \in R_S \rangle$ and S is a semisimple part of a Levy subgroup in G. One can identify $S_{\mathbb{C}}$ with the subgroup $\langle \mathcal{X}_\alpha \mid \alpha \in R_S \rangle \subset G_{\mathbb{C}}$. Let $x_S \in S$ and $z \in S_{\mathbb{C}}$ be regular unipotent elements. The definition of a labelled Dynkin diagram (see comments in the Introduction) yields that such diagrams coincide for x_S and z. One easily concludes that the analog of Lemma 2.25 holds for classical groups over $\mathbb{C}$ as well. Using the weight identifications mentioned above, one can assume that for some maximal torus $T_{\mathbb{C}} \subset G_{\mathbb{C}}$ the subgroup $T_S = T_{\mathbb{C}} \cap S_{\mathbb{C}}$ is a maximal torus in $S_{\mathbb{C}}$ and the restriction of weights from $T_{\mathbb{C}}$ to T_S induces the homomorphism θ (regarded as a mapping from $\mathbf{X}(G_{\mathbb{C}})$ to $\mathbf{X}(S_{\mathbb{C}})$). Next, one can choose z and a subgroup $A \cong A_1(\mathbb{C})$ such that $z \in A \subset S_{\mathbb{C}}$, $T_A = T_{\mathbb{C}} \cap A$ is a maximal torus in A, and the restriction of weights from $T_{\mathbb{C}}$ to T_A induces the Dynkin homomorphism $\mathcal{D}_z$. Using Lemma 2.25 2) and taking into account the construction of S, now we can find $\mathcal{D}_z(\alpha_{r-1})$ and $\mathcal{D}_z(\alpha_r)$ and conclude that $x_S \in C_i$ if $x \in C_i$ for $i = 1, 2$. Hence $x_S \in \mathrm{cl}(x)$ as required. The proposition is proved. □

In what follows we can and shall assume that $x \in S(x)$. Then we can write $x = \prod_{j=1}^{c} x_j$ where x_j is a regular unipotent element in S_j and consider $d_\mu(x)$ for $\mu \in \operatorname{Irr} S$.

COROLLARY 2.28. *Let x and u_1 be as in Proposition 2.27. Then $\overline{\mathrm{cl}(x)}$ contains an element x_1 with $J(x_1) = (u_1, 1, \dots, 1)$ if u_1 is proper for G and $J(x_1) = (u_1, u_1, 1, \dots, 1)$ if u_1 is improper for G (the number of one-dimensional blocks in $J(x_1)$ can be zero).*

PROOF. Keep the notation introduced in Proposition 2.27 and just after it. The construction of the group S_1 and the information on the form of root elements of G in the base (2.2) given at the beginning of this section yield that $J(x_1)$ is such as required. As x is a regular unipotent element in S, [**Car85**, Proposition 5.1.2]) forces that each unipotent element in S (in particular, x_1) lies in $\overline{\mathrm{cl}(x)}$. □

REMARK 2.29. Corollary 2.28 could be deduced from the description of the inclusion relations between the conjugacy classes of G in [**Spa82**, Theorems I.2.4 and II.8.2], easy arguments above are given for the reader's convenience since anyway we have to state all details of Proposition 2.27.

COROLLARY 2.30. *Let x and S be as in Proposition 2.27, $x \in S$, $\varphi \in \operatorname{Irr}$, and $\chi = \varphi(\theta(\omega(\varphi))) \in \operatorname{Irr} S$. Then $d_\varphi(x) \geq d_\chi(x)$.*

PROOF. Let M be a module affording φ and $v \in M$ be a nonzero highest weight vector. The construction of S yields that one can set $U^+(S) = S \cap U^+$. Hence v generates an indecomposable S-module with highest weight $\omega_S(v)$. By Proposition 2.27, $\omega_S(v) = \theta(\omega(\varphi))$. This yields that χ is a composition factor of $\varphi|S$ and completes the proof. □

LEMMA 2.31. *Let $G = A_r(K)$, $B_r(K)$ or $C_r(K)$ and $x \in G$ be a regular unipotent element. Then $S = G$ and hence Theorem 1.10 is valid for x if Theorem 1.7 holds for x.*

PROOF. Recall that x has a single Jordan block on V. The construction of the groups $H(x)$ and $S(x)$ and the homomorphism θ in the Introduction and Proposition 2.27 and the choice of the elements h_f implies that $H(x) = G_{\mathbb{C}}$, θ is the canonical identification of the weight systems $\mathbf{X}$ and $\mathbf{X}(G_{\mathbb{C}})$, $S = G$, and $J(h_f) = J(z_f)$ for $0 \leq f \leq s$. Hence $\psi = \varphi_{\mathbb{C}}$. This completes the proof. □

THEOREM 2.32. *Theorems* 1.1, 1.3, 1.7, *and* 1.10 *hold for elements of order* p.

PROOF. For elements of order p Theorem 1.7 follows immediately from Theorem 1.14. Let $|x| = p$ and $C = \mathrm{cl}(x) \subset G$. Assume that $J(x) = (k_1, \dots, k_t)$. Then all $k_i \leq p$. Theorem 1.3 follows directly from Theorem 1.14 and Algorithm 1.6. So suppose that $k_1 < p$. Fix a representation $\varphi \in \mathrm{Irr}$ with highest weight ω. Construct the groups H_j, $1 \leq j \leq c$, and $H = H(x)$, the mapping $\theta = \theta(x)$, and the representation $\psi \in \mathrm{Irr}\, H$ as in the Introduction. Denote by C' the image of C under the bijection f of Theorem 1.14. Let C_H be the regular unipotent class in H. By Theorem 1.14, to prove Theorem 1.10 for the class C, it suffices to verify that $d_{\varphi_{\mathbb{C}}}(C') = d_\psi(C_H)$. Let $\mathcal{D}$, $\mathcal{D}_H$, and $\mathcal{D}_j$, $1 \leq j \leq c$, be the Dynkin homomorphisms for C', C_H, and for regular unipotent classes in H_j, respectively. Formula (2.9) reduces the question to proving that $\mathcal{D}(\bar{\omega}) = \mathcal{D}_H(\theta(\bar{\omega}))$. Since θ is a homomorphism, now it suffices to deduce that $\mathcal{D}(\varepsilon) = \mathcal{D}_H(\theta(\varepsilon))$ for $\varepsilon \in \mathcal{E}$.

Next, consider the sequence $N(C')$ associated with the class C' in Algorithm 1.6. As the restriction of the standard $G_{\mathbb{C}}$-module to an A_1-type subgroup containing a representative of C' is a sum of irreducible $A_1(\mathbb{C})$-modules of dimensions $k_1, \dots, k_t$, well-known facts on representations of the latter group (see, for instance, [**Bou75**, Ch. VIII, §1]) yield that $N(C')$ consists of the values of $\mathcal{D}$ on the weights of this standard module. Since $\mathcal{D}(\alpha) \geq 0$ for all $\alpha \in \Pi$, this implies that $\mathcal{D}(\varepsilon_1) \geq \mathcal{D}(\varepsilon_2) \geq \dots \geq \mathcal{D}(\varepsilon_{r+1})$ for $G = A_r(K)$, $\mathcal{D}(\varepsilon_1) \geq \mathcal{D}(\varepsilon_2) \geq \dots \geq \mathcal{D}(\varepsilon_r) \geq 0$ for $G = B_r(K)$ or $C_r(K)$, and $\mathcal{D}(\varepsilon_1) \geq \mathcal{D}(\varepsilon_2) \geq \dots \geq \mathcal{D}(\varepsilon_{r-1}) \geq |\mathcal{D}(\varepsilon_r)|$ for $G = D_r(K)$. Furthermore, if $G = D_r(K)$ and some of k_i is odd, there are at least two such values k_j and hence at least two zeros in the collection $N(C')$. So in this case $D(\varepsilon_r) = 0$. If $G = D_r(K)$ and all k_i are even, there are two classes C' with the same $N(C')$ by Lemma 1.4 ii) and the values of $\mathcal{D}(\varepsilon_r)$ for them differ by the sign. These arguments imply that $\mathcal{D}(\varepsilon_i)$ is the ith maximal integer in the collection $N(C')$ if $G \neq D_r(K)$, or $i < r$, or some k_i is odd; in the exceptional case $\mathcal{D}(\varepsilon_r)$ is such integer for $C' = C_1$ and its opposite for $C' = C_2$ where C_1 and C_2 are the relevant classes in Algorithm 1.6 and Lemma 1.4 ii).

Set $d(r) = \mathcal{D}_H(\theta'(\varepsilon_r))$ if $G = D_r(K)$ and all k_i are even, and $d(i) = \mathcal{D}_H(\theta(\varepsilon_i))$ otherwise (here θ' is such as in the Introduction). Now our goal is to show that $d(i)$ is equal to the ith maximal integer in $N(C')$. Apply induction on i. Observe that $\mathcal{D}_H(\alpha) = 2$ and $\mathcal{D}_j(\beta) = 2$ for $\alpha \in \Pi(H)$ and $\beta \in \Pi(H_j)$ as these Dynkin homomorphisms are associated with regular unipotent classes. Recall that a regular unipotent element of H_j has a single Jordan block in the standard realization of H_j. This implies that for $\varepsilon_q^j \in \mathcal{E}(H^j)$ one has $\mathcal{D}_j(\varepsilon_q^j) = u_j - 2q + 1 = \nu(\varepsilon_q^j)$ where ν is the function defined in the course of constructing θ. Define the collection N^+ as follows: let N' be the number of the integers k_i equal to 1, for each $j \leq c$ collect all the nonnegative values $\nu(\varepsilon)$ with $\varepsilon \in \mathcal{E}(H_j)$, call this sequence $N^+(j)$, write down every $N^+(j)$ once if u_j is proper and twice if u_j is improper for G, then add N' zeros and take the collection obtained for N^+. Recall the construction of the group H and observe that every $z \in \mathbb{Z}^+$ occurs in $N(C')$ and in N^+ with the same multiplicity. It is clear that if $\mathbf{j}(i) = j$, then $d(i) = \mathcal{D}_j(\theta_j(\varepsilon_i))$. Now Rule 8 in the Introduction forces that $d(1) = k_1 - 1$ and is equal to the maximal member of $N(C')$. For $i > 1$ denote by N_i the collection obtained by deleting from $N(C')$ the $i - 1$ integers equal to $d(k)$, $k < i$. We assume that our claim on $d(k)$ holds for $k < i$ and want to show that $d(i)$ is equal to the maximal integer in N_i. Since integers a and $-a$ occur in $N(C')$ with the same multiplicity and Rules 4 and 9 in the Introduction yield that $d(i) = d(n+1-i)$ for $G = A_r(K)$, we can assume that $i \leq (n+1)/2$ for this group. Notice that Rules 6 and 7 are irrelevant for elements of order p. First assume that $\mathbf{j}(i) = j$ and $\theta_j(\varepsilon_i) = \varepsilon^j_q$. Set $\nu = \nu(\varepsilon^j_q)$. Rules 2, 3, 5, and 9 imply that $\nu \geq 0$, integers greater than ν do not occur in N_i, and ν does occur in N_i. This yields our claim. If $\theta_j(\varepsilon_i) = -\varepsilon^j_q$, the dimension u_j is improper for G and Rule 2 yields that $\theta_j(\varepsilon_{i-1}) = \varepsilon^j_{u_j+1-q}$. Applying Rules 2, 3, 5, and 9 to $\theta(\varepsilon_{i-1})$ and arguing as above, one can prove the claim for this case as well. Finally, if $\theta(\varepsilon_i) = 0$, Rules 2, 3, 5, and 9 imply that N_i contains no positive integers and 0 occurs in N_i. This completes the proof of our claim on $d(i)$ and that of Theorem 1.10 for elements of order p.

It remains to consider Theorem 1.1. By Lemma 1.4 iii), the standard realizations of the groups $B_r(K)$ and $D_r(K)$ do not contain elements x with $J(x) = (2, 1, \ldots, 1)$ (transvections). Algorithm 1.6 implies that $d_{\pi_i}(C') - 1$ can be less than 2 in the following cases only: $G = A_r(K)$ or $C_r(K)$; $G = B_r(K)$, $i = 1$ or r and $G = D_r(K)$, $i \in \{1, r-1, r\}$. Now Proposition 1.5 and Formulae (2.4) given in the Introduction yield that $d_{\varphi_{\mathbb{C}}}(C') > p$ if φ is p-large. Hence, by Theorem 1.14, $d_\varphi(C) = p$. All the claims of the theorem are proved. □

Now we present some information on the restrictions of representations to certain special subgroups.

THEOREM 2.33. (Smith [**Smi82**] and Jantzen [**Jan73**, Satz 1.5]) *Let $\mathcal{G}$ be a semisimple algebraic group over a field $\mathcal{F}$, $I \subset \{1, \ldots, r(\mathcal{G})\}$ and $\mathcal{G}_I = \mathcal{G}(i \mid i \in I)$. Assume that M is an irreducible $\mathcal{G}$-module and $v \in M$ is a nonzero highest weight vector. Then $\mathcal{F}\mathcal{G}_I v$ is an irreducible direct summand of $M|\mathcal{G}_I$ with highest weight $\omega_{\mathcal{G}_I}(v)$.*

For $G = B_r(K)$ or $C_r(K)$ set $G^+ = A_{n-1}(K)$. Then G can be naturally mapped into G^+ and regular unipotent elements of G go to such elements of G^+. By the abuse of notation we use the symbol $\rho|G$ when consider restrictions of representations of G^+ to the image of this mapping. Let $\varphi \in \operatorname{Irr}$ and $\omega(\varphi) = \sum_{i=1}^r a_i\omega_i$. Denote by φ^+ the representation in $\operatorname{Irr} G^+$ with highest weight $\sum_{i=1}^r a_i\omega_i$ (here ω_i are fundamental weights of G^+).

PROPOSITION 2.34. (Seitz [**Sei87**, 8.1]) *Let $G = B_r(K)$ and $\omega(\varphi) = \omega_i$ with $1 \leq i < r$ or $G = C_r(K)$ and $\omega(\varphi) = a\omega_1$ with $a < p$ or $a\omega_i + (p-1-a)\omega_{i+1}$ with $1 \leq i < r$ and $a \neq 0$ for $i = r-1$. Then $\varphi \cong \varphi^+|G$.*

In what follows $\bigwedge^i M$ is the ith exterior power of a module M. In Lemma 2.35 and Proposition 2.36 $\widehat{G} = A_r(\tilde{K})$ or $A_r(\mathbb{C})$, $n = r+1$, by the abuse of notation, V is the standard $\widehat{G}$-module, vectors $v_i \in V$, $1 \leq i \leq n$, form a base of this module analogous to the base (2.2).

LEMMA 2.35. ([**Jan03**, Part II, Item 2.15]) *Let $\widehat{G} = A_r(K)$ or $A_r(\mathbb{C})$. Then the representation $\varphi(\omega_i) \in \operatorname{Irr}\widehat{G}$ can be realized in the $\widehat{G}$-module $\bigwedge^i V$ for $1 \leq i \leq r$.*

Proposition 2.36 and Lemma 2.37 are well known, but we failed to find explicit references including all details we need. For $\Delta = A_t(K)$ set $\omega_0 = \omega_{t+1} = 0$ and denote by φ_i^{t+1} the representation $\varphi(\omega_i) \in \operatorname{Irr}\Delta$ with $0 \leq i \leq t+1$. So both φ_0^{t+1} and φ_{t+1}^{t+1} denote the trivial representation of Δ. The assertions of Proposition 2.36 and Lemma 2.37 justify this unusual notation. In Proposition 2.36 we write $\varphi|F =$

$\bigoplus a_j\lambda_j$ if the restriction of a representation $\varphi \in \operatorname{Irr}\widehat{G}$ to a subgroup F is a direct sum of representations each of which is a direct sum of a_j copies of a certain representation $\lambda_j \in \operatorname{Irr} F$, here we do not care that λ_j and λ_k may be isomorphic. For a subset $I \subset \{1, 2, \dots, n\}$ denote by $\widehat{G}(I)$ the subgroup in $\widehat{G}$ that consists of all elements fixing each vector v_j with $j \notin I$.

PROPOSITION 2.36. *Let $\widehat{G} = A_r(K)$ or $A_r(\mathbb{C})$ and $\varphi = \varphi(\omega_i) \in \operatorname{Irr}\widehat{G}$, $1 \le i \le r$. Assume that $I_j \subset \{1, 2, \dots, n\}$, $1 \le j \le l+1$, $l \ge 1$, $I_j \cap I_k = \varnothing$ for $j \ne k$, and $\bigcup_{j=1}^{l+1} I_j = \{1, 2, \dots, n\}$. Put $|I_j| = n_j$. Assume that $n_j > 1$ for $j \le l$ and set $F = \prod_{j=1}^{l} \widehat{G}(I_j)$. Then*

$$\varphi|F \cong \bigoplus_{\sigma} \binom{n_{l+1}}{i_{l+1}} \varphi_{i_1}^{n_1} \otimes \varphi_{i_2}^{n_2} \dots \varphi_{i_l}^{n_l}$$

where the sum is taken over all $l+1$-tuples σ of nonnegative integers $i_1, i_2, \dots, i_l, i_{l+1}$ with $i_j \le n_j$ and $\sum_{j+1}^{l+1} i_j = i$.

PROOF. Set $V_j = \langle v_i \mid i \in I_j \rangle$ for $1 \le j \le l+1$ (if $I_{l+1} = \varnothing$, we have $V_{l+1} = 0$). Put $\wedge(c_1, \dots, c_i) = v_{c_1} \wedge \dots \wedge v_{c_i}$. For each admissible $l+1$-tuple $\sigma = (i_1, \dots, i_{l+1})$ set $s_0 = 0$, $s_t = s_t(\sigma) = \sum_{j=1}^{t} i_t$ for $1 \le t \le l+1$ and

$$\bigwedge(\sigma) = \langle \wedge(c_1, \dots, c_i) \mid c_m \in I_t \text{ for } s_{t-1} < m \le s_t,\ c_k < c_m \text{ if } s_{t-1} < k < m \le s_t \rangle.$$

Since $V = \bigoplus_{j=1}^{l+1} V_j$, one easily observes that $\bigwedge^i V = \bigoplus_\sigma \bigwedge(\sigma)$ where the sum is taken over all admissible $l+1$-tuples σ. Furthermore, it is clear that each $\bigwedge(\sigma)$ is an F-module isomorphic to a direct sum of $\binom{n_{l+1}}{i_{l+1}} = \dim(\bigwedge^{i_{l+1}} V_{l+1})$ copies of the F-module $\bigwedge^{i_1} V_1 \otimes \dots \otimes \bigwedge^{i_l} V_l$. To complete the proof, apply Lemma 2.35 and [**Ste68**, Corollary a) of Lemma 68]. □

For $\Delta = B_g(K)$ denote by $\xi(\Delta)$ the representation $\varphi(\omega_g) \in \operatorname{Irr}\Delta$ and for $\Delta = D_g(K)$ let $\xi_-(\Delta)$ and $\xi_+(\Delta)$ denote the representations $\varphi(\omega_{g-1})$ and $\varphi(\omega_g) \in \operatorname{Irr}\Delta$, respectively. If Δ is fixed, we sometimes omit the indication of Δ in this notation. If $\Sigma = A_{t-1}(K)$ with $t = 2l$, set $\mu_i^t = \varphi_{l+i}^t \in \operatorname{Irr}\Sigma$ for $-l \le i \le l$.

LEMMA 2.37. *Let $\Delta = D_t(K)$ and $t = 2l$. Denote by $I_1 = I_1(t)$ the set of all odd integers i with $-l \le i \le l$ and by $I_2 = I_2(t)$ the set of all even integers in this interval. Let Σ_1 and $\Sigma_2 \subset \Delta$ be the subgroups constructed in Lemma 2.25.2. Then $\xi_-(\Delta)|\Sigma_1 \cong \xi_+(\Delta)|\Sigma_2 \cong \bigoplus_{i \in I_1} \mu_i^t$ and $\xi_+(\Delta)|\Sigma_1 \cong \xi_-(\Delta)|\Sigma_2 \cong \bigoplus_{i \in I_2} \mu_i^t$.*

PROOF. Set $\Omega = \{(\pm\varepsilon_1 \pm \varepsilon_2 \pm \dots \pm \varepsilon_t)/2\} \subset \mathbf{X}(\Delta)$. Let Ω_o and $\Omega_e \subset \Omega$ be the sets of all such linear combinations with an odd and even number of "minus" signs, respectively. It is well known that $\mathbf{X}(\xi_+) = \Omega_e$ and $\mathbf{X}(\xi_-) = \Omega_o$ and that ξ_+ and ξ_- are miniscule representations. By the abuse of notation, set $\varepsilon_i = \varepsilon_{i,\Delta}$. Arguing as in the proof of Corollary 2.26, one can take $\Pi(\Sigma_1) = \{\varepsilon_1 - \varepsilon_3, \dots, \varepsilon_{2l-3} - \varepsilon_{2l-1}, \varepsilon_{2l-1} + \varepsilon_{2l}, \varepsilon_{2l-2} - \varepsilon_{2l}, \dots, \varepsilon_2 - \varepsilon_4\}$ and obtain $\Pi(\Sigma_2)$ from this set interchanging ε_{2l} and $-\varepsilon_{2l}$. Considering the action of Σ_1 and Σ_2 on the Δ-modules $M(\omega_{t-1})$ and $M(\omega_t)$, we obtain the required decomposition.

Let M_i, $-l \le i \le l$, be the $A_{t-1}(K)$-module affording μ_i^t. Then a nonzero highest weight vector in $M(\omega_t)$ generates a Σ_1-module isomorphic to M_0 and a Σ_2-module isomorphic to M_1. Furthermore, for $1 \le i \le l$ set $c_i = l-i+1$, $\beta_i = (\varepsilon_1 + \varepsilon_2 + \dots + \varepsilon_{2i-3} + \varepsilon_{2i-2} - \varepsilon_{2i-1} + \varepsilon_{2i} - \dots - \varepsilon_{2l-1} + \varepsilon_{2l})/2$ (the signs "minus" just for ε_{2k-1}

with $k \geq i$), $\gamma_i = \beta_i - \varepsilon_{2l}$, $\delta_i = (\varepsilon_1 + \varepsilon_2 + \ldots + \varepsilon_{2i-2} + \varepsilon_{2i-1} - \varepsilon_{2i} + \ldots + \varepsilon_{2l-1} - \varepsilon_{2l})/2$ (the signs "minus" just for ε_{2k} with $k \geq i$), and $\lambda_i = \delta_i + \varepsilon_{2l}$. Then β_i and $\delta_i \in \Omega_e$ if c_i is even and lie in Ω_o if c_i is odd; γ_i and $\lambda_i \in \Omega_o$ if c_i is even and lie in Ω_e if c_i is odd. Let $m_{1,i}, m_{2,i}, m_{3,i}$, and $m_{4,i}$ be nonzero vectors of weights $\beta_i, \gamma_i, \delta_i$, and λ_i, respectively, in the Δ-module $M(\omega_{t-1}) \oplus M(\omega_t)$. One can conclude that $m_{1,i}$ and $m_{2,i}$ generate a Σ_1-module and a Σ_2-module, respectively, isomorphic to M_{-c_i}, $m_{3,i}$ and $m_{4,i}$ generate such modules isomorphic to M_{c_i}. This implies the presence of required factors in the restrictions $\xi_\pm|\Sigma_1$ and $\xi_\pm|\Sigma_2$. The dimension arguments show that no other factors occur. As μ_i^t are miniscule representations, we get direct sums in the restrictions considered. □

Next, we prove a series of lemmas that enable us to compare the minimal polynomials of some elements in certain representations of G and $G_\mathbb{C}$. Denote by $\mathbb{F}_{\tilde{p}}$ the field of $\tilde{p}$ elements.

LEMMA 2.38. *Let $\beta_1, \ldots, \beta_k \in R^+(\tilde{G})$, $t_1, \ldots, t_k \in \mathbb{Z}$, and let $\overline{t_j}$ be the image of t_j under the natural mapping $\mathbb{Z} \to \mathbb{F}_{\tilde{p}}$. Let $x = \prod_{j=1}^k x_{\beta_j}(\overline{t_j}) \in \tilde{G}$ and $x_\mathbb{C} = \prod_{j=1}^k x_{\beta_j}(t_j) \in \tilde{G}_\mathbb{C}$. Assume that $\varphi \in \operatorname{Irr} \tilde{G}$. Then $d_\varphi(x) \leq d_{\varphi_\mathbb{C}}(x_\mathbb{C})$.*

PROOF. First assume that $\varphi \in \operatorname{Irr}_{\tilde{p}} \tilde{G}$. Then $\omega(\varphi) = \omega(\varphi_\mathbb{C})$. We use the construction of the module $M(\omega)$ described in [**Ste68**, §3 and §12]. (Strictly speaking, the construction in [**Ste68**] starts with a module for the Lie algebra $L(\tilde{G}_\mathbb{C})$, but it is well known that these modules can be identified.) Let $M_\mathbb{C}$ be the irreducible $\tilde{G}_\mathbb{C}$-module with highest weight $\omega(\varphi)$. Denote by $\mathcal{U}$ the universal envelopping algebra of $L(\tilde{G}_\mathbb{C})$ and by $\mathcal{U}_\mathbb{Z}$ and $\mathcal{U}_\mathbb{Z}^-$ the $\mathbb{Z}$-subalgebras in $\mathcal{U}$ generated by the operators $X_\alpha^i/i!$, $\alpha \in R(\tilde{G})$ or $R^-(\tilde{G})$, respectively. Set $M_\mathbb{Z} = \mathcal{U}_\mathbb{Z}^- v_\mathbb{C}$. By [**Ste68**, Theorem 2 and its Corollary 1], M is $\mathcal{U}_\mathbb{Z}$-invariant and contains a base $m_1, \ldots, m_l$ of $M_\mathbb{C}$ which consists of weight vectors. We can assume that m_1 is a highest weight vector. Next, construct a $\tilde{G}$-module M' as in [**Ste68**, §3 and §12]. Set $M' = M_\mathbb{Z} \otimes_\mathbb{Z} \tilde{K}$. The operators $X_\alpha^i/i!$, $\alpha \in R(\tilde{G})$, act on M' and are zero for i large enough. Hence for $t \in \tilde{K}$ and $\alpha \in R(\tilde{G})$ one can define an operator $\exp(tX_\alpha) = \sum_{i=0}^\infty t^i X_\alpha^i/i! \in \operatorname{End} M'$ in the standard way. By the arguments in the proof of [**Ste68**, Theorem 39], we can define a rational homomorphism $\tilde{G} \to GL(M')$ (i.e., make M' a $\tilde{G}$-module) mapping $x_\alpha(t)$ into $\exp(tX_\alpha) \in \operatorname{End} M'$. Then M' is an indecomposable $\tilde{G}$-module generated by $m_1' = m_1 \otimes 1$ as $M_\mathbb{Z} = \mathcal{U}_\mathbb{Z}^- m_1$. One easily observes that m_1' is a highest weight vector in M'. Hence $M(\omega)$ can be identified with a quotient of M' and $d_{M'}(x) \geq d_\varphi(x)$. It is clear that all the elements $x_{\beta_j}(t_j)$ and hence $x_\mathbb{C}$ have integer matrices in the base $(m_1, \ldots, m_l)$ and the matrix of x in the base $(m_1 \otimes 1, \ldots, m_l \otimes 1)$ can be obtained from that of $x_\mathbb{C}$ in the base $(m_1, \ldots, m_l)$ by reduction modulo p. Now apply Lemma 2.4 to complete the proof for $\varphi \in \operatorname{Irr}_{\tilde{p}} \tilde{G}$.

Now, let φ be arbitrary. By Theorem 2.2, $\varphi = \otimes_{j=0}^a \varphi_j \operatorname{Fr}^j$ where $\varphi_j \in \operatorname{Irr}_{\tilde{p}} \tilde{G}$. We have $\omega(\varphi_\mathbb{C}) = \sum_{j=0}^a \omega(\varphi_j)$. Obviously, Fr does not influence the degree of the minimal polynomial. Several applications of Corollary 2.13 yield that $d_\varphi(x) \leq 1 + \sum_{j=0}^a (d_{\varphi_j}(x) - 1)$. We have just shown that $d_{\varphi_j}(x) \leq d_{(\varphi_j)_\mathbb{C}}(x_\mathbb{C})$. Now Proposition 1.5 implies that $d_\varphi(x) \leq d_{\varphi_\mathbb{C}}(x_\mathbb{C})$. □

The following lemma shows the possibility of applying Lemma 2.38 to representatives of each unipotent conjugacy class. Probably, it is well known, but we cannot find a reference for it.

LEMMA 2.39. *Let $C \subset G$ and $C' \subset G_{\mathbb{C}}$ be nonidentity unipotent conjugacy classes with the same labelled Dynkin diagram. Then there exist roots $\beta_1, \dots, \beta_f \in R^+$ and integers $l_1, \dots, l_f = \pm 1$ such that the elements of G and $G_{\mathbb{C}}$ equal to $\prod_{j=1}^{f} x_{\beta_j}(l_j)$ lie in C and C', respectively.*

PROOF. In this proof we call an element x special if $x = \prod_{j=1}^{a} x_{\beta_j}(l_j)$ with $\beta_j \in R^+$ and $l_j = \pm 1$. For special x denote by $x_{\mathbb{C}}$ the element $\prod_{j=1}^{a} x_{\beta_j}(l_j) \in G_{\mathbb{C}}$. If C consists of regular unipotent elements, set $x = \prod_{i=1}^{r} x_i(1)$. Then $x \in C$ and $x_{\mathbb{C}} \in C'$ by [**Car85**, Proposition 5.13]. If $G = D_r(K)$ and $J(C) = (u_1, u_2)$ with $u_1 = 2f + 1$ and $u_2 = 2g + 1$, set $x = x_1 x_2$ with x_1 and x_2 as in Lemma 2.24. As C and C' are determined by the canonical Jordan form of their elements, the arguments of Lemmas 2.23 and 2.24 applied to the action of x on V and to that of $x_{\mathbb{C}}$ on the standard $G_{\mathbb{C}}$-module imply that $x \in C$ and $x_{\mathbb{C}} \in C'$. In other cases the proof will be heavily based on the arguments in that of Proposition 2.27. Keep the notation of Proposition 2.27. Until the end of the proof S is the subgroup of Proposition 2.27 whose regular unipotent elements lie in C, the integers u_j are determined by $J(C)$. Next, let $G = D_r(K)$ and all u_j be even. Then $R(S)$ can be identified with a subset of R. Set $r' = r(S)$ and, using Corollary 2.26, fix roots $\gamma_1, \gamma_2, \dots, \gamma_r' \in R$ that form a base of $R(S)$. Put $x = \prod_{i=1}^{r'} x_{\gamma_i}(1)$. Actually it has been shown in the proof of Proposition 2.27 that $x \in C$ and $x_{\mathbb{C}} \in C'$. If $G = C_r(K)$, r is odd and $J(C) = (r, r)$, the construction of S and Corollary 2.26 yield that $S = G(\beta_1, \dots, \beta_{r-1})$ with $\Pi(S) = \{\beta_1, \dots, \beta_{r-1}\}$ and $S \cong A_{r-1}(K)$. Then C and C' contain regular unipotent elements of the groups S and $G_{\mathbb{C}}(\beta_1, \dots, \beta_{r-1})$. Arguing as for regular unipotent elements of G, set $x = \prod_{i=1}^{r-1} x_{\beta_i}(1)$ and conclude that $x \in C$ and $x_{\mathbb{C}} \in C'$. Now we can exclude these cases.

Observe that the arguments above are valid for the groups $B_1(K)$, $B_2(K)$, $C_1(K)$, $D_2(K)$, and $D_3(K)$ as well.

So until the end of the proof we suppose that C is not regular, that for $G = C_r(K)$ either some u_j is even or a representative of C has more than 2 Jordan blocks, and for $G = D_r(K)$ some u_j is odd and a representative of C has more than 2 Jordan blocks. Hence all $u_j < n$ and for $G = D_r(K)$ all $u_j < n - 1$. Apply induction on r and $\dim G$ assuming that the assertion of the lemma holds for classical groups of smaller rank or dimension than G. We claim that there exist proper subgroups G_1 and $G_2 \subset G$ such that G_1 is a classical algebraic group, G_2 is such group or $G_2 = 1$, $G_1 \cap G_2 = 1$, G_1 and G_2 commute, $G_1 = G(\delta_1, \dots, \delta_k)$ with $\delta_i \in R^+$, $G_2 = G(\mu_1, \dots, \mu_t)$ with $\mu_i \in R^+$ if $G_2 \neq 1$, and $S \subset G_1 G_2$; furthermore, there exist $V_1, V_2 \subset V$ such that $V = V_1 \oplus V_2$, G_i acts trivially on V_j if $\{i, j\} = \{1, 2\}$, V_1 is a standard G_1-module, and V_2 is a standard G_2-module if $G_2 \neq 1$. If our claim holds, for $x \in C$ write $x = g_1 g_2$ with $g_i \in G_i$. It is clear that C is determined by the conjugacy classes of g_i in G_i. Denote these classes by C_i. Let G_1' and $G_2' \subset G_{\mathbb{C}}$ be the subgroups generated by the "same" root subgroups as G_1 and G_2, respectively (we have $G_2' = 1$ if $G_2 = 1$). Applying the induction conjecture to the pairs G_i and G_i' with $i = 1, 2$, we can choose such special elements $g_i \in C_i$ that $(g_i)_{\mathbb{C}}$ belongs to the conjugacy class of G_i' with the same labelled Dynkin diagrams as C_i. Then $x = g_1 g_2$ and $x_{\mathbb{C}}$ yield required elements for C and C'. Observe that g_2 can be equal to 1 for $G_2 \neq 1$. This does not influence our arguments.

So it remains to prove the claim. Set $\Omega = \{1, 2, \dots, |\mathcal{E}|\}$. In this paragraph for $N \subset \Omega$ the subgroups G_N and G_N^1 are defined as in the proof of Proposition 2.27

(under the same restrictions). Set $\widehat{G_N} = 1$ if $N = \varnothing$ or $|N| = 1$ and $G = A_r(K)$ or $D_r(K)$, and $\widehat{G_N} = G_N$ otherwise. The construction of S implies that under our assumptions there exist subsets N_1 and $N_2 \subset \Omega$ such that $\Omega = N_1 \cup N_2$, $N_1 \cap N_2 = \varnothing$, $|N_1| > 1$ for $G \neq C_r(K)$, $N_2 \neq \varnothing$ for $G \neq B_r(K)$, $S \subset G^1_{N_1}\widehat{G_{N_2}}$ for $G = B_r(K)$, and $S \subset \widehat{G_{N_1}}\widehat{G_{N_2}}$ otherwise. Put $G_1 = G^1_{N_1}$ and $V_2 = \langle v_0, v_{\pm j} \mid j \in N_2\rangle$ for $G = B_r(K)$. In all other cases set $G_i = \widehat{G_{N_i}}$ and $V_i = \langle v_{\pm j} \mid j \in N_i\rangle$ for $i = 1$ or 2. One easily observes that the groups G_1 and G_2 and the subspaces V_1 and V_2 satisfy our claim. This completes the proof. □

COROLLARY 2.40. *Let $C \subset G$ and $C' \subset G_{\mathbb{C}}$ be unipotent conjugacy classes with the same labelled Dynkin diagram. Assume that $x \in C$, $x_{\mathbb{C}} \in C'$, and $\varphi \in \mathrm{Irr}$. Then $d_\varphi(x) \leq d_{\varphi_{\mathbb{C}}}(x_{\mathbb{C}})$.*

PROOF. This follows immediately from Lemmas 2.38 and 2.39. □

COROLLARY 2.41. *Let $x \in G$ be a regular unipotent element and $\varphi \in \mathrm{Irr}$. Assume that $|x| = p^{s+1}$. Then $d_\varphi(x) \leq \min\{p^{s+1}, p^i d_{\varphi_{\mathbb{C}}}(z_i) \mid 0 \leq i \leq s\}$ where the elements $z_i \in G_{\mathbb{C}}$ are such as in Theorem 1.7.*

PROOF. Observe that the conjugacy classes of the elements x^{p^i}, $0 \leq i < s$, are completely determined by their canonical Jordan forms. Now the corollary follows from Proposition 2.5 a) and Lemmas 2.38 and 2.39. □

LEMMA 2.42. *Let Δ be a semisimple algebraic group over $\mathbb{C}$ or $\tilde{K}$, $I \subset \Pi(\Delta)$ be a proper subset, and M be a Δ-module. Denote by Σ_I the set of integer linear combinations of the simple roots in I. Set $R_I = R(\Delta) \cap \Sigma_I$ and $R' = R^+(\Delta) \setminus (R_I \cap R^+(\Delta))$. Let $m = m_1 + \ldots + m_k \in M$ and m_j, $1 \leq j \leq k$, be the weight components of m. If $k > 1$, assume that $\omega(m_i) - \omega(m_j) \in \Sigma_I$ for $1 \leq i < j \leq k$. Suppose that $x' \in \langle \mathcal{X}_\alpha \mid \alpha \in R'\rangle$, $x_I \in \langle \mathcal{X}_\alpha \mid \alpha \in R_I \cap R^+(\Delta)\rangle$, $x = x'x_I$, and $(x_I - 1)^a m \neq 0$. Then $(x-1)^a m \neq 0$. In particular, $d_M(x) \geq d_M(x_I)$.*

PROOF. Set

$$\Omega = \{\omega(m_j) \mid 1 \leq j \leq k\}, \quad \mathbf{X}_1 = \{\mu \in \mathbf{X}(\Delta) \mid \mu = \omega + \sigma, \quad \omega \in \Omega,\ \sigma \in \Sigma_I\},$$
$$\mathbf{X}_2 = \mathbf{X}(\Delta) \setminus \mathbf{X}_1, \quad M_i = \langle M_\lambda \mid \lambda \in \mathbf{X}_i\rangle$$

for $i = 1, 2$. Using [**Ste68**, Lemma 72], one easily concludes that $(x_I - 1)^a m \in M_1$ and $(x-1)^a m = (x_I - 1)^a m + m'$ with $m' \in M_2$. Since $M_1 \cap M_2 = 0$ and $(x_I - 1)^a m \neq 0$, this yields the lemma. □

To apply induction, Proposition 2.5, and Lemma 2.42, we need more details concerning $C_G(y)$ for an element $y = x^{p^s}$. They are given in Proposition 2.43 below; there we set $A_0(K) = B_0(K) = C_0(K) = A_{-1}(K) = B_{-1}(K) = C_{-1}(K) = 1$. In Proposition 2.43 it is inconvenient to identify the groups $B_1(K)$ and $C_1(K)$ with $A_1(K)$ and $B_2(K)$ with $C_2(K)$, so we keep our assumptions on r, but do not identify such subgroups of G. In particular, here we consider the group $B_1(K)$ with the fundamental weight $\omega_1 = \varepsilon_{1,B}/2$.

PROPOSITION 2.43. *Let $G = A_r(K)$, $B_r(K)$, or $C_r(K)$, and let $x \in G$ be a regular unipotent element. Assume that $s, a, b \in \mathbb{Z}^+$, $s \geq 1$, $p^s < n = ap^s + b \leq p^{s+1}$, and $0 < b \leq p^s$. Set $y = x^{p^s}$ and $e = p^s - b$. Then $C_G(y)$ contains a semisimple group Γ with the following properties.*

1). Γ *is a central product of its simple components* Γ_1 *and* Γ_2*;*

$$\Gamma_1 = A_{b-1}(K), \quad \Gamma_2 = A_{e-1}(K) \quad \text{for} \quad G = A_r(K);$$
$$\Gamma_1 = C_l(K), \quad \Gamma_2 = B_{(e-1)/2}(K) \quad \text{for} \quad G \neq A_r(K), \quad b = 2l;$$
$$\Gamma_1 = B_l(K), \quad \Gamma_2 = C_{e/2}(K) \quad \text{for} \quad G \neq A_r(K), \quad b = 2l+1.$$

2). Write weights $\mu \in \mathbf{X}(\Gamma)$ *in the form* $\mu = (\mu_1, \mu_2)$ *where* $\mu_j \in \mathbf{X}(\Gamma_j)$ *is the restriction of the weight* μ *to some fixed maximal torus of* Γ_j*. Assume that* $\mathbf{X}(\Gamma_j) = 0$ *if* $\Gamma_j = 1$*. Denote by* ε_i^j*,* $j = 1, 2$*, the weights in* $\mathcal{E}(\Gamma_j)$ *with the standard labelling. There exist maximal tori* $T \subset G$ *and* $T_\Gamma = \Gamma \cap T \subset \Gamma$ *such that the restriction of weights from* T *to* T_Γ *determines the homomorphism* $\rho : \mathbf{X} \to \mathbf{X}(\Gamma)$ *with the following formulae for* $\rho(\varepsilon_i)$ *with* $i \leq r+1$ *for* $G = A_r(K)$ *and* $i \leq r$ *otherwise: if* $i = cp^s + d$ *and* $0 < d \leq p^s$*, then:*

(2.10)

$$\rho(\varepsilon_i) = \begin{cases} (\varepsilon_d^1, 0) & \text{for } 1 < b \leq p^s, \ d \leq b \text{ if } G = A_r(K) \\ & \text{and } d \leq [b/2] \text{ otherwise;} \\ (0, \varepsilon_{d-b}^2) & \text{for } b < d \leq p^s \text{ if } G = A_r(K) \\ & \text{and } b < d \leq [(p^s + b)/2] \text{ otherwise;} \\ (0, 0) & \text{if } b = d = 1, \text{ or } e = 1, \ d = p^s, \text{ or } G \neq A_r(K) \\ & \text{and either } b = 2l, \ d = (p^s + b + 1)/2 \\ & \text{or } b = 2l+1, \ d = l+1; \\ (-\varepsilon_{b+1-d}^1, 0) & \text{for } G \neq A_r(K), \ b = 2l \text{ and } l < d \leq b \\ & \text{or } b = 2l+1 > 1 \text{ and } l+1 < d \leq b; \\ (0, -\varepsilon_{p^s+1-d}^2) & \text{for } G \neq A_r(K), \ b = 2l \text{ and } d > (p^s + b + 1)/2 \\ & \text{or } b = 2l+1 < p^s \text{ and } d > (p^s + b)/2. \end{cases}$$

3). Set $J_1 = \{j \mid 1 \leq j \leq r, \ j \equiv b \text{ or } j \equiv 0(\mod p^s)\}$*,* $J = \{1, 2, \ldots, r \setminus J_1\}$*,* $G_J = G(i \mid i \in J)$*. One can assume that* $U^{\pm}(\Gamma) \subset U^{\pm}(G_J)$ *(naturally, the same sign in both parts of the formula) and* $x = x_\Gamma x'$ *where* $x_\Gamma = \prod_{\alpha \in R^+(G_J)} x_\alpha(t_\alpha)$ *is a regular unipotent element of* Γ *and* $x' = \prod_{\beta \notin R^+(G_J)} x_\beta(t_\beta)$*.*

4) Let $|y| = p$*. Then* y *can be embedded into a Zariski closed subgroup* $A \subset G$ *with the following properties:* A *is a homomorphic image of* $A_1(K)$*;* $\Gamma \subset C_G(A)$*; if* T *is the torus mentioned in Item 2, then* $T_1 = T \cap A$ *is a maximal torus of* A*;* $y \in U^+(A) \subset U^+(G)$ *and* $U^-(A) \subset U^-(G)$*. Furthermore, let* $\zeta : \mathbf{X} \to \mathbb{Z}$ *be the natural homomorphism determined by restricting weights from* T *to* T_1*. Then one can assume that for* $i = lp^s + i_1$ *with* $0 \leq l \leq a$ *and* $1 \leq i_1 \leq p^s$ *the value* $\zeta(\omega(v_i)) = a - 2l$ *if* $i_1 \leq b$ *and* $a - 1 - 2l$ *otherwise. In particular,* $\zeta(\omega(v_i)) \geq \zeta(\omega(v_j))$ *for* $i < j$*.*

PROOF. Item 1) is well known (see, for instance, [**SS70**, Ch. IV, 2.25] and [**Sei00**, Proposition 2.7]), but we need more details on the embedding of Γ into G. Recall that V is the standard G-module; if $G \neq A_r(K)$, then Φ is a nondegenerate bilinear form on V preserved by G (Φ is uniquely determined up to a scalar multiple). Set $I(V) = SL(V)$ for $G = A_r(K)$, $SO(V)$ for $G = B_r(K)$, and $Sp(V)$ for $G = C_r(K)$; $I_l = SL_l(K)$ for $G = A_r(K)$, $Sp_l(K)$ for $G \neq A_r(K)$, l even, and $SO_l(K)$ for $G \neq A_r(K)$, l odd. (Here $SO_1(K) = 1$.) Obviously, $I(V) = G$ for $G = A_r(K)$ or $C_r(K)$. Observe that for $G = B_r(K)$ the unipotent elements of G

are completely determined by their action on V, so in all cases we can consider them as elements of $SL(V)$ when it is convenient. In this proof we shall also need similar groups $I(U)$ for some other spaces U. If $G = A_r(K)$, set $I(U) = SL(U)$. Otherwise in all cases a space U will be equipped by a nondegenerate bilinear form symmetric if $\dim U$ is odd and antisymmetric if $\dim U$ is even, and we take for $I(U)$ the intersection of the isometry group of this form with $SL(U)$. Hence $I(U) = Sp(U)$ or $SO(U)$.

Recall that x has a single Jordan block on V. Then Proposition 2.5 b) forces that y has b blocks of size $a+1$ and e blocks of size a on V. Obviously, $a+1 \not\equiv b(\mod 2)$ if n is even and $a+1 \equiv b(\mod 2)$ for odd n. Observe also that $a > 0$ in our assumptions. Lemma 1.4i) implies that the conjugacy class of y is completely determined by its canonical Jordan form. Now we proceed to construct subspaces V_1, $V_2 \subset V$ and subgroups G_1, $G_2 \subset SL(V)$ with the following properties: $V = V_1 \oplus V_2$; $\dim V_1 = (a+1)b$, $\dim V_2 = ae$; if $G \neq A_r(K)$, then V_1 is nondegenerate with respect to Φ and V_2 is nondegenerate or zero; G_i stabilizes the subspaces V_j, $i, j = 1, 2$, G_i acts trivially on V_j if $i \neq j$; $G_1|V_1 \cong I_{a+1} \otimes I_b$, $G_2|V_2 \cong I_a \otimes I_e$ if $e > 0$ and $G_2 = 1$ for $e = 0$; $y \in G_1G_2$, $y|V_1 \in I_{a+1} \otimes 1$, $y|V_2 \in I_a \otimes 1$ for $e > 0$. It is clear that the subgroup of G_1G_2 consisting of matrices u with $u|V_1 \in 1 \otimes I_b$ and $u|V_2 \in 1 \otimes I_e$ for $e > 0$ centralizes y. We take this subgroup for Γ for $G = A_r(K)$ or $C_r(K)$ and take for Γ the commutator subgroup of the inverse image of this subgroup in G if $G = B_r(K)$.

We shall call a base $u_1, \ldots, u_l$ of a space U equipped with a nondegenerate bilinear form Ψ almost hyperbolic if $\Psi(u_i, u_j) \neq 0$ just for $i + j = l + 1$. One easily concludes that a standard hyperbolic base can be obtained from an almost hyperbolic one with the help of a diagonal transformation. So it is clear that the group of diagonal matrices in an almost hyperbolic base that have determinant 1 and preserve Ψ is a maximal torus in $I(U)$ and the group of upper (lower) triangular matrices preserving Ψ is a maximal unipotent subgroup in $I(U)$. Furthermore, one can assume that the base vectors have such weights as relevant vectors of a standard hyperbolic base in [**Bou75**, §13].

Now let U_1, W_1, U_2, and W_2 be linear spaces over K of dimensions $a+1$, b, a, and e, respectively. Let

$$\{u_1^1, \ldots, u_{a+1}^1\},\ \{w_1^1, \ldots, w_b^1\},\ \{u_1^2, \ldots, u_a^2\},\ \{w_1^2, \ldots, w_e^2\} \tag{2.11}$$

be bases of the spaces U_1, W_1, U_2, and W_2, respectively (the base of W_2 is considered only for $e > 0$). Set $n_i = \dim W_i$, $i = 1, 2$. If $G \neq A_r(K)$, assume also that our spaces are equipped with nondegenerate bilinear forms Φ_U^1, Φ_W^1, Φ_U^2, and Φ_W^2, respectively, that the forms are antisymmetric for even dimensional and symmetric for odd dimensional spaces, and that the bases fixed before are almost hyperbolic for these forms. Here and until the end of the proof we consider all objects connected with W_2 only for $e > 0$. Next, set $V_1' = U_1 \otimes W_1$, $V_2' = U_2 \otimes W_2$ if $e > 0$ and $V_2' = 0$ for $e = 0$; $V' = V_1' \oplus V_2'$. Fix a base in V' as follows. For $1 \leq i \leq n$, write an integer i in the form $i = cp^s + d$ with $c, d \in \mathbb{Z}^+$ and $0 < d \leq p^s$; observe that $c < a$ if $d > b$. Set $v_i' = u_{c+1}^1 \otimes w_d^1$ for $d \leq b$ and $v_i' = u_{c+1}^2 \otimes w_{d-b}^2$ if $d > b$. If $G \neq A_r(K)$, define bilinear forms Φ_i on V_i' setting

$$\Phi_i(u_k^i \otimes w_l^i, u_m^i \otimes w_s^i) = \Phi_U^i(u_k^i, u_m^i)\Phi_W^i(w_l^i, w_s^i).$$

Using the well-known standard embeddings $Sp_k \otimes Sp_l \to SO_{kl}$, $SO_k \otimes SO_l \to SO_{kl}$, and $Sp_k \otimes SO_l \to Sp_{kl}$, one deduces that the forms Φ_i are always nondegenerate and that both Φ_1 and Φ_2 are antisymmetric for even and symmetric for odd n. Next, define a bilinear form Φ' on V' setting $\Phi'|V'_i = \Phi_i$ and assuming that V'_1 and V'_2 are mutually orthogonal.

Now it is clear that there exists a vector space isomorphism $\nu : V' \to V$ that is an isometry if $G \neq A_r(K)$. Set $\widetilde{v_i} = \nu(v'_i)$. The mapping ν determines the isomorphism $\mu : I(V') \to I(V)$ for which the matrices of $g \in I(V')$ and $\mu(g)$ in the bases $(v'_1, \dots, v'_n)$ and $(\widetilde{v_1}, \dots, \widetilde{v_n})$ coincide. Denote by St the stabilizer of the subspaces V'_1 and V'_2 in $I(V')$. We write elements $\sigma \in St$ in the form (σ_1, σ_2) where $\sigma_i = \sigma|V'_i$. Let $y_i \in I(U_i)$, $i = 1, 2$, be regular unipotent elements that have upper unitriangular matrices in the bases (2.11), and let $y' = (y_1 \otimes 1, y_2 \otimes 1) \in St$. Recall that y_i has a single Jordan block on U_i. One can assume that $y = \mu(y')$. Put $\Gamma'_1 = (1 \otimes I_b, 1)$, $\Gamma'_2 = (1, 1 \otimes I_e) \subset St$, and $\overline{\Gamma_i} = \mu(\Gamma'_i)$. Let Γ_i be the commutator subgroup of the inverse image of $\overline{\Gamma_i}$ in G for $G = B_r(K)$ and $\Gamma_i = \overline{\Gamma_i}$ otherwise. Set $\Gamma = \Gamma_1\Gamma_2$. Then $\Gamma \subset C_G(y)$ and, obviously, 1) holds.

Put $n_1 = b$ and $n_2 = e$. Observe that the base $(\widetilde{v_1}, \dots, \widetilde{v_n})$ is almost hyperbolic. It follows from the previous arguments on almost hyperbolic bases that one can assume that $\omega(\widetilde{v_i}) = \varepsilon_i$ for $1 \leq i \leq n$, $\omega(w^j_k) = \varepsilon^j_k$ for $1 \leq k \leq n_j$ if $G = A_r(K)$ and otherwise the first equality holds for $i \leq r$ and the second one for $k \leq r(\Gamma_j)$, $j = 1, 2$. Hence actually $\langle \widetilde{v_i} \rangle = \langle v_i \rangle$ where v_i are the vectors of the base (2.2). Observe that our assumption determines the weights of all v_i and w^j_k as we deal with almost hyperbolic bases. In particular, we have $\omega(w^j_k) = -\varepsilon^j_{n_j+1-k}$ for $k > r(\Gamma_j)$ and $k \neq (n_j + 1)/2$ and $\omega(w^j_k) = 0$ if $k = (n_j + 1)/2$. This yields 2).

Now proceed to prove the claims of Item 3). Set

$$L_i = \langle v_{(i-1)p^s+1}, \dots, v_{(i-1)p^s+b} \rangle, \ 1 \leq i \leq a+1,$$
$$M_i = \langle v_{(i-1)p^s+b+1}, \dots, v_{ip^s} \rangle, \ 1 \leq i \leq a,$$

$$N_i = L_i \oplus M_i, \ Q_i = \bigoplus_{j=1}^{i} N_j, \ R_i = L_i \oplus Q_{i-1}, \ 1 \leq i \leq a$$

(we set $Q_0 = 0$ and have $M_i = 0$ if $e = 0$, and $R_1 = L_1$ in all cases). All arguments on M_i until the end of the proof concern the case where $e > 0$; if $e = 0$, we deal with the subspaces L_i only. Denote by $\overline{L_i}$ and $\overline{N_i}$ the images of the relevant subspaces under the canonical surjection $V \to V/Q_{i-1}$ and by $\overline{M_i}$ the image of M_i under such surjection $V \to V/R_i$; similarly define the vectors $\overline{v}_l$ for $\widetilde{v}_l \in N_i$. In fact, $\overline{L_1} = L_1$, but it is convenient to have a unified notation. It is clear that $L_1 = (y-1)^a V$ and N_1 is the fixed subspace of y. Hence $C_G(y)$ (that contains x) preserves L_1 and N_1. Using induction on i, we deduce that $C_G(y)$ preserves Q_i and R_i, $1 \leq i \leq a$. Indeed, assume that $i > 1$ and $C_G(y)$ preserves Q_{i-1}. It is clear that $\overline{N_i}$ is the fixed subspace of y acting on V/Q_{i-1} and $\overline{L_i} = (y-1)^{a-i+1}(V/Q_{i-1})$. Hence the image of $C_G(y)$ in $GL(V/Q_{i-1})$ preserves $\overline{L_i}$ and $\overline{N_i}$ and $C_G(y)$ and x preserve Q_i and R_i. So one can consider the action of x on $\overline{L_i}$ and $\overline{M_i}$. Set $l^i_j = \overline{v}_{(i-1)p^s+j}$ for $1 \leq i \leq a+1$, $1 \leq j \leq b$ and $m^i_k = \overline{v}_{(i-1)p^s+k+b}$ for $1 \leq i \leq a$, $1 \leq k \leq e$. Denote by x_{i1} and x_{i2} the transformations induced by x on $\overline{L_i}$ and $\overline{M_i}$, respectively, and by $\mathbf{x}_{i1}$ and $\mathbf{x}_{i2}$ the matrices of the transformations x_{i1} and x_{i2} in the bases $(l^i_1, \dots, l^i_b)$ and $(m^i_1, \dots, m^i_e)$. As $x \in C_G(y)$, we get that $\mathbf{x}_{i1} = \mathbf{x}_{11}$ and $\mathbf{x}_{i2} = \mathbf{x}_{12}$ for all i. If $G \neq A_r(K)$, define nondegenerate bilinear forms Φ_L on L_1 and Φ_M on $\overline{M_1}$ setting $\Phi_L(\widetilde{v_s}, \widetilde{v_t}) = \Phi^1_W(w_s, w_t)$ and $\Phi_M(\overline{v}_s, \overline{v}_t) = \Phi^1_W(w_{s-b}, w_{t-b})$. Since $x \in I(V)$, we deduce that $x_{11} \in I(L_1)$ and $x_{12} \in I(\overline{M_1})$. Conjugating x_{11} and x_{12} by suitable elements of $I(L_1)$ and $I(\overline{M_1})$, respectively, one can make the matrices $\mathbf{x}_{11}$ and $\mathbf{x}_{12}$ upper unitriangular. This implies that conjugating by a suitable element of $C_G(y)$ makes x upper unitriangular in the base $(v_1, \dots, v_n)$ (for $G = A_r(K)$ this is obvious). So we can and shall assume that x is such.

Define J and G_J as in Item 3) of the assertion of our Proposition. The construction of Γ implies that $\Gamma L_i = L_i$ and $\Gamma M_i = M_i$. Analyzing the action of the root subgroups of G on V, we deduce that $\Gamma \subset G_J$ and one can assume that $U^{\pm}(\Gamma) \subset U^{\pm}(G_J)$ (naturally, with the same sign "plus" or "minus" in both parts of this formula). Furthermore, one easily observes that a unipotent element $g \in C_G(y)$ such that $gL_i = L_i$, $gM_i = M_i$, $g|L_1 \in I(L_1)$, and $g|M_1 \in I(M_1)$, lies in Γ. By [**Ste68**, Corollary 2 of Lemma 18], x can be written in the form $x_\Gamma x'$ where $x_\Gamma = \prod_{\alpha \in R^+(G_J)} x_\alpha(t_\alpha)$, $x' = \prod_{\beta \notin R^+(G_J)} x_\beta(t_\beta)$, $t_\alpha, t_\beta \in K$. Observe that the action of x on $\overline{L_i}$ and $\overline{M_i}$ is completely determined by x_Γ. Now the arguments above on this action yield that $x_\Gamma \in \Gamma$. Since x is regular, $x_\Gamma|L_1$ and $x_\Gamma|M_1$ are regular unipotent elements in $I(L_1)$ and $I(M_1)$, respectively. This implies that x_Γ is a regular unipotent element of Γ.

Finally, assume that $|y| = p$. Then $a < p$. Set $\Delta = A_1(K)$, fix a maximal torus $T_\Delta \subset \Delta$. It follows from well-known facts of the representation theory of groups of type A_1 that there exist homomorphisms $\xi_i : \Delta \to I(U_i)$, $i = 1, 2$, such that ξ_1 and ξ_2, respectively, realize the representations $\varphi(a)$ and $\varphi(a-1) \in \mathrm{Irr}_\Delta$, the elements of $\xi_i(U^+(\Delta))$ and $\xi_i(U^-(\Delta))$ have upper and lower unitriangular matrices, respectively, in the bases (2.11) for U_1 and U_2, $\xi_i(T_\Delta)$ has the diagonal form in these bases, and $g_i = \xi_i(x_1(1))$. Then the analysis of the weight systems $\mathbf{X}(\varphi(a))$ and $\mathbf{X}(\varphi(a+1))$ shows that $\omega_\Delta(u_i^1) = a - 2(i-1)$ and $\omega_\Delta(u_i^2) = a - 1 - 2(i-1)$. Set $A' \subset St = \{(\xi_1(\delta) \otimes 1, \xi_2(\delta) \otimes 1) \mid \delta \in \Delta\}$ and $A = \mu(A')$. Take

$$T_1 = \mu(\{(\xi_1(t) \otimes 1, \xi_2(t) \otimes 1) \mid t \in T_\Delta\}).$$

Now the construction of the bases (2.11) and the subgroup Γ above implies that all assertions of Item 4) hold.

Now all our claims are proved. □

Throughout the text Γ, Γ_1, Γ_2, G_J, A, ρ, and ζ are such as in Proposition 2.43. We write weights $\lambda \in \mathbf{X}(\Gamma)$ in the form (λ_1, λ_2) with $\lambda_i \in \mathbf{X}(\Gamma_i)$, $i = 1, 2$.

COROLLARY 2.44. *In the assumptions of Proposition* 2.43 *suppose also that M is a G-module, $d_M(y) = g + 1$, $M_y = (y-1)^g M$, the subspace M_y contains a Γ-submodule N generated by a weight vector, and $d_N(x_\Gamma) = h$. Then $d_M(x) \geq gp^s + h$.*

PROOF. First we want to apply Lemma 2.42 and to get a required estimate for $d_{M_y}(x)$. In the assumptions of Lemma 2.42 put $\tilde{G} = G$ and $I = J$. Define Σ_J as in Lemma 2.42. Let $N = K\Gamma u$ for a weight vector u. Set $\mathbf{X}_1 = \{\omega(u) + \mu \mid \mu \in \Sigma_J\}$. Starting with this set $\mathbf{X}_1$, define $\mathbf{X}_2$, M_1, and M_2 as in the proof of Lemma 2.42. Since $\Gamma \subset G_J$, it is clear that $N \subset M_1$. Now one easily concludes that N contains a vector m with $(x_\Gamma - 1)^{h-1} m \neq 0$ that satisfies the assumptions of Lemma 2.42 with respect to G_J. By Proposition 2.43 3), $x = x_\Gamma x'$ where x' is a product of root elements associated with positive roots that do not lie in Σ_J. So Lemma 2.42 forces that $(x-1)^{h-1} m \neq 0$ and hence $d_{M_y}(x) \geq h$. Now apply Proposition 2.5 b). □

REMARK 2.45. In the assumptions of Proposition 2.43 let $G = C_r(K)$ and Γ_i be of type B_l for $i = 1$ or 2. Then $\Gamma_i \cong SO_{2l+1}(K)$. Hence $\mathbf{X}(\Gamma_i)$ consists of integer linear combinations of weights in $\mathcal{E}(\Gamma_i)$.

This follows at once from Formulae (2.10).

In the definition and Lemmas 2.46–2.49 below M is an indecomposable G-module with highest weight $\omega = \sum_{i=1}^{r} a_i\omega_i$, $v \in M$ is a nonzero highest weight vector.

DEFINITION 2.2. Let $1 \leq i, j \leq r$ and let all the roots α_t with t in the interval with the ends i and j form a chain. Assume that $0 < a_j < p$. Set $b_k = -\langle\alpha_{k+1}, \alpha_k\rangle$ and $c_k = -\langle\alpha_{k-1}, \alpha_k\rangle$. For an integer d with $0 < d \leq a_j$ define the vector $v(i, j, d)$ as follows. Put $d_j = d$. If $i < j$, set $d_k = a_k + d_{k+1}b_k$ for $i \leq k < j$. If $i > j$, put $d_k = a_k + d_{k-1}c_k$ for $i \geq k > j$. Now take $v(i, j, d) = v(i \cdot d_i, \dots, k \cdot d_k, \dots, j \cdot d)$. For $i = j$ put $v(i, j, d) = X_{-i,d}v$.

LEMMA 2.46. *We have $v(i, j, d) \neq 0$ and $X_{l,b}v(i, j, d) = 0$ for positive $l \neq i$ and $b > 0$. Hence $\mathcal{X}_l$ fixes $v(i, j, d)$.*

PROOF. One can obtain the inequality applying Lemma 2.1 several times. Other assertions of the lemma are contained in [**Sup97**, Lemma 2.9]. In that lemma only irreducible modules are considered, but the relevant arguments hold for indecomposable ones as well. □

LEMMA 2.47. ([**Sup97**, Lemma 2.10]) *Let M be irreducible and p-restricted. Put $m = v(i, j, d)$. If $i < j$, suppose that $i > 1$ and $\langle\alpha_k, \alpha_{k-1}\rangle = -1$ for $i \leq k \leq j$. If $i > j$, assume that $i < r$ and $\langle\alpha_k, \alpha_{k+1}\rangle = -1$ for $j \leq k \leq i$. Put $l = i - 1$ for $i < j$ and $l = i + 1$ for $i > j$. If $i = j$, suppose that $l \in \{i - 1, i + 1\}$ and $\langle\alpha_i, \alpha_l\rangle = -1$. Let $\omega_l(m) = p$. Then $X_{-l}m \neq 0$.*

LEMMA 2.48. *Let $1 \leq i < j < k \leq r$. Assume that $f_i, f_k \in \mathbb{Z}^+$, $0 < f_i \leq a_i$, $0 < f_k \leq a_k$, and that $a_i < p$ if $f_i < a_i$ and $a_k < p$ if $f_k < a_k$. Set $c_s = \langle\alpha_{s-1}, \alpha_s\rangle$ for $i < s \leq j$, $f_s = a_s - c_s f_{s-1}$ for $i < s < j$, $d_s = \langle\alpha_{s+1}, \alpha_s\rangle$ for $j \leq s \leq k$, $f_s = a_s - d_s f_{s+1}$ for $j < s < k$, and $f_j = a_j - c_j f_{j-1} - d_j f_{j+1}$. Put $m = v(j \cdot f_j, (j-1) \cdot f_{j-1}, \dots, i \cdot f_i, (j+1) \cdot f_{j+1}, \dots, k \cdot f_k)$. Then $m \neq 0$ and is fixed by $\mathcal{X}_s$ for $s \neq j$.*

PROOF. By Formula (2.1), $\mathcal{X}_s$ fixes m if and only if $X_{s,b}m = 0$ for all $b > 0$. Set $m_1 = v(j + 1, k, f_k)$ and $G_1 = G(1, 2 \dots, j)$. By Lemma 2.46, $m_1 \neq 0$ and is fixed by $U^+(G_1)$. One easily concludes that $M' = KG_1m_1$ is an indecomposable G_1-module with highest weight $\omega^1 = (\sum_{i=1}^{j-1} a_i\omega_i) + (a_j - f_{j+1}d_j)\omega_j$. We have $m = m_1(j, i, f_i)$ (in M'). Applying Lemma 2.46 to M', we deduce that $m \neq 0$ and is fixed by $\mathcal{X}_s$ for $s < j$.

Next, set $m^l = X_{j+1,l}m$ with $l > 0$. Assume that $m^l \neq 0$. Then $\omega_j(m^l) = -f_j + ld_j < -f_j$ and $\omega(w_j m^l) \notin \mathbf{X}(M)$ which yields a contradiction. Hence $m^l = 0$ as required.

Now let $j + 1 < s \leq r$. Using the commutator relations in Lemma 2.1 (ii), one easily observes that it suffices to show that $X_{s,l}m_1 = 0$ for $l > 0$. But this follows from Lemma 2.46 applied to m_1. This completes the proof. □

One easily observes that Lemmas 2.46–2.48 are valid in characteristic 2 as well.

LEMMA 2.49. *Let* $G = C_2(K)$ *and* ω *be* p*-restricted. Assume that* $m_1 = X_{-1,a_1+2a_2}X_{-2,a_2+c_1}X_{-1,c_1}v$ *with* $c_1 \leq a_1$, $m_2 = X_{-2,a_1+a_2}X_{-1,a_1+2c_2}X_{-2,c_2}v$ *with* $c_2 \leq a_2$, *and* $m_3 = X_{-1,a_1+2c_3-1}X_{-2,c_3}v$ *with* $0 < c_3 \leq a_2$. *Then* $m_i \neq 0$ *for* $i = 1, 2, 3$, $\mathcal{X}_2$ *fixes* m_1 *and* m_3, *and* $\mathcal{X}_1$ *fixes* m_2.

PROOF. Several applications of Lemma 2.1 (ii) yield that m_1 and $m_2 \neq 0$. We claim that $l_k = X_{2,k}m_1 = 0$ for $k > 0$. Assume this is false for some k. Then $l'_k = w_2w_1l_k \in M$ and $\omega(l'_k) \in \mathbf{X}(M)$. One can deduce that $\omega_1(l_k) = -(a_1 + 2a_2 + 2k)$, $\omega(w_1l_k) = \omega(l_k) + (a_1 + 2a_2 + 2k)\alpha_1$, and $\omega_2(w_1l_k) = -(a_2 + c_1)$. Hence $\omega(l'_k) = \omega(l_k) + (a_1 + 2a_2 + 2k)\alpha_1 + (a_2 + c_1)\alpha_2$ which yields a contradiction since $\omega(l_k) = \omega - (c_1 + a_1 + 2a_2)\alpha_1 + (a_2 + c_1 - k)\alpha_2$. So $l_k = 0$ as required and by Formula (2.1), $\mathcal{X}_2$ fixes m_1.

Similarly we prove that $\mathcal{X}_1$ fixes m_2: set $l_k = X_{1,k}m_2$ and show that $l_k = 0$ for $k > 0$ since otherwise $\omega(w_1w_2l_k) \notin \mathbf{X}(M)$.

Now consider m_3. Set $d = a_1 + 2c_3 - 1$. Lemma 2.1 (i) and (ii) yields that $X_2X_1m_3 = tX_{-1,d-1}X_{-2,c_3-1}v$ with $t \in K^*$ and so $X_2X_1m_3 \neq 0$. (Notice that $d - 1 = a_1 + 2c_3 - 2$.) Hence $m_3 \neq 0$. Set $X_{2,k}m_3 = u_k$ for $k > 0$. If $u_k \neq 0$ for some k, we have $\omega_1(u_k) = -2k - d + 1 < -d$. This yields a contradiction as $\omega(w_1u_k) \in \mathbf{X}(M)$. Hence $u_k = 0$ for all $k > 0$. This completes the proof. □

REMARK 2.50. Lemmas 2.46–2.49 do not depend upon our restrictions on r. They will be applied to subgroups $G(i_1, \ldots, i_t) \subset G$ as well. This was already done in the proof of Lemma 2.48.

The following lemma and proposition yield a weakened version of Proposition 1.5 for characteristic p.

LEMMA 2.51. *Let* λ_1 *and* $\lambda_2 \in \mathbf{X}^+(\tilde{G})$, $\omega = \lambda_1 + \lambda_2$, $M_j = M(\lambda_j)$, *and* $M = M(\omega)$. *Assume that* $x \in \tilde{G}$ *is a unipotent element. Then* $d_M(x) \leq d_{M_1}(x) + d_{M_2}(x) - 1$.

PROOF. Set $M' = M_1 \otimes M_2$. It is well known that M' has a composition factor isomorphic to M. Hence $d_M(x) \leq d_{M'}(x)$. To complete the proof, apply Corollary 2.13. □

PROPOSITION 2.52. ([**Sup96**, Proposition 2.15]) *Let* ω, λ_j, M_j $(j = 1, 2)$, *and* M *be as in Lemma* 2.51. *Suppose that* $x \in U^-(\tilde{G})$. *For each* $a \in \mathbb{Z}^+$ *set*

$$\mathbf{X}_a = \{\mu \in \mathbf{X}(M) \mid \omega - \mu = \sum_{i=1}^{\tilde{r}} c_i\tilde{\alpha}_i, \quad \sum_{i=1}^{\tilde{r}} c_i \geq a\}.$$

Let $v_j \in M_j$ *and* $v \in M$ *be nonzero highest weight vectors, and let* $(x-1)^{f_j}v_j \neq 0$ *for some* $f_j \in \mathbb{Z}^+$, $j = 1, 2$. *Assume that* $f = f_1 + f_2$, $\binom{f}{f_1} \not\equiv 0 (\mod \tilde{p})$, *the vectors* $(x-1)^{f_j}v_j$ *have nonzero weight components of weights* μ_j, *and* $\dim V(\omega)_\mu = \dim M_\mu$ *for* $\mu = \mu_1 + \mu_2$. *Then* $(x-1)^f v \neq 0$. *In particular, this inequality holds if* $\dim V(\omega)_\mu = \dim M_\mu$ *for all* $\mu \in \mathbf{X}_f$ *(for instance, if* $\mathbf{X}_f$ *consists of the lowest weight of* M *or* $V(\omega)$ *is irreducible).*

Notice that though [**Sup96**] is devoted to computing the minimal polynomials of elements of order p, this proposition was proved there for arbitrary unipotent elements.

PROPOSITION 2.53. *Let Δ_1 and Δ_2 be semisimple algebraic groups over $\tilde{K}$, $\Delta = \Delta_1 \times \Delta_2$, and let U be an indecomposable Δ-module generated by a nonzero highest weight vector u. Set $\mu = \omega(u)$, $\mu_i = \omega_{\Delta_i}(u)$ for $i = 1, 2$, and $U_1 = \tilde{K}\Delta_1 u$. Assume that $U_1 \cong V(\mu_1)$ (as a Δ_1-module). Then U has a quotient isomorphic to $V(\mu_1) \otimes M(\mu_2)$.*

PROOF. Put $V_\Delta = V(\mu)$, $V_i = V(\mu_i)$, and $M = M(\mu_2)$. By the universal property of the Weyl modules [**Jan03**, Part II, Lemma 2.13], $U = V_\Delta/N$ where $N \subset V$ is a Δ-submodule. Using [**Ste68**, §12, Corollary a) of Lemma 68] and dimension arguments, one also concludes that $V_\Delta \cong V_1 \otimes V_2$ and a nonzero highest weight vector $v \in V_\Delta$ can be identified with $m_1 \otimes m_2$ where $m_i \in V_i$ are nonzero highest weight vectors. We have $M = V_2/N_2$ where $N_2 \subset V_2$ is the maximal Δ_2-submodule. Set $V_1' = V_1 \otimes m_2$. It is clear that $U_1 \cong V_1'/V_1' \cap N$ (as Δ_1-modules). Since $U_1 \cong V_1$, we get $V_1' \cap N = 0$. It suffices to prove that $N \subset V_1 \otimes N_2$. Indeed, in this case $V_1 \otimes M \cong V_1 \otimes V_2/V_1 \otimes N_2 \cong (V_1 \otimes V_2/N)/(V_1 \otimes N_2/N)$. Suppose $N \not\subset V_1 \otimes N_2$. Let $x_1, \dots, x_d$ and $y_1, \dots, y_f$ be bases of V_1 and V_2, respectively, that consist of weight vectors. Assume also that for all $\mu \in \mathbf{X}(V_2)$ the sequence $y_1, \dots, y_f$ contains bases of $(N_2)_\mu$. Then there exists a vector $z \in N$ such that $z = \sum_{i,j} b_{ij} x_i \otimes y_j$ where $b_{ij} \in K$ and for some k, l the coefficient $b_{kl} \neq 0$ and $y_l \notin N_2$. Let

$$\Omega_t = \{\nu \in \mathbf{X}(V_2) \mid \mu_2 - \nu \quad \text{is a sum of } t \text{ roots of } \quad \Pi(\Delta_2)\}, \; J_t = \{j \mid \omega(y_j) \in \Omega_t\}.$$

Choose maximal t such that for some $l \in J_t$ and some k the coefficient $b_{kl} \neq 0$ and $y_l \notin N_2$. Fix these k and l. Observe that N contains weight components of its vectors. In particular, if a linear combination $a \otimes m_2 + \sum_j u_j \otimes w_j \in N$, $a, u_j \in V_1$, $w_j \in V_2$ are weight vectors and $\omega(w_j) < \mu_2$, then $a \otimes m_2 \in N$. Hence $t > 0$ since $V_1' \cap N = 0$. Set $\Theta = \{q \mid \omega(y_q) = \omega(y_l), b_{kq} \neq 0\}$ and $u = \sum_{q \in \Theta} b_{kq} y_q$. Let $\overline{u}$ and $\overline{m_2}$ be the images of u and m_2 under the canonical homomorphism $V_2 \to M$. Then $\overline{u} \neq 0$ due to our assumptions on the base $(y_1, \dots, y_f)$. Since only highest weight vectors of M are fixed by the group $U^+(\Delta_2)$, there exists an operator $X = X_{\beta_1, d_1} \dots X_{\beta_s, d_s}$ where $\beta_1, \dots, \beta_l \in R^+(\Delta_2)$ such that $X\overline{u} = cm_2$ where $c \in \tilde{K}^*$. As the operator X sends distinct weight subspaces to distinct ones, it is not difficult to deduce that Xz has a nonzero weight component collinear to $x_k \otimes m_2$. This yields a contradiction since $Xz \in N$. Hence $N \subset V_1 \otimes N_2$ as required. □

Now we collect some facts on representations of the group $A_1(\tilde{K})$ we need and complete this section of preliminary results.

LEMMA 2.54. ([**Sup97**, Lemma 2.6]) *Let $\Delta = A_1(\tilde{K})$, $a \in \mathbb{Z}^+$, $M = M(a)$, and $V_a = V(a)$.*

i) We have $V_a = \langle X_{\alpha,d} v \mid 0 \leq d \leq a \rangle$ where $\{\alpha\} = R^-(\Delta)$ and $v \in V_a$ is a nonzero highest weight vector; $\dim V_a = a + 1$. In particular, $X_{\alpha,d} v \neq 0$ for $0 \leq d \leq a$.

ii) Let $a < \tilde{p}$, $x \in \Delta$ be a nonidentity positive root element, and $m \in M$ be a nonzero highest weight vector. Then $\dim M = a + 1$ and $(x-1)^a M = \langle m \rangle$.

iii) Let $a = \sum_{j=0}^k a_j \tilde{p}^j$ with all $a_j < \tilde{p}$. Then $\dim M = \prod_{j=0}^k (a_j + 1)$ and

$$d_M(x) = \min\{\tilde{p}, \; 1 + \sum_{j=0}^k a_j\}.$$

PROOF. The formula for $d_M(x)$ in iii) is the unique assertion of the lemma which was not formulated explicitly in [**Sup97**, Lemma 2.6]. This formula follows from Theorem 2.2 and Lemma 2.8. □

LEMMA 2.55. *Let $\Delta = A_1(\tilde{K})$ and let N be an indecomposable Δ-module of highest weight $\tilde{p}+b$ with $0 \le b < \tilde{p}-1$. Assume that $v \in N$ is a highest weight vector and $X_\alpha^{b+1}v \neq 0$ for $\alpha \in R^-(\Delta)$. Then $N \cong V(\tilde{p}+b)$.*

PROOF. Set $b_1 = \tilde{p}+b$ and $b_2 = p - \tilde{b} - 2$. By the universal property of the Weyl module [**Jan03**, Part II, Lemma 2.13 b)], N is a quotient of $V(b_1)$. It follows from [**CC76**](and can be easily deduced from the weight structure of $V(b_1)$) that $V(b_1)$ has two composition factors: $M(b_1)$ and $M(b_2)$. Theorem 2.2 forces that $b_2 \notin \mathbf{X}(M(b_1))$. However, $b_2 \in \mathbf{X}(N)$ as $X_\alpha^{b+1}v \neq 0$. This implies that $N \ncong M(b_1)$ and completes the proof. □

LEMMA 2.56. *Let $\Delta = A_1(\tilde{K})$, $x \in U^+(\Delta)$, $x \neq 1$, $b \in \mathbb{Z}^+$, and $b < \tilde{p}-2$. Set $N = V(\tilde{p}+b)$. Then $d_N(x) = \tilde{p}$ and $(x-1)^{\tilde{p}-1}N = N_{\tilde{p}+b-2}$.*

PROOF. Put $M = M(\tilde{p}+b)$ and $N_x = (x-1)^{\tilde{p}-1}N$. By Lemma 2.54 iii), $d_M(x) = b+2 < \tilde{p}$. It follows from the arguments in the proof of Lemma 2.55 that $M \cong N/M_1$ where $M_1 \cong M(\tilde{p}-b-2)$. Hence $N_x \subset M_1$. Put $N^+ = \langle N_a \mid a \geq \tilde{p}-b-2\rangle$. Let $v \in N$ be a nonzero highest weight vector and α be the negative root of Δ. By Lemma 2.54 i), $N = \langle X_{\alpha,d}v \mid 0 \le d \le \tilde{p}+b\rangle$ and $X_{\alpha,d}v \neq 0$ for these d. Lemma 2.1 i) forces that $X_{-\alpha}^{\tilde{p}-1}X_{\alpha,\tilde{p}+b}v = (\tilde{p}-1)!X_{\alpha,b+1}v \neq 0$ and that $N_x \subset N^+$. This yields that $N_x \neq 0$ and hence $N_x = N_{\tilde{p}+b-2}$ as $M_1 \cap N^+ = N_{\tilde{p}+b-2}$. □

3. The general scheme of the proof of the main results

This section contains a discussion of the general scheme mentioned above and several lemmas that are heavily used for treating unipotent elements with a single Jordan block. We proceed to prove Theorems 1.1, 1.3, 1.7, 1.9, and 1.10 simultaneously using induction on the group rank and the order of an element. For $G \neq A_r(K)$ we also use the relevant results for $A_i(K)$. More exactly, we prove these results for elements of order $p^{s+1} > p$ in G assuming that the following Conjecture (r,s) holds.

Conjecture (r,s). Theorems 1.1, 1.3, 1.7, 1.9, and 1.10 are valid for classical groups of rank $< r$ and for unipotent elements in G of order $\le p^s$. If $G \neq A_r(K)$, we also assume that these results hold for unipotent elements of order $\le p^{s+1}$ in $A_{n-1}(K)$.

The induction base is given by Theorem 2.32 which yields Conjecture $(r,1)$. Notice that Theorem 1.9 does not include the case where $|x| = p$.

The additional assumption for $G \neq A_r(K)$ in Conjecture (r,s) causes no confusion. Notice that the proofs of the main results for $G = A_r(K)$ do not depend upon groups of other series. Actually, we could first write these proofs and then consider other groups, but we find more convenient to unify approaches and required technical lemmas when it is possible.

Now we fix some notation that will be used throughout the text or its major part. In what follows $x \in G$ is unipotent, $|x| = p^{s+1} > p$ for fixed s, except Lemma 3.1 (see comments before that lemma), $y = x^{p^s}$, $\varphi \in \mathrm{Irr}$, $\omega = \omega(\varphi) = \sum_{i=1}^r a_i\omega_i$, $M = M(\omega)$, $v \in M$ is a nonzero highest weight vector. If $d_M(y) = d$, set $M_y = (y-1)^{d-1}M$. Naturally, we can assume that $\omega \neq 0$. For $1 \leq f \leq g \leq r$ set $\omega(f,g) = \sum_{i=f}^{g} a_i\omega_i$. For $h \leq r-g$ put also $\omega_+(f,g,h) = \sum_{i=f}^{g} a_{i+h}\omega_i$. Throughout the text $\omega(f,g)$ and $\omega_+(f,g,h)$ are often considered as formal expressions for weights of some explicitly determined subgroup $\Delta \subset G$, i.e. some weight in $\mathbf{X}(\Delta)$ is a linear combination of the fundamental weights with the coefficients connected with a_i in such way. This subgroup is always clear from the context. For a unipotent element u or a unipotent conjugacy class C of a simple classical group $\mathcal{G}$ over K and a representation $\lambda \in \operatorname{Irr}\mathcal{G}$ denote by $d(u,\lambda)$ or $d(C,\lambda)$, respectively, the value of the degree of the minimal polynomial of $\lambda(u)$ ($\lambda(C)$) given by Theorems 1.1, 1.3, 1.7, 1.9, and 1.10. In fact, these theorems hold for $\mathcal{G}$ if and only if $d_\lambda(u) = d(u,\lambda)$ for each unipotent $u \in \mathcal{G}$. To prove them for G, we shall show that in all cases $d_\varphi(x) = d(x,\varphi)$. If x is regular, then $z_j \in G_{\mathbb{C}}$ are unipotent elements with $J(z_j) = J(x^{p^j})$, and we set $d_j = d_j(\varphi) = d_{\varphi_{\mathbb{C}}}(z_j)$, $0 \leq j \leq s$. Recall that for regular x Theorems 1.7 and 1.10 give the same value of $d(x,\varphi)$ according to Lemma 2.31. In what follows the groups Γ, Γ_1, Γ_2 and A and the homomorphisms ρ and ζ are such as in Proposition 2.43. Recall that x_Γ is a regular unipotent element of Γ. Write $x_\Gamma = g_1g_2$ where $g_i \in \Gamma_i$ are regular unipotent elements, $i = 1,2$. Our unified proof is subdivided into the following cases.

1. p-large representations.

2. p-restricted representations and elements with a single Jordan block on V (those are regular unipotent elements for the groups of types A_r, B_r, and C_r).

3. The general case ($J(x) \neq (n)$ or $\varphi \notin \operatorname{Irr}_p$).

Cases 2 and 3 have their own subdivisions almost common for both of them. These subdivisions are connected with relations between the size k_1 of the maximal block in $J(x)$ and $|x|$ and basically are as follows:

a) $k_1 = p^s + a$ with $0 < a < p$;

b) $k_1 = p^s + b$, $p \leq b < 2p$, $G = B_r(K)$ or $D_r(K)$, $a_r \neq 0$ for $G = B_r(K)$, and $a_{r-1} + a_r \neq 0$ for $G = D_r(K)$;

c) $k_1 = p^{s+1}$;

d) the exceptional cases in Theorems 1.7 and 1.10;

e) cases covered by Theorem 1.9;

f) the general case in Theorems 1.7 and 1.10.

In many situations where we need to prove that $d_\varphi(x) = |x|$ for a unipotent element $x \in G$ and $\varphi \in \operatorname{Irr}$ the approach is as follows: we find an element $g \in \overline{\operatorname{cl}(x)}$ such that $|g| = |x|$ and g lies in a subgroup $\Delta \subset G$ which is a classical group of smaller rank, then we construct an irreducible quotient π of $\varphi|\Delta$ such that Conjecture (r,s) yields that $d_\pi(g) = |g|$ and apply Lemma 2.19 to complete the argument.

In particular, arguments of such nature allow us to reduce the analysis of p-restricted p-large representations to the following cases: $n = k_1 = p^s+1$ or p^s+2, and $k_1 = k_2 = r = p^s+1$, $G = D_r(K)$ (here k_2 is the size of the second Jordan block in $J(x)$).

By Corollary 2.41, $d_\varphi(x) \leq d(x,\varphi)$ for regular x. So for this element it suffices to show that $d_\varphi(x) \geq d(x,\varphi)$. Arguments of the following kind are typically used both in Case 1 and in Case 2. Let $x \in G$ be a regular unipotent element of order p^{s+1}. We need to prove that $d_\varphi(x) = ap^s + b$ with $a, b \in \mathbb{Z}^+$, $0 < b \leq p^s$, and we know that $d_\varphi(x) \leq ap^s + b$. Using Theorem 1.14, we get $d_\varphi(y) = a+1$ (other value of $d_\varphi(y)$ would yield a contradiction by Proposition 2.5 a)). Since $\Gamma \subset C_G(y)$, the subspace M_y is a $K\Gamma$-module. Using Proposition 2.43, one readily concludes that $M_y|\Gamma$ has a composition factor ψ with highest weight $\rho(\omega)$. Obviously, the ranks of simple components of Γ are less than r. Using Conjecture (r,s) and Theorem 2.9, we find out that $d_\psi(x_\Gamma) = b$. Then Corollary 2.44 is applied to complete the proof.

Case 2 appears to be the central one. Here we deal with unipotent elements that do not lie in some naturally embedded proper semisimple subgroups. So Conjecture (r,s) can be applied only indirectly as in the arguments described above that are connected with using Lemma 2.19 and Proposition 2.43.

The following results are used heavily in Case 2. In Lemma 3.1, to get a required fact in all cases we need it and to avoid additionary complicated notation, we include groups of ranks 1 and 2 for all types considered. So here a regular unipotent element may have order p and then $s = 0$, contrary to the general assumption made at the beginning of this section.

LEMMA 3.1. *Let $G \neq D_r(K)$, $x \in G$ be a regular unipotent element, and $\varphi \in \mathrm{Irr}_p$. Assume that $r \geq 1$ for all types considered. Put $n_0 = 3$ for $G = B_1(K)$, 5 for $G = B_2(K)$, and $n_0 = n$ otherwise. Set $\Omega = \{0, \omega_1, \omega_r\}$ for $G = A_r(K)$ or $G = B_r(K)$ with $r = 2, 3$, $\Omega = \{0, \omega_1, 2\omega_1\}$ for $G = B_1(K)$, and $\Omega = \{0, \omega_1\}$ otherwise. Let $\omega \notin \Omega$. Then $d(x,\varphi) > n_0$. If $s > 0$, $p^{s+1} - p < n_0 \leq p^{s+1}$, and $\omega \notin \Omega$, one has $d(x,\varphi) = p^{s+1}$.*

PROOF. Proposition 1.5 implies that it suffices to consider the fundamental weights $\omega \notin \Omega$, the cases where $\omega = 2\omega_i \notin \Omega$ with $\omega_i \in \Omega$, and the weight $\omega = 3\omega_1$ for $G = B_1(K)$ and $p > 3$. Set $s + 1 = t$. Then $t \geq 1$ and $p^{t-1} < n_0 \leq p^t$. Write the p-adic expansion $n_0 = \sum_{l=0}^{u} b_l p^l$ with $u = t-1$ or t and $b_u \neq 0$. Observe that for $u = t$ one has $b_u = 1$ and $b_j = 0$ if $j < u$. For $0 < j \leq u$ set $n_j^- = \sum_{l=0}^{j-1} b_l p^l$ and $n_j^+ = \sum_{l=j}^{u} b_l p^{l-j}$. By Proposition 2.5 b), if $j > 0$, then z_j has n_j^- Jordan blocks of size $n_j^+ + 1$ and $p^j - n_j^-$ blocks of size n_j^+ on V. Next, apply Algorithm 1.6 and Proposition 1.5 and obtain the following:

a) $t = 1$ and $d_0 = 4$ for $G = B_1(K)$ and $\omega = 3\omega_1$ with $p > 3$;

b) if $G = B_2(K)$ and $\omega = 2\omega_2$, then $d_0 = 7$, for $p > 3$ one has $t = 1$, and for $p = 3$ we get $t = 2$ and $d_1 = 3$;

c) if $G = B_3(K)$ and $\omega = 2\omega_1$, then $d_0 = 13$, for $p > 5$ one has $t = 1$, if $p \leq 5$, then $t = 2$, $d_1 = 3$ for $p = 5$ and 5 for $p = 3$;

d) in other cases where $\omega = 2\omega_i$ with $\omega_i \in \Omega$ we have $d_0 = 2n_0 - 1$; $d_j = 2p^{t-j} - 1$ for $j > 0$ and $n_0 = p^t$; if $n_0 < p^t$ and $j > 0$, then either $d_j \geq 2b_{t-1}p^{t-1-j} + 1$, or $b_{t-1} > 1$ and $d_j \geq 2b_{t-1}p^{t-1-j} - 1$;

e) if $G = B_r(K)$ with $r \geq 4$ and $\omega = \omega_r$, then $d_0 > 3r$ for $r > 4$ and $d_0 = 11$ for $r = 4$; for $p = 3$ we have $d_1 \geq 2b_{t-1}3^{t-2} + 4$ if $n_0 \neq 3^t$ or $b_{t-1}3^{t-1} + b_0$, $d_1 \geq 2 \times 3^{t-2}$ if $n_0 = 3^{t-1} + 2$, $d_1 \geq 4 \times 3^{t-2} - 1$ if $n_0 = 2 \times 3^{t-1} + 1$, and $d_1 \geq 2 \times 3^{t-1} - 2$ for $n_0 = 3^t$; in other cases for $j > 0$ one has $d_j \geq 2b_u p^{u-j}$ (to check this, consider the cases $p \geq 5$ and $p = 3$, $j > 1$ separately);

f) in all other situations where $\omega = \omega_i \notin \Omega$ one gets $d_0 \geq 2n - 3 > n_0$, for $j > 0$ we have $d_j \geq 2b_u p^{u-j} - 1$ and $d_j \geq 2b_u p^{u-j}$ if $n_0 \neq b_u p^u$.

Observe that $n_0 > p^j$ if $n_0 = b_u p^u$ and $j < t$. If $n_0 > p^t - p$ with $t > 1$, then $b_j = p - 1$ for $j > 0$. Furthermore, if in this case $n_0 = 2r + 1$, then $3r > p^t$ or $p = 3$ and $n_0 = 7$. Now the assertion of the lemma follows from a)–f) since $p > 2$. $\square$

LEMMA 3.2. *Let G and x be as in Lemma 3.1 and $\varphi \in \mathrm{Irr}$. For $G = A_r(K)$ assume that $a_i = 0$ if $p \leq i \leq n - p$, otherwise suppose that $a_i = 0$ for $i \geq p$. Assume also that $|x| > p$. Then $p^l(d_l - 1) \leq p^k(d_k - 1)$ for $k < l$.*

PROOF. Keep the notation introduced in the proof of the previous lemma for the p-adic expansion of n_0. Set $D_j = d_j - 1$ for $0 \leq j < t$. First assume that $\omega = \omega_i$ with $i < p$. Let $i < r$ or $G \neq B_r(K)$. Algorithm 1.6 and Proposition 2.5 b) yield that $D_0 = i(n_0 - i)$; if $n_0 = p^t$, then $D_j = i(p^{t-j} - 1)$ for $0 < j < t$; otherwise we have $D_j = in_j^+$ for $n_j^- \geq i$ and $D_j = b_0 n_j^+ + (i - b_0)(n_j^+ - 1)$ if $n_j^- < i$. Observe that the latter formula holds if and only if $n_j^- = b_0 < i$. Obviously, $n_k^- \leq n_l^-$ for $k < l$. Now consider different possibilities for D_k and D_l in turn and check directly that in all cases $p^l D_l \leq p^k D_k$ as desired.

As $d_j(\varphi) = d_j(\varphi^*)$, the result just proved yields the assertion of the lemma for $G = A_r(K)$ and $\omega = \omega_m$ with $m > n - p$.

Next, let $i = r$ and $G = B_r(K)$. Then $n_0 = p + b$ with $0 < b = 2c \leq p - 1$ and $r = c + (p-1)/2$. Hence $D_0 = (c + (p-1)/2)(c + (p+1)/2)/2 > pc = pD_1$ since $c \leq (p-1)/2$. Now all the possibilities for fundamental representations have been considered.

Apply Proposition 1.5 to complete the proof. □

COROLLARY 3.3. *In the assumptions of Lemma 3.2 let $d_{t-1} > p$. Then $d(x, \varphi) = |x|$.*

PROOF. By Lemma 3.2, in this case $p^j d_j > |x|$ for all j. This yields the corollary. □

Now return to Case 3. Here we also start with $\varphi \in \mathrm{Irr}_p$. If $k_1 = p^{s+1}$, we prove Theorem 1.3 using the fact that it holds in Case 2 and Lemma 2.19. Next, assume that $k_1 < p^{s+1}$. Construct the groups H_j^p, H_j, H^p, and $H = H(x)$ and the homomorphism $\theta : \mathbf{X} \to \mathbf{X}(H)$ just as in the Introduction. Set $\lambda = \theta(\omega(\varphi))$. In Proposition 2.27 we have constructed a subgroup $S = S(x) \subset G$ such that S is isomorphic to a quotient of H^p by a central subgroup and regular unipotent elements of S lie in $\mathrm{cl}(x)$. By Corollary 2.30, $\varphi|S$ has a composition factor $\chi = \varphi(\lambda)$. Using Conjecture (r, s) and Theorem 2.9, we deduce that when the pair (x, φ) does not satisfy the assumptions of Theorem 1.9, then almost always $d_\chi(x) = d(x, \varphi)$. (Obviously, the ranks of simple components of S are less than r since now x has more than one Jordan block.) In exceptional cases special arguments are used.

It remains to show that $d_\varphi(x) \leq d(x, \varphi)$. Here the arguments are based on the analysis of the explicit realizations of the fundamental representations and their restrictions to S and on Proposition 1.5 and Lemmas 2.38 and 2.39. The treatment of all cases where $\varphi \notin \mathrm{Irr}_p$ is based on Theorem 2.2 and Results 2.8–2.18.

Proposition 2.5, Lemma 2.19, reduction to groups of smaller ranks, arguments on centralizers similar to those of Proposition 2.43, Theorem 2.2 and Results 2.8–2.18 can be applied for computing the minimal polynomials of unipotent elements in representations of the classical groups in characteristic 2 and of the exceptional groups as well.

4. p-large representations

This section is devoted to Theorem 1.1. We assume that Conjecture (r, s) holds and prove Theorem 1.1 for elements of order $p^{s+1} > p$ in G.

Some comments on the scheme of the proof. In fact, this proof is based on reduction to special elements and ranks and Lemma 2.19 as we mentioned in Section 3. Actually we reduce the task to considering regular elements in groups of type A_r, B_r, and C_r with $n = p^s + 1$ or $p^s + 2$. For this purpose we observe that in all other cases for each element x of order p^{s+1} the set $\overline{\mathrm{cl}(x)}$ contains some specific element w of the same order that lies in a naturally embedded subgroup Δ of rank $r-1$. Furthermore, we have $\Delta \cong A_{r-1}(K)$ for $G = A_r(K)$, $B_{r-1}(K)$ or $A_{r-1}(K)$ for $G = B_r(K)$, $C_{r-1}(K)$ for $G = C_r(K)$, and $A_{r-1}(K)$ or $B_{r-1}(K)$ for $G = D_r(K)$. This will be done in Lemma 4.1. Then we deal with the exceptional values of n and p-restricted representations (Results 4.4–4.8). Here x is a regular unipotent element of G and y is a root element. By Theorem 1.9, $d_M(y) = p$ for a p-large module M. In the majority of cases the arguments for these r and x are based on applying Propositions 2.5 and 2.43 and Corollary 2.44. It is crucial here that for the values of n under consideration the group Γ constructed in Proposition 2.43 is isomorphic to $SL_{p^s-1}(K)$ or $Sp_{p^s-1}(K)$ or has a normal subgroup isomorphic to $Spin_{p^s-2}(K)$ and that for almost all representations of the latter three groups their regular unipotent elements have the minimal polynomials of degree p^s in these representations (see Lemmas 4.2 and 4.3). So in the notation of Proposition 2.43 we have $a = p - 1$ and almost always we can show that $b = p^s$ using Lemma 2.42. However, these arguments fail in several specific cases and we have to apply Proposition 2.52 and to elaborate some special approaches. Next, we pass to bigger ranks and to the group $D_r(K)$, show that $\varphi|H$ has a p-large composition factor for all p-large $\varphi \in \mathrm{Irr}_p$ and complete the proof for such φ applying Lemma 2.19 as we planned. This is done in Proposition 4.9. Finally, in Lemma 4.10 we pass to arbitrary p-large representations using the Steinberg tensor product theorem.

LEMMA 4.1. *Let $x \in G$ be an element of order p^{s+1}. Then $\overline{\mathrm{cl}(x)}$ contains an element w with $J(w) = (p^s + 1, 1, \dots, 1)$ for $G = A_r(K)$ or $C_r(K)$ and $J(w) = (p^s + 2, 1, \dots, 1)$ or $(p^s + 1, p^s + 1, 1, \dots, 1)$ for $G = B_r(K)$ or $D_r(K)$ (here in all cases the number of one-dimensional blocks in $J(w)$ can be zero). Furthermore, set $G_0 = G(2, \dots, r)$ for $G = A_r(K)$, $B_r(K)$, and $C_r(K)$; $G_1 = G(1, \dots, r-1)$ for $G = B_r(K)$ and $D_r(K)$; $G_2 = G(1, \dots, r-2, r)$ for $G = D_r(K)$ and denote by G_3 a naturally embedded subgroup $B_{r-1}(K)$ in $D_r(K)$. Then one can choose w in G_0 if $n > p^s + 1$ and $G = A_r(K)$ or $C_r(K)$, in G_0 or G_1 if $n > p^s + 2$ and $G = B_r(K)$, and in G_1, G_2, or G_3 for $G = D_r(K)$.*

PROOF. The first claim could be deduced from the description of the inclusion relations between the conjugacy classes of G in [**Spa82**, Theorems I.2.4 and II.8.2], but the whole assertion of the lemma requires some comments. Let $J(x) = (k_1, k_2, \dots)$. By Corollary 2.28, $\overline{\mathrm{cl}(x)}$ contains an element x_1 with $J(x_1) = (k_1, 1, \dots, 1)$ if k_1 is proper, and $J(x_1) = (k_1, k_2, 1, \dots, 1)$ if k_1 is improper for G (the number of ones in $J(x_1)$ can be zero). Set $\Sigma = G(r + 2 - k_1, \dots, r)$ for $G = A_r(K)$; $\Sigma = G(r + 1 - m, \dots, r)$ for $G = B_r(K)$, $k_1 = 2m + 1$ and for $G = C_r(K)$, $k_1 = 2m$; $\Sigma = G(1, 2, \dots, k_1 - 1)$ for $G = B_r(K)$ or $C_r(K)$ and improper k_1 and for $G = D_r(K)$ and improper $k_1 < r$; and $\Sigma = G(r - m, \dots, r)$ for $G = D_r(K)$, $k_1 = 2m + 1$. If $G = D_r(K)$ and r is even, the arguments in [**Sup96**, Item 2.1] yield that regular unipotent elements of the groups G_1 and G_2 are not conjugate in G. If in this case $k_1 = r$, Lemma 1.4 ii) implies that $\mathrm{cl}(x_1)$ contains regular unipotent elements of G_i for $i = 1$ or 2; denote the relevant group G_i by Σ. In all other situations $\mathrm{cl}(x_1)$ is completely determined by $J(x_1)$ and hence, analyzing the action of regular unipotent elements of Σ on V, we can conclude that they lie in $\overline{\mathrm{cl}(x_1)}$. Now [**Car85**, Proposition 5.1.2] implies that in all cases $\overline{\mathrm{cl}(x_1)}$ contains all unipotent elements of Σ.

Using Formulae (2.3), it is not difficult to deduce that the group Σ contains an element w with required $J(w)$. To complete the proof of the first claim of the lemma, it suffices to notice that $\overline{\mathrm{cl}(x)} \supset \overline{\mathrm{cl}(x_1)}$. Now assume that $n > p^s + 1$ for $G = A_r(K)$ or $C_r(K)$ and $n > p^s + 2$ for $G = B_r(K)$. Using Lemma 1.4, one easily concludes that an element w can be chosen inside G_0, G_1, G_2, or G_3 as required in the assertion of the lemma for the relevant case. □

To consider the special cases where $n = p^s + 1$ or $p^s + 2$, we need to know the minimal polynomials of regular unipotent elements in irreducible representations of the groups $SL_{p^s-1}(K)$, $Sp_{p^s-1}(K)$, and $Spin_{p^s-2}(K)$.

LEMMA 4.2. *Let $l_1 = p^s - 2$, $l_2 = (p^s - 1)/2$, $\mathcal{G} = A_{l_1}(K)$ or $C_{l_2}(K)$, and let $g \in \mathcal{G}$ be a regular unipotent element. Assume that $\mu \in \mathrm{Irr}\,\mathcal{G}$ and that Conjecture (r, s) holds. Then $d_\mu(g) = p^s$ unless $\omega(\mu) \in \{0, p^a\omega_1, p^a\omega_{l_1}\}$ for $\mathcal{G} = A_{l_1}(K)$ and $\omega(\mu) \in \{0, p^a\omega_1\}$ for $\mathcal{G} = C_{l_2}(K)$.*

PROOF. It is well known that $J(g) = (p^s - 1)$. Proposition 2.5 b) implies that for $0 < i < s$ the element g^{p^i} has at least $p - 1 \geq 2$ Jordan blocks of size p^{s-i} in the standard realization of $\mathcal{G}$. Now Proposition 1.5, Algorithm 1.6, and Formula (1.4) imply that if μ is not an exceptional representation mentioned in the assertion of the lemma, then $d(g, \mu) = p^s$. It remains to apply Conjecture (r, s). □

LEMMA 4.3. *Let $l = (p^s - 3)/2 \geq 2$, $\mathcal{G} = B_l(K)$, and let $g \in \mathcal{G}$ be a regular unipotent element. Assume that $\mu \in \mathrm{Irr}\,\mathcal{G}$ and that Conjecture (r, s) holds. Then $d_\mu(g) = p^s$ unless $\omega(\mu) \in \{0, p^a\omega_1\}$, or $l = 2$ and $\omega(\mu) = 7^a\omega_2$, or $l = 3$ and $\omega(\mu) = 3^a\omega_3$.*

PROOF. Observe that now $J(g) = (p^s - 2)$ and apply Propositions 1.5 and 2.5, Algorithm 1.6, and Formula (1.4) just as in the proof of Lemma 4.2. □

The following lemma deals with a special case arising in the proof of Proposition 4.5.

LEMMA 4.4. *Let $G = A_r(K)$, $n = p^s + 1$, $s > 0$. Assume that $\varphi \in \mathrm{Irr}_p$ and $\omega = a_1\omega_1 + a_r\omega_r$ with $a_1 + a_r = p$. Then $d_\varphi(x) = p^{s+1}$ for a regular unipotent element $x \in G$.*

PROOF. Naturally, we can suppose that $x \in U^-$. The proof is based on several applications of Proposition 2.52. First set $\lambda = a\omega_1$ with $a < p$, consider the module $M_0 = M(\lambda)$, and prove that $d_{M_0}(x) = ap^s + 1$ and $(x - 1)^{ap^s} v_0 \neq 0$ for a nonzero highest weight vector $v_0 \in M_0$. Apply induction on a. It is clear that the claim holds for $a = 0$ or 1. Suppose that $a > 1$ and the claim holds for $a - 1$. In the assumptions of Proposition 2.52 set $\lambda_1 = (a - 1)\omega_1$, $\lambda_2 = \omega_1$, $f_1 = (a - 1)p^s$, and $f_2 = p^s$. Then $f = f_1 + f_2 = ap^s$. Using Lemma 2.7, one can conclude that $\binom{bp^s}{dp^s} \not\equiv 0 (\mod p)$ for $0 < d \leq b < p$, $d, b \in \mathbb{Z}^+$. It is well known (and can be easily deduced from [**Sei87**, 1.14]) that $V(a\omega_1)$ is irreducible for $a < p$. Put $M_1 = M((a - 1)\omega_1)$ and $M_2 = M(\omega_1)$. Let $m_i \in M_i$ be nonzero highest weight vectors. Then $(x - 1)^{(a-1)p^s} m_1 \neq 0$ and $(x - 1)^{p^s} m_2 \neq 0$ by our assumptions. Now

Proposition 2.52 forces that $(x-1)^{ap^s}v_0 \neq 0$. As M_0 is a quotient of $M_1 \otimes M_2$, we get $d_{M_0}(x) \leq ap^s+1$ by Corollary 2.13. Hence $d_{M_0}(x) = ap^s+1$ as required.

Next, apply Proposition 2.52 to another pair λ_1 and λ_2. Now set

$$\lambda_1 = a_1\omega_1,\ \lambda_2 = a_r\omega_r,\ M_i = M(\lambda_i),\ f_1 = a_1p^s,\ f_2 = a_rp^s - 1$$

and keep the notation m_i. We have $f = f_1 + f_2 = p^{s+1} - 1$. Actually it has been proved in the previous paragraph that $(x-1)^{f_1}m_1 \neq 0$. Passing to the dual module and using the facts proved above, one easily deduces that $d_{M_2}(x) = a_rp^s+1 = f_2+2$. For a G-module N let $d(N)$ be as in Definition 2.1. It is not difficult to observe that $d(M_2) = a_rp^s = f_2+1$. Hence by Lemma 2.20, $(x-1)^{f_2+1}m_2 \neq 0$ and so $(x-1)^{f_2}m_2 \neq 0$. Set $\mathbf{X}_a = \{\mu \in \mathbf{X} \mid \omega - \mu = \sum_{i=1}^r b_i\alpha_i,\ \sum_{i=1}^r b_i = a\}$. Since $f = d(M) - 1$, it is clear that $\dim M_\mu = \dim V(\omega)_\mu$ for all $\mu \in \mathbf{X}_a$ with $a \geq f$. By Lemma 2.7, $\binom{f}{f_1} \not\equiv 0 (\mod p)$. Hence all assumptions of Proposition 2.52 hold and that proposition forces $(x-1)^f v \neq 0$ as required. □

The proofs of Propositions 4.5, 4.7, and 4.8 and Lemma 4.4 below are based on Propositions 2.5 and 2.43. In these proofs β is the maximal root of G.

Proposition 4.5. *Let $n = p^s + 1$ and $G = A_r(K)$ or $C_r(K)$. Assume that $\varphi \in \mathrm{Irr}_p$ and is p-large. Then $d_\varphi(x) = p^{s+1}$ for all $x \in G$ of order p^{s+1}.*

Proof. Observe that in our situation only regular unipotent elements have order p^{s+1}. By Proposition 2.5 b), we have $J(y) = (2,1,\dots,1)$. Hence $\Gamma_1 = 1$, $\Gamma_2 = A_{p^s-2}(K)$ for $G = A_r(K)$ and $C_{(p^s-1)/2}(K)$ for $G = C_r(K)$. Assume that $x \in U^+$. Then $y \in \mathcal{X}_\beta$. Recall that $\beta = \sum_{i=1}^r \alpha_i$ for $G = A_r(K)$ and $\beta = (2\sum_{i=1}^{r-1}\alpha_i) + \alpha_r$ for $G = C_r(K)$. Observe that $\langle\alpha_1,\beta\rangle = 1$ for both choices of G, $\langle\alpha_r,\beta\rangle = 1$ for $G = A_r(K)$, and in all other cases $\langle\alpha_i,\beta\rangle = 0$. One easily concludes that under our assumptions $\Gamma = G(2,\dots,r-1)$ for $G = A_r(K)$ and $\Gamma = G(2,\dots,r)$ for $G = C_r(K)$. Since φ is p-large and p-restricted, there exist $j < r$ and $d \in \mathbb{Z}^+$ with $0 < d \leq a_j$ such that

$$a_j - d + \sum_{i=j+1}^{r} a_i = p - 1. \tag{4.1}$$

Put $m = v(1,j,d)$ and $\mu = \omega_\Gamma(m)$. Lemma 2.46 shows that the vector $m \neq 0$ and generates an indecomposable Γ-module $N = K\Gamma m$ with highest weight μ. Set $f = r-2$ for $G = A_r(K)$ and $f = r-1$ for $G = C_r(K)$. We have $r \geq 3$ for $G = A_r(K)$. One easily deduces that

$$\mu = (a_2 + d)\omega_1$$

for $G = A_3(K)$, $j = 1$ and for $G = C_2(K)$,

$$\mu = (a_2 + d)\omega_1 + \omega_+(2, f, 1)$$

in other cases with $j = 1$,

$$\mu = (a_1 + a_2 - d)\omega_1$$

for $G = A_3(K)$, $j = 2$, and

$$\mu = \omega(1, j-2) + (a_{j-1} + a_j - d)\omega_{j-1} + (d + a_{j+1})\omega_j + \omega_+(j+1, f, 1)$$

otherwise (here for $G = A_r(K)$ and $j = r-1$ the last summand is $(a_{r-2} + a_{r-1} - d)\omega_{r-2}$). It is clear that $\mathcal{X}_\beta$ preserves m, and (4.1) yields that $\langle\omega(m),\beta\rangle = p-1$. We can take $A = G(\beta)$. Put $\nu = \varphi(\mu) \in \mathrm{Irr}\,\Gamma$. Applying Lemma 2.54 to the group

A, we obtain that $d_M(y) = p$ and $m \in M_y$. Hence $N \subset M_y$ as $\Gamma \subset C_G(y)$. By Proposition 2.5 b), it suffices to show that $d_{M_y}(x) = p^s$. Corollary 2.44 implies that $d_{M_y}(x) \geq d_{M_y}(x_\Gamma) \geq d_N(\Gamma)$. Obviously, N has a composition factor ν. Hence $d_\varphi(x) = p^{s+1}$ if $d_\nu(x_\Gamma) = p^s$. Lemma 4.2 implies that $d_\nu(x_\Gamma) = p^s$, except the following cases:

1) $G = A_r(K)$, $\omega = a_1\omega_1 + a_r\omega_r$, $a_1 + a_r = p$;
2) $G = A_r(K)$, $\omega = a_1\omega_1 + a_2\omega_2 + a_r\omega_r$, $j = 1$, $a_2 + d = p$;
3) $G = C_r(K)$, $\omega = a_1\omega_1 + a_2\omega_2$, $j = 1$, $a_2 + d = p$;
4) $G = A_r(K)$, $\omega = \omega_{r-2} + (p-1)\omega_r$, $r > 3$;
5) $G = A_r(K)$, $\omega = a_{r-2}\omega_{r-2} + a_{r-1}\omega_{r-1} + a_r\omega_r$, $a_{r-2} + a_{r-1} = p$, $a_{r-1} + a_r = p - 1$;
6) $G = A_r(K)$, $\omega = a_1\omega_1 + a_{r-1}\omega_{r-1} + (p-1)\omega_r$, $a_1 \in \{0, 1\}$, $a_{r-1} \neq 0$;
7) $G = A_r(K)$, $\omega = a_{r-1}\omega_{r-1} + (p-2)\omega_r$;
8) $G = A_r(K)$, $\omega = a_{r-2}\omega_{r-2} + a_{r-1}\omega_{r-1} + a_r\omega_r$, $j = r-1$, $a_{r-2} + a_{r-1} - d = p$, $a_{r-2} = a_r + 1$.

Now we conclude that $d_\varphi(x) = p^{s+1}$ or one of Cases 1)–8) holds. It is clear that $d_\varphi(x) = d_{\varphi^*}(x)$. Observe that for $G = A_r(K)$ we have $r \geq 3$ and $p^s = 3$ if $r = 3$. Hence passing to the dual representation allows us to eliminate Case 4) and Cases 5) and 8) for $r > 3$ and to make some other reductions. As $\varphi \in \mathrm{Irr}_p$, in Case 8) there is the unique possibility for $r = 3$: $\omega = 2\omega_1 + 2\omega_2 + \omega_3$.

Case 1 was handled in Lemma 4.4.

Consider Cases 2) and 3) and Case 5) with $r = 3$. Observe that $a_2 \neq 0$. Set $m_1 = X_{-2}m$. Lemmas 2.1 and 2.47 imply that $m_1 \neq 0$ and $X_2 m_1 = 0$. Hence $U^+(\Gamma)$ fixes m_1. First assume that $r(\Gamma) = 1$. Then $p = 3$, $G = A_3(K)$ or $C_2(K)$, $\Gamma \cong A_1(K)$, and N is an indecomposable Γ-module with highest weight 3. By Lemma 2.55, $N \cong V(3)$. Set $g = x_{-2}(1)$. It follows from Lemma 2.54 i) that $X_{-2}^2 m \neq 0$. Hence $(g-1)^2 m \neq 0$ and $d_N(g) = 3 = p$. As $g \in \mathrm{cl}(x_\Gamma)$, this forces $d_N(x_\Gamma) = 3 = p$ and $d_\varphi(x) = 9$ as required.

Mow let $r(\Gamma) > 1$. Then m_1 generates an indecomposable Γ-module with highest weight $\mu = (p-2)\omega_1 + \omega_2$ and so N has a composition factor $F = M(\mu)$. Now Lemma 4.2 implies that $d_N(x_\Gamma) = d_F(x_\Gamma) = p^s$ as desired.

Next, let $G = A_3(K)$ and $\omega = 2\omega_1 + 2\omega_2 + \omega_3$. Observe that $p = 3$ in this case. Set $m' = X_{-1}^2 X_{-3} v$ and $N' = K\Gamma m'$. Applying Lemmas 2.46 and 2.54 and the arguments used for the vector m constructed before, we conclude that N' is an indecomposable Γ-module with highest weight 5 and $N' \subset M_y$. Since N' has a composition factor $M(5)$, Lemma 4.2 implies that $d_{N'}(x_\Gamma) = 3$ as desired. This settles Case 8).

Finally, let Case 6) or 7) hold. Set $\omega^* = \omega(\varphi^*)$. We have $\omega^* = (p-1)\omega_1 + a_2\omega_2$, or $(p-1)\omega_1 + a_2\omega_2 + \omega_r$, or $(p-2)\omega_1 + a_2\omega_2$ with $a_2 \neq 0$ in all cases. Now it is clear that φ^* does not satisfy the assumptions of Cases 6) and 7) and hence $d_{\varphi^*}(x) = p^{s+1}$ by previous arguments. So $d_\varphi(x) = p^{s+1}$ as required. This completes the proof. □

LEMMA 4.6. *Let $G = B_r(K)$, $n = p^s + 2$, and $x \in U^+$ be a regular unipotent element. Assume that $\varphi \in \mathrm{Irr}_p$ is p-large. Then the group Γ constructed in Proposition 2.43 equals $G(1, 3, \ldots, r)$ and $d_\varphi(x) = p^{s+1}$ if $d_{M_y}(x_\Gamma) = p^s$.*

PROOF. Using the arguments in the proof of Proposition 2.5 b), one concludes that $y \in \mathcal{X}_\beta$ and $J(y) = (2, 2, 1, \dots, 1)$. Now the construction of Γ in Proposition 2.43 implies that

$$\Gamma_1 = G(1) \cong A_1(K), \quad \Gamma_2 = G(3, 4, \dots, r) \cong B_{r-2}(K), \quad \text{and} \quad \Gamma = G(1, 3, \dots, r).$$

Applying Corollary 2.44, we deduce the second assertion of the lemma. □

For the group G satisfying the assumptions of Lemma 4.6 we shall write a weight $\mu \in \mathbf{X}(\Gamma)$ in the form (a, μ_2) where $a = \langle \mu, \alpha_1 \rangle$ and μ_2 is the restriction of μ to Γ_2. By [**Ste68**, Corollary a) of Lemma 68], a representation $\chi \in \operatorname{Irr}\Gamma$ with highest weight $\mu = (a, \mu_2)$ is equivalent to the tensor product of the representations $\varphi(a) \in \operatorname{Irr}\Gamma_1$ and $\varphi(\mu_2) \in \operatorname{Irr}\Gamma_2$. Put $\chi_1 = \varphi(a) \in \operatorname{Irr}\Gamma_1$ and $\chi_2 = \varphi(\mu_2) \in \operatorname{Irr}\Gamma_2$. For a weight vector $m \in M$ set $\omega_\beta(m) = \langle \omega(m), \beta \rangle$. Recall that $\beta = \alpha_1 + 2\sum_{i=2}^{r} \alpha_i$ and $\langle \omega, \beta \rangle = a_1 + 2(\sum_{i=2}^{r-1} a_i) + a_r$. This notation will be used until the end of the section.

PROPOSITION 4.7. *Let $G = B_r(K)$, $n = p^s + 2$ and $\omega = (p-1)\omega_1 + a_2\omega_2$ with $a_2 \neq 0$. Then $d_\varphi(x) = p^{s+1}$ for a regular unipotent element $x \in G$.*

PROOF. For $a_2 > 1$ put $t_1 = v(r, 2, a_2 - 1)$ and $t = X_{-2}^{a_2} X_{-3}^{a_2-1} \dots X_{-(r-1)}^{a_2-1} t_1$. (We have $t = X_{-2}^{a_2} t_1$ for $r = 3$.) If $a_2 = 1$, set $t = X_{-2}v$. Lemmas 2.1 and 2.46 and [**Sup97**, Lemma 4.2] yield that $t \neq 0$ and the groups $\mathcal{X}_i$ with $i > 0$ and $i \neq 2$ fix t. Observe that $\omega_\beta(t) = p$, $\omega_\Gamma(t) = (p - 2 + 2a_2, \omega_1)$ for $r > 3$ and $\omega_\Gamma(t) = (p - 2 + 2a_2, 2\omega_1)$ if $r = 3$. We claim that $X_{-\beta}t \neq 0$.

Until the end of the proof for vectors $l, m \in M$ or endomorphisms l, m of M we write $l = sc \cdot m$ if $l = cm$ for some $c \in K^*$. We need this notation (connected with the phrase "some constant") to avoid introducing many different constants whose exact values are not needed.

For $1 \leq i \leq r$ let s_i be an involution in the group $G(i)$ that normalizes T and maps onto the reflection w_i under the canonical homomorphism $N_G(T) \to W$. Write $w = s_2 \dots s_{r-1} s_r s_{r-1} \dots s_2$, and let $w' \in W$ be the image of w under the homomorphism mentioned above. One can directly check that $w = w^{-1}$ and $w'\alpha_1 = \beta$. Using [**Ste68**, Lemma 19], one gets $X_{-\beta} = sc \cdot wX_{-1}w$ and $s_2 X_{-1} = sc \cdot X_{-(\alpha_1+\alpha_2)} s_2$ (as endomorphisms of M). We shall need the following assertion.

(*) Let $m \in M \backslash \{0\}$, $\alpha \in R$, $\mathcal{X}_\alpha$ fixes m, and $\langle \omega(m), \alpha \rangle = a > 0$. Then $s_\alpha m = sc \cdot X_{-\alpha,a} m$.

Indeed, by the universal property of the Weyl module [**Jan03**, Part II, Lemma 2.13b)], the $G(\alpha)$-module generated by m is a quotient of the Weyl module $V(a)$. One may assume that m is the image of a highest weight vector $m' \in V(a)$ under the canonical homomorphism onto this module. Using the construction of the Chevalley groups in [**Ste68**, § 3] and [**Ste68**, Theorem 39], we get the required equality for m' and then transfer it to m.

Observe that our vector t is obtained from $t' = X_{-2,a_2-1}v$ in the result of a series of the following steps: one has a vector l such that $\mathcal{X}_i$ fixes l for some i and $\omega_i(l) = a$ and then puts $l_1 = X_{-i,a}l$. Using this observation and the assertion (*), we deduce that $wt = sc \cdot s_2 t'$. By [**Ste68**, Lemma 19], $\omega(s_2 t') = \omega - \alpha_2$. Since $\dim M_{\omega-\alpha_2} = 1$ and $t' \neq 0$, we get $s_2 t' = sc \cdot X_{-2}v$. Now it is clear that $X_{-\beta}t = sc \cdot wX_{-1}X_{-2}v \neq 0$ by Lemma 2.47.

Set $X_{-\beta}t = m$. As $\langle \beta, \alpha_i \rangle = 0$ for $i \neq 2$, the weight $\omega_\Gamma(m) = \omega_\Gamma(t)$ and $X_{-\beta}$ commutes with Γ. Hence the groups $\mathcal{X}_i$, $i \neq 2$, fix m. It is clear that $\mathcal{X}_\beta$

fixes t. Now Lemmas 2.55 and 2.56 imply that the vector t generates the Weyl module $V(p)$ with respect to the group $G(\beta)$ and $m \in M_y$. Put $N = K\Gamma m$. Then $N \subset M_y$ as $\Gamma \subset C_G(y)$. Set $\mu = \omega_\Gamma(t)$ and $\chi = \varphi(\mu) \in \operatorname{Irr}\Gamma$. It is clear that $d_{\chi_2}(g_2) = p^s - 2$. If $d_{\chi_1}(g_1) > 2$, Corollary 2.13 forces $d_\chi(x_\Gamma) = p^s$. Since $p - 2 + 2a_2 < 3p \leq p^2$, Lemma 2.54 yields that $d_{\chi_1}(g_1) > 2$ if $a_2 > 1$. Let $a_2 = 1$. Then $t = X_{-2}v$. Put $q = X_{-1}m = X_{-1}X_{-\beta}t$. We claim that $q \neq 0$. Indeed, put $q_1 = s_2X_{-1}X_{-2}v$. As $a_2 = 1$, we get $s_2v = sc \cdot X_{-2}v$. So the arguments above show that $q_1 = sc \cdot X_{-(\alpha_1+\alpha_2)}v$. Then

$$q = sc \cdot X_{-1}wX_{-1}X_{-2}v = sc \cdot X_{-1}X_{-2}^2X_{-3}\dots X_{-(r-1)}X_{-r}^2X_{-(r-1)}\dots X_{-3}q_1$$

for $r > 3$ and $q = scX_{-1}X_{-2}^2X_{-3}^2q_1$ if $r = 3$. Using commutator relations in L, we first get that

$$X_2q = sc \cdot X_{-1}X_{-2}X_{-3}\dots X_{-(r-1)}X_{-r}^2X_{-(r-1)}\dots X_{-3}q_1$$

for $r > 3$ and $X_2q = sc \cdot X_{-1}X_{-2}X_{-3}^2q_1$ if $r = 3$, then conclude that

$$X_3\dots X_{r-1}X_r^2X_{r-1}\dots X_3X_2q = sc \cdot X_{-1}X_{-2}q_1$$

for $r > 3$ and $X_3^2X_2q = scX_{-1}X_{-2}q_1$ for $r = 3$. Hence it suffices to show that $X_{-1}X_{-2}q_1 \neq 0$. Put $q_2 = X_{-2}q_1$. We claim that $q_2 \neq 0$. Indeed, using commutator relations in L and taking into account that $\omega_2(q_1) = 0$, we get that $X_2q_2 = sc \cdot X_{-2}X_{-1}v \neq 0$ by Lemma 2.1. Therefore $q_2 \neq 0$. Set $\Delta = G(1,2)$. By Theorem 2.33, v generates an irreducible Δ-module M_Δ with highest weight $(p-1)\omega_1 + \omega_2$. It is clear that $q_2 \in M_\Delta$ and is not a lowest weight vector in M_Δ. As M_Δ is p-restricted, [**Bor70**, Lemma 6.2 and Theorem 6.4] yield that $X_{-\alpha}q_2 \neq 0$ for some $\alpha \in R^+(\Delta)$ ([**Bor70**, Lemma 6.2] deals with $R^+(\Delta)$ and highest weight vectors, but one can replace $R^+(\Delta)$ by $R^-(\Delta)$ and the highest weight by the lowest one). Now it follows from the commutator relations in $L(\Delta)$ that $X_{-i}q_2 \neq 0$ for $i = 1$ or 2. One easily observes that $\omega(X_{-2}q_2) \notin \mathbf{X}(\varphi)$. Hence $X_{-2}q_2 = 0$ and $X_{-1}q_2 \neq 0$. Thus we have deduced the desired inequality $X_{-1}X_{-2}q_1 \neq 0$ which forces $q \neq 0$. By Lemma 2.55, $K\Gamma_1m \cong V(p)$ (as a Γ_1-module). Now Proposition 2.53 yields that N has a quotient module N_1 isomorphic to $V(p)\otimes M(\lambda)$ where $\lambda \in \mathbf{X}(\Gamma_2)$, $\lambda = \omega_1$ if $r > 3$ and $\lambda = 2\omega_1$ for $r = 3$. It follows from Lemma 2.56 that g_1 has the minimal polynomial of degree p in the module $V(p)$. Then Corollary 2.13 forces $d_{N_1}(x_\Gamma) = p^s$ as $x_\Gamma = g_1g_2$. Hence in all cases $d_N(x_\Gamma) = p^s$. Thus $d_{M_y}(x_\Gamma) = p^s$. Now Lemma 4.6 yields that $d_\varphi(x) = p^{s+1}$. □

PROPOSITION 4.8. *Let $G = B_r(K)$, $n = p^s + 2$, and let $\varphi \in Irr_p$ be p-large. Assume that Conjecture (r,s) holds. Then $d_\varphi(x) = |x|$ for each $x \in G$ of order p^{s+1}.*

PROOF. Lemma 1.4 iii) implies that it suffices to consider regular unipotent elements since only they have order p^{s+1}. Obviously, one can assume that $x \in U^+$. Below the arguments in the proof of Lemma 4.6 and the notation introduced just after that lemma are used. Our goal is to prove that $d_\varphi(x) = p^{s+1}$. Recall that $y \in \mathcal{X}_\beta$. We shall construct a nonzero weight vector $m \in M$ such that the groups $\mathcal{X}_i$ with $i > 0$ and $i \neq 2$ and $\mathcal{X}_\beta$ fix m and $\omega_\beta(m) = p - 1$. By Lemma 2.54, $m \in M_y$. Set $N = K\Gamma m$ and $N_2 = K\Gamma_2 m$. We have $N \subset M_y$ as $\Gamma \subset C_G(y)$. Lemma 4.6 implies that it suffices to show that $d_N(x_\Gamma) = p^s$. It is clear that N is an indecomposable Γ-module with highest weight $\mu = \omega_\Gamma(m)$. Set $\chi = \varphi(\mu) \in \operatorname{Irr}\Gamma$.

For the weight μ and the representation χ define a, μ_2, χ_1 and χ_2 as in the comments after Lemma 4.6. Obviously, $d_N(x_\Gamma) \geq d_\chi(x_\Gamma)$. Observe that $d_\chi(x_\Gamma) \geq d_{\chi_2}(g_2)$.

Our construction of the vector m can be naturally subdivided into 3 subcases.

I. Let $\sum_{i=1}^{r-1} a_i > p-1$. Then there exist $j, b \in \mathbf{Z}^+$, $1 < j < r$, $0 < b \leq a_j$ such that

$$a_j - b + \sum_{i=1}^{j-1} a_i = p - 1. \tag{4.2}$$

Put $w = v(r,j,b)$, $c_{r-1} = \omega_{r-1}(w)$, $c_l = \omega_l(w) + c_{l+1}$ for $2 \leq l < r-1$, $m = X_{-2,c_2} \dots X_{-(r-1),c_{r-1}} w$. Applying [**Sup97**, Lemma 4.2], one can conclude that $m \neq 0$ and the groups $\mathcal{X}_\beta$ and $\mathcal{X}_i$ with $i > 0$ and $i \neq 2$ fix m, verify that $\omega_\beta(m) = p-1$, and find $\omega_\Gamma(m)$. We have

$$a = b + a_r + \sum_{i=1}^{j} a_i + 2\sum_{i=j+1}^{r-1} a_i; \tag{4.3}$$

(4.4)

$$\mu_2 = \omega_+(1, j-3, 1) + (a_{j-1} + b)\omega_{j-2} + (a_j - b + a_{j+1})\omega_{j-1} + \omega_+(j, r-2, 2)$$
$$\text{for} \quad 2 < j < r-1;$$
$$\mu_2 = \omega_+(1, r-4, 1) + (a_{r-2} + b)\omega_{r-3} + (2a_{r-1} - 2b + a_r)\omega_{r-2}$$
$$\text{for} \quad 2 < j = r-1;$$
$$\mu_2 = (a_2 - b + a_3)\omega_1 + \omega_+(2, r-2, 2) \quad \text{for} \quad j = 2 < r-1;$$

and $\mu_2 = (2a_2 - 2b + a_3)\omega_1$ for $r = 3$.

Recall that $r \geq 3$. It is clear that $a_{j-1} + b < 2p$ and $a_j - b + a_{j+1} < 2p$. If $\mu_2 = 0$, Formulae (4.2) and (4.4) yield that $j = 2$ and $\omega = (p-1)\omega_1 + a_2\omega_2$. This case has been handled in Proposition 4.7. So we assume that $\mu_2 \neq 0$. Observe that $p = 5$ and $s = 1$ if $r = 3$.

First let $r > 3$. Then Formulae (4.2) and (4.4) and Lemmas 4.2 and 4.3 imply that $d_{\chi_2}(g_2) = p^s$ unless one of the following holds:

1) $\omega = (p-1)\omega_1 + a_2\omega_2 + \omega_3$;
2) $\omega = a_1\omega_1 + a_2\omega_2 + a_3\omega_3$, $a_1 + a_2 = p-1$, $a_2 + a_3 = p$;
3) $\omega = (p-2)\omega_1 + a_2\omega_2$, $a_2 > 1$;
4) $\omega = a_1\omega_1 + a_2\omega_2 + (a_1+1)\omega_3$, $j = 2$, $a_2 - b + a_1 = p-1$;
5) $p = 3$, $r = 5$, $\omega = 2\omega_1 + a_2\omega_2 + \omega_5$, $a_2 \neq 0$;
6) $p = 7$, $r = 4$, $\omega = 6\omega_1 + a_2\omega_2 + \omega_4$, $a_2 \neq 0$.

Using Conjecture (r,s) and Algorithm 1.6, we get that $d_{\chi_2}(g_2) \geq p^s - 2$ in Cases 1)–5) and $d_{\chi_2}(g_2) = 4$ in Case 6). If $d_{\chi_1}(g_1) > 2$ in Cases 1)–5) and > 3 in Case 6), Corollary 2.13 yields that $d_\chi(x_\Gamma) = p^s$ as desired. Formula (4.3) implies that $a = p+1+2a_2$, $a_1 + a_3 + p$, $p - 3 + 2a_2$, $3a_1 + 2 + a_2 + b$, $2a_2 + 3$, and $2a_2 + 7$ in Cases 1), 2), 3), 4), 5), and 6), respectively. It is clear that $a < p^2$ in Cases 1)–3), 5), 6) and in Case 4) with $p > 3$. If $p = 3$ in Case 4), we get $a_1 = b = 1$, $a_2 = 2$, and $a = 8$ as φ is p-restricted. Then Lemma 2.54 forces that in all our cases $d_{\chi_1}(g_1) > 2$ and $d_{\chi_1}(g_1) > 3$ in Case 6).

Now let $r = 3$. Corollary 2.13 implies that $d_\chi(x_\Gamma) = 5$ if $d_{\chi_i}(g_i) \geq 3$ for $i = 1, 2$. Observe that a and $2a_2 - 2b + a_3 < p^2$ (take into account Formula (4.3) for a). By (4.2), we have $a_1 + a_2 - b = 4$. Hence $a > 5$. Then Formula (4.4) and Lemma 2.54

imply that it suffices to consider the situations where $2a_2 - 2b + a_3 = 1$ or 5 (recall that we have assumed that $\mu_2 \neq 0$). This yields the following cases:

i) $\omega = 4\omega_1 + a_2\omega_2 + \omega_3$;

ii) $\omega = 3\omega_1 + a_2\omega_2 + 3\omega_3$;

iii) $\omega = 2\omega_1 + a_2\omega_2 + \omega_3$.

Recall that Case i) with $a_3 = 0$ was handled in Proposition 4.7. We have $a = 4 + 2b + a_3$, so a is odd in all these cases as a_3 is odd. Observe that $a \geq 7$ in Cases i) and iii) and $a \geq 9$ in Case ii). Then Lemma 2.54 implies that in all these situations $d_\chi(x_\Gamma) = 5$ and completes the analysis of Case I).

II. Now let $\sum_{i=1}^{r-1} a_i \leq p-1$, but $\sum_{i=1}^{r} a_i \geq p$. Then we have $p - 1 = a_r - b + \sum_{i=1}^{r-1} a_i$ where $0 < b \leq a_r$. Take $m = v(2, r, b)$. Lemma 2.46 implies that $m \neq 0$ and the groups $\mathcal{X}_i$ fix m for $i > 0$ and $i \neq 2$. It is clear that $\mathcal{X}_\beta$ fixes m. One easily deduces that $\omega_\beta(m) = p - 1$, $a = b + \sum_{i=1}^{r-1} a_i$, and

$$\mu_2 = \omega_+(1, r-3, 1) + (2a_{r-1} + a_r)\omega_{r-2}. \tag{4.5}$$

Arguing as in Part I of the proof, we conclude that $d_\chi(x_\Gamma) = p^s$ if $d_{\chi_1}(g_1) + d_{\chi_2}(g_2) > p^s$. So we can exclude such situations. It is clear that $a_r \neq 0$, $a < 2p - 1$, and $2a_{r-1} + a_r < p^2$. As φ is p-large, one has $a > 1$. Lemma 4.3, Algorithm 1.6, and Conjecture (r, s) imply that $d_{\chi_2}(g_2) = p^s$ unless $r = 4$ or 5 with $2a_{r-1} + a_r = 1$ or p and $\omega = a_1\omega_1 + a_{r-1}\omega_{r-1} + a_r\omega_r$, or $r = 3$. Furthermore, one has $p^s = 7$ and $d_{\chi_2}(g_2) = 4$ in the exceptional case with $r = 4$ and $p^s = 9$ and $d_{\chi_2}(g_2) = 7$ in that with $r = 5$.

First let $r = 3$. Applying Corollary 2.13 as in Part I, we conclude that it suffices to consider the situations where $d_{\chi_i}(g_i) < 3$ for $i = 1$ or 2. Now Lemma 2.54 implies that it remains to consider the cases where either $a = 5$ or $2a_2 + a_3 = 5$ (if $2a_2 + a_3 = 1$, one gets $a_2 = 0$, $a_3 = b = 1$, $a_1 = 4$, and $a = 5$). If $a = 5$, Lemma 2.47 yields that $X_{-1}m \neq 0$. Then $K\Gamma_1 m \cong V(p)$ by Lemma 2.55. Applying Proposition 2.53 and [**Sup97**, Lemma 2.11] and arguing as in the proof of Proposition 4.7, we conclude that $d_N(g_1) = 5 = p^s$. Hence $d_N(x_\Gamma) = p^s$ as desired. Next, let $2a_2 + a_3 = 5 \neq a$. Then a_3 is odd. As $a_1 + a_2 + a_3 - b = 4$ and is even, $a_1 + a_2 - b$ and a are odd. Now Lemma 2.54 forces that $d_{\chi_1}(g_1) \geq 4$ and so $d_\chi(x_\Gamma) = 5$ by Corollary 2.13.

Next, let $r = 4$ or 5, $2a_{r-1} + a_r = 1$ or p, and $\omega = a_1\omega_1 + a_{r-1}\omega_{r-1} + a_r\omega_r$. Arguing as for $r = 3$, we conclude that a_r, $a_1 + a_{r-1} - b$ and a are odd. Hence $a \neq 2$ or 8. If $a = p$, applying Lemmas 2.47 and 2.55 and Proposition 2.53 and arguing as for $r = 3$, one can deduce that $K\Gamma_1 m \cong V(p)$ and $d_N(g_1) = d_N(x_\Gamma) = p^s$ in this case.

Now assume that $a \neq p$. Since $1 < a < 2p - 1$, Lemma 2.54 and Corollary 2.13 yield that $d_\chi(x_\Gamma) = p^s$. This completes Part II of the proof.

III. Finally, let $\sum_{i=1}^{r} a_i \leq p - 1$. Then we have

$$p - 1 = \left(\sum_{i=1}^{j} a_i\right) + a_j - b + 2\left(\sum_{i=j+1}^{r-1} a_i\right) + a_r \tag{4.6}$$

for some j, $2 \leq j < r$ and some $b \in \mathbf{Z}^+$, $0 < b \leq a_j$. Put $m = v(2, j, b)$. By Lemma 2.46, $m \neq 0$ and the groups $\mathcal{X}_i$ fix m for $i > 0$ and $i \neq 2$. It is clear that $\mathcal{X}_\beta$ also fixes m. One easily observes that $\omega_\beta(m) = p - 1$,

$$a = b + \sum_{i=1}^{j} a_i, \tag{4.7}$$

(4.8)

$$\mu_2 = \omega_+(1, j-3, 1) + (a_{j-1} + a_j - b)\omega_{j-2} + (a_{j+1} + b)\omega_{j-1} + \omega_+(j, r-2, 2) \quad \text{for} \quad 3 \leq j < r-1,$$

$$\mu_2 = (a_3 + b)\omega_1 + \omega_+(2, r-2, 2) \quad \text{for} \quad j = 2, \quad r > 3,$$

$$\mu_2 = \omega_+(1, r-4, 1) + (a_{r-2} + a_{r-1} - b)\omega_{r-3} + (a_r + 2b)\omega_{r-2} \quad \text{for} \quad 2 < j = r-1,$$

and $\mu_2 = (a_3 + 2b)\omega_1$ for $r = 3$.

As before, we can exclude the situations where $d_{\chi_1}(g_1) + d_{\chi_2}(g_2) > p^s$. First assume that $r > 3$. Formula (4.6) implies that $a_3 + b < p$ if $j = 2$. Then Formulae (4.8), Lemma 4.3, Algorithm 1.6, and Conjecture (r, s) yield that $d_{\chi_2}(g_2) = p^s$ unless one of the following holds:

1) $\omega = a_1\omega_1 + a_2\omega_2$, $a_1 + 2a_2 = p$;

2) $r = 4$, $p^s = 7$, $a_2 = 0$, $2a_3 + a_4 = 7$, $a_1 + a_3 + a_4 = 6$;

3) $r = 5$, $p^s = 9$, $a_2 = a_3 = 0$, $2a_4 + a_5 = 3$, $a_1 + a_4 + a_5 = 2$.

We have $d_{\chi_2}(g_2) = p^s - 2$ in Cases 1) and 3) and 4 in Case 2). By (4.7), $a = a_1 + b$ in Case 1), $a_1 + a_3$ in Case 2), and $a_1 + a_4$ in Case 3). Hence $a < p$ in all these cases. In Case 1) $a_1 > 0$ as a_1 is odd and so $a > 1$ and $d_{\chi_1}(g_1) > 2$ by Lemma 2.54. This allows us to eliminate Case 1). In Case 2) a_4 is odd and so either $a \geq 3$ and $d_{\chi_1}(g_1) > 3$ by Lemma 2.54, or $\omega = \omega_3 + 5\omega_4$. The first possibility yields $d_N(x_\Gamma) = 7$. In Case 3) we get $\omega = \omega_4 + \omega_5$ as $\varphi \in \mathrm{Irr}_p$.

Now let $r = 4$ and $\omega = \omega_3 + 5\omega_4$. Recall that $m = X_{-2}X_{-3}v$. Put $m_1 = X_{-4}m$. Using commutator relations in L and Lemmas 2.1 and 2.46, one can deduce that $X_3X_2m_1 \neq 0$ and the groups $\mathcal{X}_i$ fix m_1 for $i = 3$ and 4. Hence m_1 generates an indecomposable Γ_2-module with highest weight $\sigma = \omega_1 + 5\omega_2$. Set $\eta = \varphi(\sigma)$. Obviously, η is a composition factor of the Γ_2-module N. Theorem 1.14 and Algorithm 1.6 yield that $d_\eta(g_2) = 7$ and hence $d_N(g_2) = 7$.

Now let $r = 5$ and $\omega = \omega_4 + \omega_5$. Here $m = X_{-2}X_{-3}X_{-4}v$. Put $m_1 = X_{-5}m$. Using commutator relations in L and Lemma 2.1, we can deduce that $X_4X_3X_2m_1 = X_{-5}v \neq 0$. Hence $m_1 \neq 0$. The same commutator relations show that $X_5m_1 = 0$ and $\mathcal{X}_5$ fixes m_1. One easily observes that $\omega(m_1) + \alpha_j \notin \mathbf{X}(M)$ for $j = 3, 4$. Hence the groups $\mathcal{X}_j$ also fix m_1 for these j. Now it is clear that m_1 generates an indecomposable Γ_2-module with highest weight $\lambda = \omega_2 + \omega_3$ and the representation $\pi = \varphi(\lambda)$ is a composition factor of the Γ_2-module N. Since π is p-large, Conjecture (r, s) yields that $d_\pi(g_2) = 9$. Then $d_N(g_2) = 9$ as desired.

Finally, let $r = 3$. Observe that $2 \leq a_3 + 2b < 2p$ and $a = a_1 + b < p$ by (4.6) and (4.7). We have $p^s = 5$. Lemma 2.54 and Formulae (4.7) and (4.8) imply that it suffices to consider the cases where $d_{\chi_i}(g_i) < 3$ for $i = 1$ or 2 which yield the following possibilities:

i) $a_1 = 0$, $2a_2 + a_3 = 5$;

ii) $2b + a_3 = 5$, $a_1 + 2a_2 = 3b - 1$.

Both in Cases i) and ii) a_3 is odd. Then, considering the possible values of a_i in Cases i) and ii) and using Lemma 2.54, we conclude that either $d_{\chi_f}(g_f) = 4$ for $f = 1$ or 2 and $d_{\chi_1}(g_1) + d_{\chi_2}(g_2) > 5$, or $\omega = \omega_2 + 3\omega_3$. In the latter case we set $m_1 = X_{-3}m$ and, applying commutator relations in L and Lemma 2.1, deduce that $X_2m_1 \neq 0$ and hence $m_1 \neq 0$. By Lemma 2.55, this forces that $K\Gamma_2m \cong V(5)$. As we have seen before, in this situation $d_N(g_2) = 5$ as desired. Now all the possibilities have been considered. This completes the proof. □

PROPOSITION 4.9. *Let $s > 0$, $n > p^s + 1$ for $G = A_r(K)$, $C_r(K)$, or $D_r(K)$, and $n > p^s + 2$ for $G = B_r(K)$. Assume that $\varphi \in \mathrm{Irr}_p$ is p-large and Conjecture (r, s) holds. Then $d_\varphi(x) = |x|$ for each element $x \in G$ of order p^{s+1}.*

PROOF. Set $G_0 = G(2, \dots, r)$ for $G = A_r(K)$, $B_r(K)$, and $C_r(K)$ and $G_1 = G(1, \dots, r-1)$ for $G = B_r(K)$ and $D_r(K)$. If $G = D_r(K)$, put also $G_2 = G(1, \dots, r-2, r)$ and

$$G_3 = \langle \mathcal{X}_i, \mathcal{X}_{-i}, x_{r-1}(t)x_r(t), x_{-(r-1)}(t)x_{-r}(t) \mid 1 \leq i \leq r-2, \quad t \in K \rangle.$$

It is well known (see, for instance, [**Sei87**, Section 8]) that $G_3 \cong B_{r-1}(K)$. Let $\Pi(G_3) = \{\beta_1, \dots, \beta_{r-1}\}$. One may assume that $X_{\beta_i} = X_i$ for $1 \leq i \leq r-2$ and $X_{\beta_{r-1}} = X_{r-1} + X_r$. The canonical homomorphism of the weight systems $\mathbf{X} \to \mathbf{X}(G_3)$ associated with the embedding $G_3 \to G$ maps the fundamental weights ω_i, $1 \leq i \leq r-2$, of G to the weights $\omega_i \in \mathbf{X}(G_3)$, the weights ω_{r-1} and ω_r to the weight ω_{r-1}, and the roots α_{r-1} and α_r to β_{r-1}. This homomorphism coincides with the homomorphism θ from the Introduction and Proposition 2.27 constructed for a regular unipotent element of G. Conjecture (r, s) and Lemmas 4.1 and 2.19 imply that it suffices to show that $\varphi|G_i$ has a p-large composition factor.

I. Let $G = A_r(K)$ or $C_r(K)$. Then $n > 4$. First assume that $\sum_{i=3}^r a_i \neq 0$. By Theorem 2.33, $\varphi|G_0$ has a composition factor $\chi = \varphi(\omega_+(1, r-1, 1))$. If $\sum_{i=2}^r a_i \geq p$, the representation χ is p-large, and we are done.

Let $\sum_{i=2}^r a_i < p$. Since φ is p-large, we have $b + \sum_{i=2}^r a_i = p$ for some $b \in \mathbb{Z}^+$, $0 < b \leq a_1$. As $\sum_{i=3}^r a_i \neq 0$, the sum $a_2 + b < p$. Set $m_1 = X_{-1,b}v$, $\lambda = (a_2 + b)\omega_1 + \omega_+(2, r-1, 1) \in \mathbf{X}(G_0)$, and $\mu = \varphi(\lambda)$. By Lemma 2.1, $m_1 \neq 0$. It is clear that m_1 generates an indecomposable G_0-module with highest weight λ that has a p-large composition factor μ.

Finally, let $\omega = a_1\omega_1 + a_2\omega_2$. Put $m_2 = v(1, 2, a_2)$. Lemma 2.46 yields that $m_2 \neq 0$ and the groups $\mathcal{X}_i$ fix m_2 for $i > 1$. So one easily observes that m_2 generates an indecomposable G_0-module with highest weight $\nu = a_1\omega_1 + a_2\omega_2$. Hence $\varphi|G_0$ has a p-large composition factor $\varphi(\nu)$. This completes the proof for $G = A_r(K)$ or $C_r(K)$.

II. Next, assume that $G = B_r(K)$ with $n > p^s + 2$. For this group in some cases we consider restrictions to naturally embedded subgroups of type D_r as well. Put $\Delta = \langle \mathcal{X}_\alpha \mid \alpha \text{ is long} \rangle$ and $\delta = \alpha_{r-1} + 2\alpha_r$. It is well known that $\Delta \cong D_r(K)$ and that one can take $\Pi(\Delta) = \{\alpha_1, \alpha_2, \dots, \alpha_{r-1}, \delta\}$.

Let $r = 3$. This case requires some special arguments. Our assumptions on n imply that $p = p^s = 3$. We have $\Delta \cong D_3(K) \cong A_3(K)$. Let $u \in \Delta$ be a regular unipotent element. Analyzing the standard embedding of $SO_6(K)$ into $SO_7(K)$, one can deduce that $J(u) = (5, 1, 1)$. Lemma 1.4 iii) implies that every element of order 9 in G is either a regular unipotent element, or lies in $\mathrm{cl}(u)$. Now it follows from [**Car85**, Proposition 5.1.2] that $u \in \overline{\mathrm{cl}(x)}$ if $x \in G$ and $|x| = 9$. Hence Lemma 2.19 and Conjecture (r, s) imply that it suffices to show that $\varphi|\Delta$ has a p-large composition factor. Since $\langle \omega, \delta \rangle = a_2 + a_3$, one easily concludes that the $K\Delta$-module $K\Delta v$ has such factor. This completes the proof for $r = 3$.

Now assume that $r > 3$. First consider the restriction $\varphi|G_0$. Let $a_2 + a_3 \geq p$. Theorem 2.33 forces that $\varphi|G_0$ has a composition factor φ_1 with highest weight $\omega_+(1, r-1, 1)$. It is clear that φ_1 is p-large.

Now let $a_2 + a_3 < p$. Set $m = v(1, 2, a_2)$. By Lemma 2.46, $m \neq 0$ and the groups $\mathcal{X}_i$ fix m for $i > 1$. Hence m generates an indecomposable G_0-module with highest weight $\lambda = a_1\omega_1 + (a_2+a_3)\omega_2 + \omega_+(3, r-1, 1)$. Lemma 2.3 implies that $\varphi(\lambda)$ is p-large.

Next, consider the restriction $\varphi|G_1$. First assume that $\sum_{i=1}^r a_i < p$. Set $m_1 = v(r, 2, a_2)$. Lemma 2.46 yields that $m_1 \neq 0$ and generates an indecomposable G_1-module with highest weight $\nu = (a_1+a_2)\omega_1 + \omega_+(2, r-2, 1) + (\sum_{i=2}^r a_i)\omega_{r-1}$. Hence $\varphi|G_1$ has a composition factor $\sigma = \varphi(\nu)$. Lemma 2.3 implies that $a_1 + 2(\sum_{i=2}^{r-1} a_i) + a_r \geq p$ and so σ is p-large.

Now let $\sum_{i=1}^r a_i \geq p$. If $\sum_{i=1}^{r-1} a_i \geq p$, Theorem 2.33 and Lemma 2.3 yield that $\varphi|G_1$ has p-large composition factor. Let $\sum_{i=1}^{r-1} a_i < p$. Then there exists $k \in \mathbb{Z}^+$ such that $0 < k \leq a_r$ and $k + \sum_{i=1}^{r-1} a_i = p$. Assume that $a_j \neq 0$ for some j with $j < r-1$. Set $m_2 = X_{-r,k}v$. Our assumptions force that $a_{r-1} + k < p$. So Lemmas 2.3 and 2.46 imply that m_2 generates an indecomposable G_1-module with p-large highest weight $\gamma = \omega(1, r-2) + (a_{r-1}+k)\omega_{r-1}$. Obviously, $\varphi(\gamma)$ is a composition factor of $\varphi|G_1$.

Finally, let $\omega = a_{r-1}\omega_{r-1} + a_r\omega_r$ with $a_{r-1} + a_r \geq p$. In this case we shall get a desired factor analyzing the restriction $\varphi|\Delta$. Set $t = p-1-a_{r-1}$, $u = a_{r-1}+a_r-t$, and $m_3 = X_{-r,t}v$. It is clear that $0 \leq t < a_r$. By Lemma 2.46, $m_3 \neq 0$ and is fixed by the subgroups $\mathcal{X}_i$ for $i < r$. One easily observes that $\mathcal{X}_\delta$ fixes m_3 as well. We have $\langle \alpha_r, \delta \rangle = 1$. Hence m_3 generates an indecomposable Δ-module with highest weight $\mu = (p-1)\omega_{r-1} + u\omega_r$ and $M|\Delta$ has a composition factor $M' = M(\mu)$. Let $v' \in M'$ be a nonzero highest weight vector. It is clear that $0 < u < 2p$. Set $m' = X_{-\delta}v'$ if $u < p$ and $m' = X_{-\delta,p}v'$ otherwise. In the first case $m' \neq 0$ by Lemma 2.1. If $u \geq p$, Theorem 2.2 implies that $M' = M^+ \otimes M(p\omega_r)$ where M^+ is a p-restricted Δ-module. Analyzing the weight structures of the modules $M(p\omega_r)$ and M', we deduce that $\omega(M') - p\delta \in \mathbf{X}(M')$. As $M' = KU^-(\Delta)v'$, Formula (2.1) yields that $m' \neq 0$ in the second case as well. Now it is clear that in all cases m' generates an indecomposable G_1-module with p-large highest weight and so $\varphi|G_1$ has a desired factor. It can occur that this factor is not p-restricted, but observe that Conjecture (r, s) concerns arbitrary irreducible representations. This completes the proof for $G = B_r(K)$.

III. Finally, let $G = D_r(K)$. First consider the group G_1. If $\sum_{i=1}^{r-1} a_i \geq p$, Theorem 2.33 yields that v generates an irreducible G_1-module with a p-large highest weight.

Let $\sum_{i=1}^{r-1} a_i < p \leq \sum_{i=1}^r a_i$. Take $m_1 = X_{-r,a_{r-2}+a_r}X_{-(r-2),a_{r-2}}v$. Lemma 2.46 implies that $m_1 \neq 0$ and $\mathcal{X}_i$ fixes m_1 for $0 < i < r$. One can easily conclude that m_1 generates an indecomposable G_1-module with highest weight

$$\mu = \omega(1, r-4) + (a_{r-3}+a_{r-2})\omega_{r-3} + a_r\omega_{r-2} + (a_{r-2}+a_{r-1})\omega_{r-1}.$$

It is clear that $\varphi(\mu)$ is p-large.

Finally, let $\sum_{i=1}^r a_i < p$. Set $a = \sum_{i=1}^{r-2} a_i$ and $m_2 = X_{-r,a_r+a}v(r-2, 1, a_1)$. By Lemma 2.46, $m_2 \neq 0$ and the groups $\mathcal{X}_i$ fix m_2 for $0 < i < r$. One easily deduces that m_2 generates an indecomposable G_1-module with highest weight $\nu = \omega_+(1, r-3, 1) + a_r\omega_{r-2} + (\sum_{i=1}^{r-1} a_i)\omega_{r-1}$. Lemma 2.3 implies that $\varphi(\nu)$ is p-large. Hence $\varphi|G_1$ always has a p-large composition factor.

Similarly we can prove the existence of such factor for G_2.

Now consider the group G_3. One easily observes that v generates an indecomposable G_3-module with highest weight $\lambda = \omega(1, r-2) + (a_{r-1}+a_r)\omega_{r-1}$. If $a_{r-1} + a_r < p$, the weight λ is p-large, and we are done.

Let $a_{r-1}+a_r\geq p$. Applying the graph automorphism if required, one can assume $a_{r-1}\geq a_r$. It is clear that $a_r>0$. For any $h\in\mathbb{Z}$ with $0<h\leq a_r$ set

$$\Omega_h=\{\lambda\in\mathbf{X}(\varphi)\mid\omega-\lambda=a\alpha_{r-1}+b\alpha_r,\quad a+b=h\},\ M_h=\langle M_\lambda\mid\lambda\in\Omega_h\rangle.$$

Since X_r and X_{r-1} commute,using Lemma 2.1, one easily concludes that $\dim M_\lambda=1$ for $\lambda\in\Omega_h$ and $|\Omega_h|=h+1$. So $\dim M_h=h+1$. It is clear that $X_{\beta_{r-1}}M_h\subset M_{h-1}$. Hence M_h contains a nonzero vector u_h which is annihilated by $X_{\beta_{r-1}}$. Now Formula (2.1) implies that $\mathcal{X}_{\beta_{r-1}}$ fixes u_h as $h<p$. Hence u_h is fixed by $U^+(G_3)$. Now it is clear that u_h generates an indecomposable G_3-module with highest weight

$$\rho_h=\omega(1,r-3)+(a_{r-2}+h)\omega_{r-2}+(a_{r-1}+a_r-2h)\omega_{r-1}.$$

One can choose h such that $a_{r-1}+a_r-2h=p-2$ or $p-1$. Then $\varphi(\rho_h)$ is p-large by Lemma 2.3.

Now all the possibilities have been considered. The proposition is proved. □

Now pass to the general case for p-large representations.

PROPOSITION 4.10. *Let $d_\varphi(x)=|x|$ for all unipotent $x\in G$ and all p-large p-restricted representations of G. Then Theorem 1.1 holds for G.*

PROOF. Let $\varphi\in\mathrm{Irr}$ be p-large. Using Theorem 2.2, write $\varphi=\otimes_{j=1}^t\varphi_j\,\mathrm{Fr}^{i_j}$ where the superscripts i_j are all distinct, $\varphi_j\in\mathrm{Irr}_p$ and $\omega(\varphi_j)\neq0$ for all j. Apply induction on t. Let $t=1$. As the Frobenius morphism does not influence the minimal polynomial, one can assume that $\varphi\in\mathrm{Irr}_p$ in the latter case. Then Theorem 1.1 holds by our assumptions.

Let $t>1$. Then we can take $\varphi=\varphi_1\otimes\varphi_2$ where $\varphi_1,\varphi_2\in\mathrm{Irr}$ are nontrivial and $\overline{\omega(\varphi)}=\overline{\omega(\varphi_1)}+\overline{\omega(\varphi_2)}$. Theorem 2.32 implies that it suffices to consider elements of order p^{s+1} with $s>0$. Let $|x|=p^{s+1}$ and put $y=x^{p^s}$. Then $|y|=p$. Let $C=f(\mathrm{cl}(y))\subset G_{\mathbb{C}}$ where f is the canonical bijection of Theorem 1.14. Set $\psi_i=(\varphi_i)_{\mathbb{C}}$, $l_i=d_{\psi_i}(C)-1$ for $i=1,2$. By Proposition 1.5, $d_{\varphi_{\mathbb{C}}}(C)=l_1+l_2+1$. The arguments in the proof of Theorem 2.32 yield that $d_{\varphi_{\mathbb{C}}}(C)>p$ whenever φ is p-large. Hence $l_1+l_2\geq p$. Set $m_i=\min\{l_i,p-1\}$, $i=1,2$. By Theorem 1.14, $d_{\varphi_i}(y)=m_i+1$. Hence $(y-1)^{m_i}=(x-1)^{p^sm_i}\neq0$ and we get $d_{\varphi_i}(x)>p^sm_i$. So $\varphi_i(x)$ has a Jordan block of size $q_i>p^sm_i$. Since φ_i is nontrivial, $l_i>0$. Now it is clear that $q_1+q_2>p^{s+1}$. Then, by Lemma 2.18, $\varphi(x)$ has a Jordan block of size p^{s+1}, i.e. $d_\varphi(x)=p^{s+1}$ as required. □

COROLLARY 4.11. *Assume that Conjecture (r,s) holds for G. Then Theorem 1.1 holds for elements of order p^{s+1} in G.*

PROOF. For p-restricted representations this follows from Propositions 4.5, 4.8, and 4.9. Next, apply Proposition 4.10. □

In what follows when proving the other main results for elements of a fixed order we assume that Theorem 1.1 is valid for these elements.

In Sections 5–10 $\varphi\in\mathrm{Irr}_p$, $x\in U^+$ is a regular unipotent element, $q_j=r(\Gamma_j)$ and $\mathbf{X}_j=\mathbf{X}(\Gamma_j)$ for $j=1,2$.

5. Regular unipotent elements for $n = p^s + b$, $0 < b < p$

This section is devoted to proving Theorems 1.7 and 1.10 for regular unipotent elements in the groups $A_r(K)$, $B_r(K)$, and $C_r(K)$ with $n = p^s + b$, $0 < b < p$, and p-restricted representations. Set $e = p^s - b$. Observe that

$$b = 2b_1 \quad \text{for} \quad G = B_r(K) \quad \text{and} \quad 2b_1 + 1 \quad \text{for} \quad G = C_r(K)$$

with $b_1 \in \mathbb{Z}^+$;

$\Gamma_1 = A_{b-1}(K)$ for $G = A_r(K)$, $C_{b_1}(K)$ for $G = B_r(K)$, $B_{b_1}(K)$ for $G = C_r(K)$;

$\Gamma_2 = A_{e-1}(K)$ for $G = A_r(K)$, $B_{(e-1)/2}(K)$ for $G = B_r(K)$, $C_{e/2}(K)$ for $G = C_r(K)$

(with the same notation for groups of small ranks as in Proposition 2.43). In particular, we have

$\Gamma_1 = 1$ for $b = 1$,

$\Gamma_1 \cong A_1(K)$ for $G = B_r(K)$, $b = 2$ and for $G = C_r(K)$, $b = 3$,

$\Gamma_1 \cong C_2(K)$ for $G = C_r(K)$, $b = 5$,

$\Gamma_2 = 1$ for $s = 1$, $b = p - 1$,

$\Gamma_2 \cong A_1(K)$ for $s = 1$, $G = B_r(K)$, $b = p - 3$ or $G = C_r(K)$, $b = p - 2$,

$\Gamma_2 \cong C_2(K)$ for $s = 1$, $G = C_r(K)$, $b = p - 5$.

Set

$$\Sigma_b = \begin{cases} \sum_{i=b+1}^{r-b} a_i & \text{for } G = A_r(K), \\ \sum_{i=b+1}^{r-1} a_i & \text{for } G = B_r(K) \text{ with } r \neq p-1, \\ \sum_{i=b+1}^{r} a_i & \text{for } G = C_r(K). \end{cases}$$

Put

(5.1)

$$N(\varphi) = \begin{cases} \sum_{i=1}^{b} a_i i + b\Sigma_b + \sum_{i=r-b+1}^{r} a_i(r+1-i) & \text{for } G = A_r(K), \\ \sum_{i=1}^{b} a_i i + b\Sigma_b + b_1 a_r & \text{for } G = B_r(K),\ n \neq 2p-1, \\ \sum_{i=1}^{p-2} a_i i + b_1 a_{p-1} & \text{for } G = B_{p-1}(K), \\ \sum_{i=1}^{b} a_i i + b\Sigma_b & \text{for } G = C_r(K). \end{cases}$$

If Δ is a classical group whose standard realization has the dimension $p^s + b'$ with $0 < b' < p$ and $\xi \in \operatorname{Irr}_p \Delta$, define a parameter $N_\Delta(\xi)$ in a similar way replacing in Formula (5.1) b and r by b' and $r(\Delta)$, respectively. Thus $N(\varphi) = N_G(\varphi)$.

Theorem 1.14 and Algorithm 1.6 imply that $d_\varphi(y) = \min\{N(\varphi) + 1,\ p\}$. The following lemma yields a base for using Corollary 2.44 in this section.

LEMMA 5.1. *Let $\nu \in \operatorname{Irr}\Gamma$ with $\omega(\nu) = (\nu_1, \nu_2)$, $\nu_j \in \mathbf{X}_j$. Assume that Conjecture (r, s) is valid and $d_\mu(x_\Gamma) < p^s$. Then one of the following holds.*

1) $\nu_2 = 0$;

2) $s = 1$, $\nu_1 = 0$;

3) $\nu_j \in \{0, p^l\omega_1, p^l\omega_{q_j}\}$ for $G = A_r(K)$ and $\nu_j \in \{0, p^l\omega_1\}$ otherwise;

4) $s = 1$, $b = p - 3$, $G = B_r(K)$, $\nu_1 = p^l\omega_1$, $\nu_2 = 2p^m\omega_1$;

5) $p = 5$, $G = B_3(K)$, $\nu_1 = 2p^l\omega_1$, $\nu_2 = p^m\omega_1$;

6) $p = 7$, $G = B_5(K)$, $\nu_1 = p^l\omega_2$, $\nu_2 = p^m\omega_1$;

7) $s = 1$, $b = p - 5$, $G = B_r(K)$, $\nu_1 = p^l\omega_1$, $\nu_2 = p^m\omega_2$;

8) $p = 7$, $G = B_4(K)$, $\nu_1 = 2p^l\omega_1$, $\nu_2 = p^m\omega_2$;

9) $G = B_r(K)$, $b = p - 7$ or $p = 3$ and $r = 5$, $\nu_1 = p^l\omega_1$, $\nu_2 = p^m\omega_3$;

10) $b = 3$, $G = C_r(K)$, $\nu_1 = 2p^l\omega_1$, $\nu_2 = p^m\omega_1$.

If $e = 1$ *and* $\nu_1 \neq 0$, *we have*

$$\nu_1 = \begin{cases} p^l\omega_1 \text{ or } p^l\omega_{p-2} & \textit{for } G = A_{2p-2}(K), \\ p^l\omega_1 & \textit{for } G = B_{p-1}(K). \end{cases} \tag{5.2}$$

PROOF. First assume that $e > 1$. Suppose that $\nu_1 \neq 0$ for $s = 1$ and $\nu_2 \neq 0$ in all cases. Set $\varphi_j = \varphi(\nu_j)$. If $s > 1$, Formula (1.5), Proposition 1.5, and Algorithm 1.6 imply that $d(\varphi_2, g_2) = p^s$ unless ν_2 is such as in Items 3), 9) with $p = 3$, or 10). Theorem 2.9 and [**Ste68**, Corollary a) of Lemma 68] yield that $d_\mu(x_\Gamma) = p^s$ if $d_{\varphi_1}(g_1) + d_{\varphi_2}(g_2) > p^s$. This reduces the lemma to the cases where either $d_{\varphi_1}(g_1) < b$, or $d_{\varphi_2}(g_2) < e$, or $d_{\varphi_1}(g_1) = b$ and $d_{\varphi_2}(g_2) = e$. Remark 2.45 forces that $\nu_j \neq p^l\omega_{q_j}$ if $G = C_r(K)$ and $\Gamma_j = B_{q_j}(K)$. Now Conjecture (r, s), Formula (1.5), Proposition 1.5, and Algorithm 1.6 yield that such situations are exhausted by Cases 1)–10) above.

Now let $e = 1$. Then $s = 1$, $n = 2p - 1$, and $G = A_{2p-2}(K)$ or $B_{p-1}(K)$. Keep the notation φ_1 from above. Formula (1.5), Proposition 1.5, and Algorithm 1.6 imply that $d_{\varphi_1}(g_1) = p$ unless ν_1 satisfies (5.2). This implies the last statement of the lemma and completes the proof. □

By Proposition 2.5 b), y has b blocks of size 2 and $p^s - b$ blocks of size 1 on V. One easily observes that $(y - 1)V = \langle v_1, \dots, v_b\rangle$. Now a well-known information on the structure of the classical groups and representations of the group $A_1(K)$ and Proposition 2.43 4) imply that $\zeta(\varepsilon_i) = 1$ for $1 \leq i \leq b$; if $G = A_r(K)$, then $\zeta(\varepsilon_i) = 0$ for $b < i \leq r + 1 - b$ and $\zeta(\varepsilon_i) = -1$ for $r + 1 - b < i$; otherwise $\zeta(\varepsilon_i) = 0$ for $b < i \leq r$.

LEMMA 5.2. *Let* $N(\varphi) < p$. *Then* $d_\varphi(x) = p^s(N(\varphi) + 1)$ *unless one of the following holds:*

1) $\omega = \omega(1, b) + \omega(p^s, r)$ *for* $G = A_r(K)$ *and* $\omega = \omega(1, b)$;

2) $n = p + b$, $\omega = \omega(b, p)$ *for* $G = A_r(K)$ *and* $\omega(b, r)$ *otherwise;*

3) $\omega = a_b\omega_b + a_{p^s}\omega_{p^s} + \mu$ *with* $b + 1 < p^s$, *and*

$$\mu \in \{\omega_{b+1}, \omega_{p^s-1}, \omega_1 + \omega_{b+1}, \omega_1 + \omega_{p^s-1}, \omega_{b-1} + \omega_{b+1}, \omega_{b-1} + \omega_{p^s-1}, \omega_{b+1} + \omega_{p^s+1}, \omega_{p^s-1} + \omega_{p^s+1}, \omega_{b+1} + \omega_r, \omega_{p^s-1} + \omega_r\}$$

for $\quad G = A_r(K) \quad$ and

$b < r, \quad \omega \in \{a_b\omega_b + \omega_{b+1}, \omega_1 + a_b\omega_b + \omega_{b+1}, \omega_{b-1} + a_b\omega_b + \omega_{b+1}\}$ otherwise;

4) $n = 2p - 3 > 3$, $G = B_r(K)$, $\omega = \omega_1 + 2\omega_{p-2}$;

5) $p = 5$, $G = B_3(K)$, $\omega = 2\omega_1 + \omega_3$;

6) $p = 7$, $G = B_5(K)$, $\omega = \omega_2 + \omega_5$;

7) $n = 2p - 5 \geq 9$, $G = B_r(K)$, $\omega = \omega_1 + a_{p-5}\omega_{p-5} + \omega_{p-3}$, $a_{p-5} < 2$ *and* $a_{p-5} = 0$ *for* $p > 11$, *or* $p = 7$ *and* $\omega = \omega_1 + 2\omega_2 + \omega_4$, *or* $p = 11$ *or* 13 *and* $\omega = \omega_{p-6} + \omega_{p-3}$;

8) $p = 7$, $G = B_4(K)$, $\omega = 2\omega_1 + a_2\omega_2 + \omega_4$, $a_2 \leq 1$;

9) $p = 3$, $G = B_5(K)$, $\omega = \omega_1 + \omega_5$.

PROOF. It is easy to check that $\zeta(\omega) = N(\varphi)$. Hence, by Lemma 2.54 ii), $v \in M_y$. Set $\lambda = \rho(\omega)$ and $M_0 = M(\lambda)$. As v is fixed by $U^+(\Gamma)$, it is clear that $K\Gamma$-module M_y has a composition factor M_0. Corollary 2.44 implies that $d_\varphi(x) = p^s(N(\varphi)+1)$ if $d_{M_0}(x_\Gamma) = p^s$. Now the construction of ρ in Proposition 2.43

and Lemma 5.1 complete the proof. Here one must take into account that ω is p-restricted and not p-large (so $a_1 + \ldots + a_r < p$) and that $N(\varphi) < p$. □

LEMMA 5.3. *Let $N(\varphi) < p$. Assume that one of the conditions* 1) – –9) *of Lemma* 5.2 *holds. Then $d_\varphi(x) = \min\{d_0, pd_1\}$.*

PROOF. Keeping the notation introduced in the proof of Lemma 5.2, set $d_\varphi(x) = d$ and $d_{M_0}(x_\Gamma) = d'$. By Corollary 2.41, it suffices to show that either $d \geq d_0$, or $d \geq pd_1$. The proof is based on Corollary 2.44 and the arguments in the proof of Lemma 5.2. In all cases we find d' using Conjecture (r, s), Formula (1.5), Proposition 1.5, and Algorithm 1.6, then conclude that $d \geq N(\varphi)p^s + d'$ by Corollary 2.44 and compare the estimate obtained for d with d_0 or pd_1. Use Formula (2.10) to find λ. As before, we write λ in the form (λ_1, λ_2) where $\lambda_j \in \mathbf{X}_j$, $j = 1, 2$. In this proof "Case i" means "Case i of Lemma 5.2".

Case 1. For all types $\lambda_2 = 0$. We have

$$\lambda_1 = \begin{cases} \omega(1, b-1) + \omega_+(1, r-p^s, p^s) & \text{for} \quad G = A_r(K), \\ \omega(1, b_1) + \sum_{i=b_1+1}^{b-1} a_i\omega_{b-i} & \text{for} \quad G = B_r(K), \\ \omega(1, b_1 - 1) + 2(a_{b_1} + a_{b_1+1})\omega_{b_1} & \\ + \sum_{i=b_1+2}^{b-1} a_i\omega_{b-i} & \text{for} \quad G = C_r(K) \text{ and } b > 1, \\ 0 & \text{for} \quad G = C_r(K) \quad \text{with} \quad b = 1. \end{cases}$$

Put $a(\varphi) = 0$ for $b = 1$. If $b > 1$, set

$$a(\varphi) = \begin{cases} \sum_{i=1}^{b-1} a_i i(b-i) + \sum_{i=p^s+1}^{r} a_i(i-p^s)(n-i) & \text{for } G = A_r(K), \\ \sum_{i=1}^{b-1} a_i i(b-i) & \text{otherwise.} \end{cases}$$

Then

$$d' = \min\{a(\varphi) + 1, p\}. \tag{5.3}$$

Observe that in our case

$$d_0 = \begin{cases} 1 + \sum_{i=1}^{b} a_i i(n-i) + \sum_{i=p^s}^{r} a_i i(n-i) & \text{for } G = A_r(K), \\ 1 + a_{p-1}p(p-1)/2 + \sum_{i=1}^{p-2} a_i i(2p-1-i) & \text{for } G = B_{p-1}(K), \\ 1 + \sum_{i=1}^{b} a_i i(n-i) & \text{otherwise.} \end{cases}$$

Now (5.1) implies that

$$d_0 = p^s N(\varphi) + a(\varphi) + 1. \tag{5.4}$$

By Proposition 2.5 b), x_1 and z_1 have b Jordan blocks of size $p^{s-1} + 1$ and $(p - b)$ blocks of size p^{s-1}. Hence

$$d_1 = \begin{cases} 1 + p^{s-1}(\sum_{i=1}^{b} a_i i + \sum_{i=p^s}^{r} a_i(n-i)) & \text{for } G = A_r(K), \\ 1 + a_{p-1}(p-1)/2 + \sum_{i=1}^{p-2} a_i i & \text{for } G = B_{p-1}(K), \\ 1 + p^{s-1}(\sum_{i=1}^{b} a_i i) & \text{otherwise.} \end{cases}$$

This forces

$$pd_1 = p^s N(\varphi) + p. \tag{5.5}$$

Now Formulae (5.3)–(5.5) yield that $d \geq \min\{d_0, pd_1\}$ and complete the proof for Case 1.

Case 2. Obviously, $\lambda_1 = 0$. One can assume that $G \neq B_{p-1}(K)$ since for this group Case 2 is a subcase of Case 1. We get $\lambda_2 = \omega_+(1, p-1-b, b)$ for $G = A_r(K)$ and $\lambda_2 = \omega_+(1, r-b, b)$ otherwise. Set

$$c(\varphi) = \begin{cases} \sum_{i=b+1}^{p-1} a_i(i-b)(p-i) & \text{for} \quad G = A_r(K), \\ a_r(r-b)(p-r)/2 + \sum_{i=b+1}^{r-1} a_i(i-b)(p-i) & \text{for} \quad G = B_r(K), \\ \sum_{i=b+1}^{r} a_i(i-b)(p-i) & \text{for} \quad G = C_r(K). \end{cases}$$

Then $d' = \min\{c(\varphi)+1, p\}$. As

$$d_0 = \begin{cases} 1 + \sum_{i=b}^{p} a_i i(p+b-i) & \text{for } G = A_r(K), \\ 1 + a_r r(p+b-r)/2 + \sum_{i=b}^{r-1} a_i i(p+b-i) & \text{for } G = B_r(K), \\ 1 + \sum_{i=b}^{r} a_i i(p+b-i) & \text{for } G = C_r(K), \end{cases}$$

we deduce from (5.1) that $d_0 = pN(\varphi) + c(\varphi) + 1$. Since $x_1 = y$, it is clear that (5.5) holds in Case 2 as well (with $s = 1$). Now complete the proof as in Case 1.

Case 3. Define the subsets $\Lambda_j \in \mathbf{X}_j$, $j = 1, 2$, as follows: $\Lambda_j = \{0, \omega_1, \omega_{q_j}\}$ for $G = A_r(K)$, $\Lambda_1 = \{0, 2\omega_1\}$ if $\Gamma_1 = B_1(K)$, and $\Lambda_j = \{0, \omega_1\}$ otherwise. Then $\lambda_j \in \Lambda_j$. It is well known (see, for instance, [**Ste68**, Corollary a) of Lemma 68]) that $M_0 \cong M(\lambda_1) \otimes M(\lambda_2)$. Put $M_j = M(\lambda_j)$. This notation will be used until the end of the proof. It is clear that $d_{M_j}(g_j) = 1$ if $\lambda_j = 0$, $d_{M_1}(g_1) = b$ if $\lambda_1 \neq 0$, $d_{M_2}(g_2) = e$ if $\lambda_2 \neq 0$ and $\Gamma_2 \neq B_1(K)$, and $d_{M_2}(g_2) = 2$ for $\lambda_2 \neq 0$ and $\Gamma_2 = B_1(K)$. Now Corollary 2.15 yields that $d' = p^s - 1$ if both λ_1 and $\lambda_2 \neq 0$ and $\Gamma_2 \neq B_1(K)$. Arguing as in the proof of Corollary 2.15, we deduce that $d' = p^s - 2$ if $\Gamma_2 = B_1(K)$ and $\lambda_j \neq 0$ for $j = 1, 2$. Analyzing all the possibilities, one easily concludes that $d_0 = p^s N(\varphi) + d'$ and hence $d = d_0$.

In Cases 4–8 both b and $e < p$. One can easily compute d' and check that $d_0 = pN(\varphi) + d'$ in all these cases. Hence $d = d_0$.

Case 9. Here $\lambda_1 = \omega_1$, $\lambda_2 = \omega_3$, $N(\varphi) = 2$, and $d_0 = 26$. We have $d_{M_1}(g_1) = 2$ and $d_{M_2}(g_2) = 7$. Hence by Corollary 2.15, $d' = 8$ and $d_0 = p^s N(\varphi) + d'$. Then $d = d_0$.

Now all the possibilities have been considered. □

PROPOSITION 5.4. *Let $N(\varphi) \geq p$. Then $d_\varphi(x) = p^{s+1} = |x|$.*

PROOF. Our assumption that φ is not p-large implies that $n > p^s + 1$ for $G = A_r(K)$ and $C_r(K)$ and $n > p^s + 2$ for $G = B_r(K)$. Hence $b > 1$ and $b > 2$ for $G = B_r(K)$. This forces $p > 3$ for $G \neq A_r(K)$. Apply induction on r supposing that our proposition holds for classical groups of smaller ranks. Theorem 1.1 for the classical groups with $n = p^s + 1$ or $p^s + 2$ provides the induction base.

Set

(5.6)
$$N_1(\varphi) = \begin{cases} \sum_{i=1}^{b-1} a_i i + (b-1)(\sum_{i=b}^{p^s} a_i) + \sum_{i=p^s+1}^{r} a_i(n-i) & \text{for } G = A_r(K), \\ (b_1 - 1)a_r + \sum_{i=1}^{b-2} a_i i + (b-2)\sum_{i=b-1}^{r-1} a_i & \text{for } G = B_r(K), \\ \sum_{i=1}^{b-2} a_i i + (b-2)\sum_{b-1}^{r} a_i & \text{for } G = C_r(K). \end{cases}$$

I. Assume that $N_1(\varphi) \geq p$. Set $G_1 = G(2, \ldots, r)$. If $a_i \neq 0$ for some i with $1 \leq i \leq b$, choose maximal such i and put $m = v(1, i, a_i)$. Otherwise take $m = v$. In all cases set $\omega^1 = \omega_{G_1}(m)$, and $\varphi_1 = \varphi(\omega_1) \in \operatorname{Irr} G_1$. Lemma 2.46 implies that $m \neq 0$ and generates an indecomposable G_1-module M_1 with highest weight ω^1. Hence φ_1 is a composition factor of $\varphi|G_1$.

We have

$$\omega^1 = \begin{cases} \omega(1, b-1) + (a_b + a_{b+1})\omega_b \\ +\omega_+(b+1, r-1, 1) & \text{if } r > b+1, \\ \omega(1, r-2) + (2a_{r-1} + a_r)\omega_{r-1} & \text{if } G = B_r(K) \text{ and } r = b \text{ or } b+1, \\ \omega(1, r-2) + (a_{r-1} + a_r)\omega_{r-1} & \text{if } G = C_r(K) \text{ and } r = b+1. \end{cases}$$

Hence φ_1 is p-restricted. Observe that $b < r-1$ for $G = A_r(K)$ as $p > 2$, and $b < r$ for $G \neq B_r(K)$. One can directly check that $N_{G_1}(\varphi_1) = N_1(\varphi)$. Let $g \in G_1$ be a regular unipotent element. Then $d_{\varphi_1}(g) = p^{s+1}$ by the induction conjecture. So Lemma 2.19 forces that $d_\varphi(x) = p^{s+1}$ as required.

II. Now assume that $N_1(\varphi) < p$. Until the end of the proof T, T_Γ, and T_1 are the maximal tori of G, Γ, and A, respectively, considered in Proposition 2.43. We construct a nonzero vector $m \in M$ such that m is a weight vector for T, T_Γ and T_1, $U^+(A)$ and $U^+(\Gamma)$ fix m, and $\omega_A(m) = p-1$. Then $m \in M_y$ by Lemma 2.54 ii). Set $M_\Gamma = K\Gamma m$ and $\lambda = \omega_\Gamma(m)$. Then M_Γ is an indecomposable Γ-module with highest weight λ and hence has a composition factor $M_0 \cong M(\lambda)$. By Corollary 2.44, it suffices to show that $d_{M_y}(x_\Gamma) = p^s$. As $M_\Gamma \subset M_y$, we are done if $d_{M_\Gamma}(x_\Gamma) = p^s$. Naturally, the latter holds if

$$d_{M_0}(x_\Gamma) = p^s. \tag{5.7}$$

Now we proceed to construct m. In the majority of the cases considered below Formula (5.7) holds. The remaining few ones will be handled with the help of special arguments. It is essential in this proof that φ is not p-large, so $\sum_{i=1}^r a_i < p$ and for $G = B_r(K)$ we have $a_1 + 2(\sum_{i=2}^{r-1} a_i) + a_r < p$ as well. Until the end of the proof $\lambda = \omega_\Gamma(m)$ and we write $\lambda = (\lambda_1, \lambda_2)$ where $\lambda_i \in \mathbf{X}_i$, $i = 1, 2$. Set $\psi_i = \varphi(\lambda_i) \in \operatorname{Irr}\Gamma_i$. Recall that $\lambda_2 = 0$ if $e = 1$. All formulae for λ_2 in Items II A)–II C) below concern the case where $e > 1$.

Set $d = N(\varphi) - p + 1$. Then Formulae (5.1) and (5.6) force that

$$d \leq \begin{cases} \sum_{i=b}^{p^s} a_i & \text{for } G = A_r(K), \\ a_{b-1} + 2(\sum_{i=b}^{r-1} a_i) + a_r & \text{for } G = B_r(K), \\ a_{b-1} + 2\sum_{i=b}^{r} a_i & \text{for } G = C_r(K). \end{cases}$$

A). Let $G = A_r(K)$. First assume that $d > \sum_{i=b}^{p^s-1} a_i$. Then $d = c + \sum_{i=b}^{p^s-1} a_i$ where $0 < c \leq a_{p^s}$. Put $m_1 = X_{-p^s,c}v$ and $m = m_1(b, p^s-1, a_{p^s-1})$ (in the $KG(1, \ldots, p^s-1)$-module generated by m_1).

If $a_b < d \leq \sum_{i=b}^{p^s-1} a_i$, there exist j and c such that $b < j \leq p^s - 1$, $0 < c \leq a_j$, and $d = c + \sum_{i=b}^{j-1} a_i$. Then set $m = v(b, j, c)$.

If $d \leq a_b$, put $m = X_{-b,d}v$. By Lemma 2.46, $m \neq 0$ in all cases and is fixed by $\mathcal{X}_i$ for $i \leq p^s - 1$, $i \neq b$. Obviously, the groups $\mathcal{X}_i$ with $p^s < i \leq r$ fix m. Hence $U^+(\Gamma)$ fixes m. One can assume that

$$U^+(A) = \{x_{\alpha_1+\ldots+\alpha_{p^s}}(t)x_{\alpha_2+\ldots+\alpha_{p^s+1}}(t)\ldots x_{\alpha_b+\ldots+\alpha_r}(t),\ t \in K\}.$$

Thus $U^+(A)$ fixes m (it is essential here that $p^s < r$ as $n > p^s + 1$). Since $\zeta(\alpha_b) = \zeta(\alpha_{p^s}) = 1$ and $\zeta(\alpha_i) = 0$ for $i \neq b$ or p^s, we deduce that $\omega_A(m) = p-1$ as required.

Using the description of the homomorphism ρ in Proposition 2.43, we obtain the following equalities.

a) Let $d > \sum_{i=b}^{p^s-1} a_i$. Then

$$(5.8)\qquad \lambda_1 = \begin{cases} (a_1 + a_{p^s+1} + c)\omega_1 + (\sum_{i=2}^{b-2}(a_i + a_{p^s+i})\omega_i) \\ +(a_{p^s+b-1} + \sum_{i=b-1}^{p^s-1} a_i)\omega_{b-1} & \text{if } b > 2, \\ (a_{p^s+1} + c + \sum_{i=1}^{p^s-1} a_i)\omega_1 & \text{for } b = 2; \end{cases}$$

and

$$(5.9)\qquad \lambda_2 = \omega_+(1, e-2, b-1) + (a_{p^s-2} + c)\omega_{e-1}.$$

b) If $a_b < d \leq \sum_{i=b}^{p^s-1} a_i$ and $d = c + \sum_{i=b}^{j-1} a_i$, we get

$$(5.10)\qquad \lambda_1 = (\sum_{i=1}^{b-2}(a_i + a_{p^s+i})\omega_i) + (a_{b-1} + d + a_{p^s+b-1})\omega_{b-1},$$

$$(5.11)\qquad \begin{aligned} \lambda_2 &= \omega_+(1, j-1-b, b-1) + (a_{j-1} + a_j - c)\omega_{j-b} + (a_{j+1} + c)\omega_{j-b+1} \\ &\quad + \omega_+(j-b+2, e-1, b) \quad \text{if} \quad j < p^s - 1, \quad \text{and} \\ \lambda_2 &= \omega_+(1, e-2, b-1) + (a_{p^s-2} + a_{p^s-1} - c)\omega_{e-1} \quad \text{for} \quad j = p^s - 1. \end{aligned}$$

c) Finally, let $d \leq a_b$. Then λ_1 is determined by Formula (5.10) and

$$(5.12)\qquad \lambda_2 = (a_{b+1} + d)\omega_1 + \omega_+(2, e-1, b).$$

B). Now let $G = B_r(K)$. Recall that $b = 2b_1 \geq 4$ and $\Gamma_1 = C_{b_1}(K)$. First assume that $b < r$ and $d > \sum_{i=b-1}^{r} a_i$. Then there exist j and c such that $b \leq j < r$, $0 < c \leq a_j$, and $d = c + (\sum_{i=b-1}^{r} a_i) + \sum_{i=j+1}^{r-1} a_i$. Set $m_1 = v(r, j, c)$. By Lemma 2.46, $m_1 \neq 0$. For $b \leq i < r$, set $f_i = \omega_i(m_1)$. If $b < r-1$, put $l_{b-1} = a_{b-1}$, $l_{r-1} = f_{r-1}$, $l_t = f_t + l_{t+1}$ for $b < t < r-1$, $l_b = l_{b-1} + f_b + l_{b+1}$, and $m = m_1(b \cdot l_b, (b-1) \cdot l_{b-1}, (b+1) \cdot l_{b+1}, \ldots, (r-1) \cdot l_{r-1})$. For $b = r-1$ we have $m_1 = X_{-r, a_r+2c} X_{-(r-1),c} v$ and set $m = X_{-(r-1), a_{r-2}+a_{r-1}+a_r} X_{-(r-2), a_{r-2}} m_1$. We claim that $m \neq 0$ and is fixed by $\mathcal{X}_i$ with $i \neq b$.

Put $\Delta = G(1, 2, \ldots, r-1)$. Lemma 2.46 implies that m_1 generates an indecomposable Δ-module with highest weight $\omega_\Delta(m_1)$. Applying Lemmas 2.48 and 2.46 to this module and vector m_1, we get that $m \neq 0$ and is fixed by $\mathcal{X}_i$ for $i < r$, $i \neq b$.

Now let $i = r$. Show that $X_{r,k} m = 0$ for $k > 0$. Using the commutator relations in Lemma 2.1, we conclude that for $r > b+1$ it suffices to prove that $u_k = X_{r,k} X_{-(r-1), l_{r-1}} m_1 = 0$. Suppose this is false for some k. It follows from the construction of m_1 that one can write $u_k = X_{r,k} X_{-(r-1), l_{r-1}} X_{-r,f} X_{-(r-1),a} u$ where $\omega(u) = \omega - \sum_{i=j}^{r-2} b_i \alpha_i$, $0 < a \leq \omega_{r-1}(u) < p$, $\omega_r(u) = a_r$, $f = a_r + 2a$, and $l_{r-1} = \omega_{r-1}(u) + a_r$. It is clear that $k \leq f$. Set $\Sigma = G(r-1, r)$. By Lemma 2.46, u generates an indecomposable Σ-module M_Σ with p-restricted highest weight $\omega_\Sigma(u)$. Applying Lemma 2.49 to M_Σ, we conclude that $u_k = 0$ as required.

Let $b = r-1$. Set $l_k = X_{r,k} m$ and assume that $l_k \neq 0$ for some $k > 0$. Then $l'_k = w_r w_{r-1} l_k \neq 0$. One can compute that $\omega_{r-1}(l_k) = -(a_{r-2} + a_{r-1} + a_r + k)$. This implies that $\omega(w_{r-1} l_k) = \omega(l_k) + (a_{r-2} + a_{r-1} + a_r + k)\alpha_{r-1}$ and hence $\omega_r(w_{r-1} l_k) = \omega_r(m_1) = -(a_r + 2c)$. But then $\omega(l'_k) \notin \mathbf{X}(M)$ which yields a contradiction. Thus all $l_k = 0$.

Hence in all cases $\mathcal{X}_r$ fixes m. This implies that $U^+(\Gamma)$ fixes m. Until the end of this proof set $g = j - b$, $h = r - b$, $a'_i = a_i + a_{b-i}$ for $i \leq b$, and $\Theta = \sum_{i=3}^{b_1-1} a'_i \omega_i$. Notice that b is a constant determined by n whenever n is fixed, but j is determined by the coefficients a_i and takes different values throughout this proof, its value will be clear from the context. One can deduce the following formulae for λ.

(5.13)

$$\lambda_1 = \begin{cases} (a_1 + d - a_{b-1})\omega_1 + (a_2' + a_{b-1})\omega_2 + \Theta + a_{b_1}\omega_{b_1} & \text{if } b > 4, \\ (a_1 + d - a_3)\omega_1 + (a_2 + a_3)\omega_2 & \text{for } b = 4; \end{cases}$$

(5.14)

$$\lambda_2 = (a_{b-1} + a_b)\omega_1 + \omega_+(2, g-1, b-1) + (a_{j-1} + c)\omega_g + (a_j - c + a_{j+1})\omega_{g+1}$$
$$+ \omega_+(g+2, h, b) \quad \text{if} \quad b+1 < j < r-1,$$

$$\lambda_2 = (a_{b-1} + a_b)\omega_1 + \omega_+(2, h-2, b-1) + (a_{r-2} + c)\omega_{h-1} + (2a_{r-1} - 2c + a_r)\omega_h$$
$$\text{if} \quad b+1 < j = r-1; \quad \text{and}$$

(5.15)

$$\lambda_2 = \begin{cases} (a_{b-1} + a_b + c)\omega_1 + (a_{b+1} - c + a_{b+2})\omega_2 & \\ +\omega_+(3, h, b) & \text{for } j = b+1 < r-1, \\ (a_{b-1} + a_b - c + a_{b+1})\omega_1 + \omega_+(2, h, b) & \text{for } j = b < r-1, \\ (a_{r-3} + a_{r-2} + c)\omega_1 + (2a_{r-1} - 2c + a_r)\omega_2 & \text{for } b+1 = j = r-1, \\ (2a_{r-2} + 2a_{r-1} - 2c + a_r)\omega_1 & \text{for } j = b = r-1. \end{cases}$$

Now let $\sum_{i=b}^r a_i < d \leq \sum_{i=b-1}^r a_i$. Then $d = c + \sum_{i=b}^r a_i$ where $0 < c \leq a_{b-1}$. If $b < r$, set $m_1 = v(b+1, r, a_r)$ and $m = X_{-b,d}X_{-(b-1),c}m_1$. For $b = r$ put $c_1 = [(c+1)/2]$ and $m = X_{-r,d}X_{-(r-1),c_1}v$. Observe that $a_r + 2c_1 \leq a_r + 2a_{r-1} < p$ as φ is not p-large. Applying Lemmas 2.46 and 2.1, one concludes that in all cases $m \neq 0$. If $b < r$ or $c = 2c_1$, Lemmas 2.48 and 2.46 force that $\mathcal{X}_i$ fixes m for $i \neq b$.

If $b = r$ and c is odd, it is clear that $\mathcal{X}_i$ fixes m for $i < r-1$. Lemma 2.49 implies that $\mathcal{X}_{r-1}$ fixes m as well. Hence $U^+(\Gamma)$ fixes m.

Recall that $\lambda_2 = 0$ if $b = r$. We have

(5.16)

$$\lambda_1 = \begin{cases} (a_1' + d - 2c)\omega_1 + (a_2' + c)\omega_2 + \Theta + a_{b_1}\omega_{b_1} & \text{if } 4 < b < r, \\ (a_1 + d + a_3 - 2c)\omega_1 + (a_2 + c)\omega_2 & \text{for } 4 = b < r, \\ (a_1' + a_r)\omega_1 + (a_2' + c_1)\omega_2 + \Theta + a_{b_1}\omega_{b_1} & \text{if } 4 < b = r \text{ and } c = 2c_1, \\ (a_1 + a_3 + a_4)\omega_1 + (a_2 + c_1)\omega_2 & \text{if } b = r = 4 \text{ and } c = 2c_1, \\ (a_1' + a_r - 1)\omega_1 + (a_2' + c_1)\omega_2 + \Theta + a_{b_1}\omega_{b_1} & \text{if } 4 < b = r \text{ and } c = 2c_1 - 1, \\ (a_1 + a_3 + a_4 - 1)\omega_1 + (a_2 + c_1)\omega_2 & \text{if } b = r = 4 \text{ and } c = 2c_1 - 1; \end{cases}$$

(5.17)

$$\lambda_2 = \begin{cases} (a_b + c)\omega_1 + \omega_+(2, r-b-1, b-1) + (2a_{r-1} + a_r)\omega_{r-b} & \text{if } b < r-1, \\ (2a_{r-1} + 2c + a_r)\omega_1 & \text{for } b = r-1. \end{cases}$$

Finally, let $d \leq \sum_{i=b}^r a_i$. If $d > a_b$, there exist j and c such that $b < j \leq r$, $0 < c \leq a_j$, and $d = \sum_{i=b}^{j-1} a_i + c$. Then put $m = v(b, j, c)$. If $d \leq a_b$, set $j = b$ and $m = X_{-b,d}$. In all cases $m \neq 0$ and $\mathcal{X}_i$ fixes m for $i \neq b$ by Lemma 2.46. So $U^+(\Gamma)$

fixes m. We have

$$\lambda_1 = (a_1' + d)\omega_1 + (\sum_{i=2}^{b_1-1} a_i'\omega_i) + a_{b_1}\omega_{b_1}; \tag{5.18}$$

$$\lambda_2 = \begin{cases} \omega_+(1, g-1, b-1) + (a_j - c + a_{j-1})\omega_g & \\ +(a_{j+1} + c)\omega_{g+1} + \omega_+(g+2, h, b) & \text{if } b < j < r-1, \\ (a_{b+1} + d)\omega_1 + \omega_+(2, h, b) & \text{for } b = j < r-1, \\ \omega_+(1, h-2, b-1) + (a_{r-1} - c + a_{r-2})\omega_{h-1} & \\ +(a_r + 2c)\omega_h & \text{for } b < j = r-1, \\ \omega_+(1, h-1, b-1) + (2a_{r-1} + a_r)\omega_h & \text{for } b < j = r, \\ (a_r + 2d)\omega_1 & \text{for } b = j = r-1. \end{cases} \tag{5.19}$$

We have $\zeta(\alpha_b) = 1$ and $\zeta(\alpha_t) = 0$ for $t \neq b$. This implies that $\omega_A(m) = p - 1$ in all cases.

For $1 \leq i \leq b_1$ put $\gamma_i = \varepsilon_i + \varepsilon_{b+1-i}$. One can assume that $U^+(A)$ consists of elements of the form $\prod_{i=1}^{b_1} x_{\gamma_i}(t_i)$. We do not need to know relations between t_i. Observe that each of the roots involved in this product is a linear combination of the simple roots with a nonzero coefficient for some α_k with $k \leq b_1$. As $b_1 < b - 1$, the construction of m implies that $U^+(A)$ fixes m.

C). Finally, let $G = C_r(K)$. Then $b = 2b_1 + 1 \geq 3$.

First assume that $d > 2a_r + \sum_{i=b-1}^{r-1} a_i$. Then there exist j and c such that $b \leq j \leq r-1$, $0 < c \leq a_j$, and $d = c + (\sum_{i=b-1}^{j} a_i) + 2\sum_{i=j+1}^{r} a_i$. Set $m_1 = v(r, j, c)$. If $b < r - 1$, put

$$f_t = \omega_t(m_1) \quad \text{for} \quad b \leq t < r,$$

$$h_{r-1} = f_{r-1},\ h_t = f_t + h_{t+1} \quad \text{for} \quad b+1 \leq t < r-1, \quad \text{and} \quad h_b = f_b + a_{b-1} + h_{b+1};$$

$$m = m_1(b \cdot h_b, (b-1) \cdot a_{b-1}, (b+1) \cdot h_{b+1}, (b+2) \cdot h_{b+2}, \dots, (r-1) \cdot h_{r-1}).$$

As the values $\langle \alpha_t, \alpha_u \rangle$ are known, one can check that $h_b = d$ if $b < j$ and $h_b = d - c$ for $j = b$. (Note that $\langle \alpha_r, \alpha_{r-1} \rangle = -2$.) If $b = r - 1$, set $m = X_{-(r-1), a_{r-2}+a_{r-1}+2a_r} X_{-(r-2), a_{r-2}} m_1$.

Applying Lemmas 2.46, 2.48, and 2.49 and arguing as in Item II of the proof, we conclude that m_1 and $m \neq 0$ and m is fixed by $\mathcal{X}_i$ for $i \neq b$. We give some comments for $i = r$. First assume that $b < r - 1$. According to the commutator relations in Lemma 2.1, it suffices to show that $u_k = X_{r,k} X_{-(r-1), f_{r-1}} m_1 = 0$ for $k > 0$. As in Item II, write $u_k = X_{r,k} X_{-(r-1), f_{r-1}} X_{-r,t} X_{-(r-1),l} u$ where $\omega(u) = \omega - \sum_{i=1}^{r-2} b_i \alpha_i$ and $l = \omega_{r-1}(u)$ or $u = v$ and $l = c$, $t = a_r + l$ in both cases. As φ is not p-large, one easily observes that $\omega_f(u) < p$ for $f = r - 1$ or r. Applying Lemma 2.49 to the indecomposable $KG(r-1, r)$-module generated by u, we conclude that $u_k = 0$.

If $b = r - 1$, we put $t = a_{r-2} + a_{r-1} + 2a_r$ and have to show that $u_k' = X_{r,k} X_{-(r-1),t} X_{-(r-2), a_{r-2}} X_{-r, a_r + c} X_{-(r-1),c} v = 0$ for $k > 0$. We assume that this is false for some k and obtain that $\omega(w_r w_{r-1} u_k') \notin \mathbf{X}(M)$ which yields a contradiction. Hence in all cases $U^+(\Gamma)$ fixes m.

Using the formulae for ρ in Proposition 2.43, we get

$$(5.20)\qquad \lambda_1 = \begin{cases} (a_1 + d - a_{b-1})\omega_1 + (a'_2 + a_{b-1})\omega_2 + \Theta + 2a'_{b_1}\omega_{b_1} & \text{if } b > 5, \\ (a_1 + d - a_4)\omega_1 + 2(a_2 + a_3 + a_4)\omega_2 & \text{for } b = 5, \\ 2(a_1 + d)\omega_1 & \text{for } b = 3; \end{cases}$$

(5.21)

$$\lambda_2 = (a_{b-1} + a_b)\omega_1 + \omega_+(2, g-1, b-1) + (a_{j-1} + c)\omega_g + (a_j - c + a_{j+1})\omega_{g+1} + \omega_+(g+2, h, b)$$

if $b + 1 < j < r - 1$,

$$\lambda_2 = (a_{b-1} + a_b)\omega_1 + \omega_+(2, h-2, b-1) + (a_{r-2} + c)\omega_{h-1} + (a_{r-1} + a_r - c)\omega_h$$

if $b + 1 < j = r - 1$,

$$\lambda_2 = (a_{b-1} + a_b + c)\omega_1 + (a_{b+1} - c + a_{b+2})\omega_2 + \omega_+(3, h, b) \quad \text{for } j = b+1 \le r-1,$$

and $\lambda_2 = (a_{b-1} + a_b - c + a_{b+1})\omega_1 + \omega_+(2, h, b)$ for $j = b \le r - 1$.

Now assume that $\sum_{i=b-1}^{r-1} a_i < d \le 2a_r + \sum_{i=b-1}^{r-1} a_i$. Set $d_- = d - \sum_{i=b-1}^{r-1} a_i$. Then $d_- = 2c$ or $2c - 1$ where $0 < c \le a_r$. Put $f_r = c$, $f_{r-1} = d_- + a_{r-1}$, and $f_i = a_i + f_{i+1}$ for $b < i < r - 1$. Set

$$m = \begin{cases} v(b \cdot d, (b-1) \cdot a_{b-1}, (b+1) \cdot f_{b+1}, \dots, \\ (r-1) \cdot f_{r-1}, r \cdot c) & \text{for } b < r - 1, \\ v((r-1) \cdot d, r \cdot c, (r-2) \cdot a_{r-2}) & \text{if } b = r-1,\ d_- = 2c \\ & \text{or } a_{r-2} = 0, \\ v((r-1) \cdot d, r \cdot c, (r-2) \cdot (a_{r-2} - 1)) & \text{if } b = r-1,\ d_- = 2c - 1, \\ & \text{and } a_{r-2} > 0. \end{cases}$$

If $d_- = 2c$, Lemma 2.48 yields that $m \neq 0$ and is fixed by $\mathcal{X}_i$ for $i \neq b$.

Let $d_- = 2c - 1$. First assume that $b < r - 1$. Put $m' = X_{-(r-1), f_{r-1}} X_{-r,c} v$. Applying Lemma 2.49 to the $G(r-1, r)$-module generated by v, we deduce that $m' \neq 0$ and $X_{r,k} m' = 0$ for $k > 0$. Now the commutator relations in Lemma 2.1 yield that $X_{r,k} m = 0$ for $k > 0$ and $\mathcal{X}_r$ fixes m. Observe that the arguments in the proof of Lemma 2.48 showing that $\mathcal{X}_i$ fixes m for relevant i require no specific assumptions for ω. Using these arguments for the $G(1, 2, \dots, r-1)$-module generated by $X_{-r,c} v$, we deduce that $\mathcal{X}_i$ fixes m for $i < r$, $i \neq b$. Next, apply Lemma 2.48 to the $G(1, 2, \dots, r-2)$-module generated by m' and conclude that $m \neq 0$.

If $b = r - 1$, we apply Lemma 2.48 for $a_{r-2} \neq 0$ and Lemma 2.49 for $a_{r-2} = 0$ and deduce that $m' \neq 0$ and is fixed by $\mathcal{X}_i$, $i \neq b$, in this case as well.

We obtain the following formulae for λ. If $b < r - 1$, or $d_- = 2c$, or $a_{r-2} = 0$, the weight λ_1 satisfies (5.20). Now assume that $b = r-1$, $d_- = 2c-1$, and $a_{r-2} > 0$. Put $\Theta' = \sum_{i=3}^{b_1-1} (a_i + a_{p-2-i})\omega_i$. Then $n = 2p - 2$, $r = p - 1$, $p \ge 5$, and

(5.22)

$$\lambda_1 = \begin{cases} (a_1 + d + 2 - a_{p-3})\omega_1 + (a_2 + a_{p-4} + a_{p-3} - 1)\omega_2 + \Theta' \\ +2(a_{b_1} + a_{b_1+1})\omega_{b_1} & \text{if } p > 7, \\ (a_1 + d + 2 - a_4)\omega_1 + 2(a_2 + a_3 + a_4 - 1)\omega_2 & \text{for } p = 7, \\ 2(a_1 + d + 1)\omega_1 & \text{for } p = 5. \end{cases}$$

We have
(5.23)

$$\lambda_2 = \begin{cases} (a_b + a_{b-1})\omega_1 + \omega_+(2, r-1-b, b-1) \\ +(a_{r-1} + a_r)\omega_{r-b} & \text{if } b+1 < r \text{ and } d_- = 2c, \\ (a_b + a_{b-1})\omega_1 + \omega_+(2, r-1-b, b-1) \\ +(a_{r-1} + a_r - 1)\omega_{r-b} & \text{if } b+1 < r \text{ and } d_- = 2c-1, \\ (a_{r-2} + a_{r-1} + a_r)\omega_1 & \text{for } b+1 = r \text{ and } d_- = 2c, \\ (a_{r-2} + a_{r-1} + a_r - 1)\omega_1 & \text{for } b+1 = r \text{ and } d_- = 2c-1. \end{cases}$$

Next, let $\sum_{i=b}^{r-1} a_i < d \leq \sum_{i=b-1}^{r-1} a_i$. Then $d = c + \sum_{i=b}^{r-1} a_i$ with $0 < c \leq a_{b-1}$. If $b = r-1$, set $m = X_{-(r-1),d} X_{-(r-2),c} v$. Otherwise put $f_{r-1} = a_{r-1}$, $f_j = a_j + f_{j+1}$ for $b < j < r-1$, and $f_b = d$, then take $m = v(b \cdot d, (b+1) \cdot f_{b+1}, \ldots, (r-1) \cdot f_{r-1}, (b-1) \cdot c)$. By Lemmas 2.46 and 2.48, $m \neq 0$ and is fixed by $\mathcal{X}_i$ for $i \neq b$. We have

$$(5.24)\quad \lambda_1 = \begin{cases} (a_1' - 2c + d)\omega_1 + (a_2' + c)\omega_2 + \Theta + 2(a_{b_1} + a_{b_1+1})\omega_{b_1} & \text{if } b > 5, \\ (a_1 + a_4 - 2c + d)\omega_1 + 2(a_2 + a_3 + c)\omega_2 & \text{for } b = 5, \\ 2(a_1 + a_2 - c + d)\omega_1 & \text{for } b = 3; \end{cases}$$

(5.25)

$$\lambda_2 = \begin{cases} (c + a_b)\omega_1 + \omega_+(2, r-1-b, b-1) + (a_{r-1} + a_r)\omega_{r-b} & \text{if } b < r-1, \\ (a_{r-1} + a_r + c)\omega_1 & \text{for } b = r-1. \end{cases}$$

Finally, let $d \leq \sum_{i=b}^{r-1} a_i$. There exist j and c such that $b \leq j \leq r-1$, $0 < c \leq a_j$, and $d = c + \sum_{i=b}^{j-1} a_i$. Set $m = v(b, j, c)$. By Lemma 2.46, $m \neq 0$ and is fixed by $\mathcal{X}_i$ for $i \neq b$. We get

$$(5.26)\qquad \lambda_1 = \begin{cases} (a_1' + d)\omega_1 + (\sum_{i=2}^{b_1-1} a_i'\omega_i) + 2a_{b_1}'\omega_{b_1} & \text{if } b > 3, \\ 2(a_1 + a_2 + d)\omega_1 & \text{for } b = 3; \end{cases}$$

(5.27)

$$\lambda_2 = \begin{cases} \omega_+(1, g-1, b-1) + (a_{j-1} + a_j - c)\omega_g + (a_{j+1} + c)\omega_{g+1} \\ +\omega_+(g+2, r-b, b) & \text{if } j > b, \\ (a_{b+1} + c)\omega_1 + \omega_+(2, r-b, b) & \text{for } j = b. \end{cases}$$

One can deduce that in all cases $\omega_A(m) = (p-1)\omega_1$.

Set $\alpha = 2\varepsilon_{b_1+1}$ and for $1 \leq i \leq b_1$ use the notation γ_i introduced at the end of Item B). We can assume that $U^+(A)$ consists of elements of the form $x_\alpha(t_\alpha)\prod_{i=1}^{b_1} x_{\gamma_i}(t_{\gamma_i})$. Observe that $\gamma_i = \alpha_i + \beta_i$ where $\beta_i \in R^+$, and $\alpha = 2(\alpha_{b_1+1} + \ldots + \alpha_{r-1}) + \alpha_r$. As $b_1 < b-1$, we conclude that $\mathcal{X}_{\gamma_i}$ fixes m for $1 \leq i \leq b_1$. If $b > 3$, then $b_1 + 1 < 2b_1 = b-1$ and hence $\mathcal{X}_\alpha$ and $U^+(A)$ fix m.

We need some special arguments for the case $b = 3$. Set $\gamma = \varepsilon_2 + \varepsilon_3$, $\delta = \varepsilon_2 + \varepsilon_4$, and $\eta = 2\varepsilon_3$. Obviously, $\mathcal{X}_\alpha$ fixes m if $m = fv(i_1 \cdot k_1, \ldots, i_l \cdot k_l)$ with $f \in K^*$ and $i_j > 2$ for $1 \leq j \leq l$. Otherwise m is one of the vectors

$$(5.28)\qquad X_{-3,k_3}X_{-2,k_2}Yv,\ X_{-3,k_3}X_{-2,k_2}YX_{-3,k_3'}v,\ X_{-3,k_3}YX_{-2,k_2}v$$

where $Y = X_{-i_1,k_{i_1}} \ldots X_{-i_t,k_{i_t}}$, $i_j > 3$ for $1 \leq j \leq t$, or $Y = 1$. As φ is not p-large, the construction of m implies that in all cases $k_l < p$. Hence $X_{\alpha,f}m = 0$ for $f \geq p$ and Formula (2.1) yields that it suffices to show that $X_\alpha m = 0$. Commutator relations in Lemma 2.1 imply that X_α commutes with $X_{-i,l}$ for $i \neq 2$, X_γ commutes with $X_{-i,l}$ for $i \neq 2$ or 3, X_δ commutes with $X_{-i,l}$ for $i \neq 2$ or 4, X_η commutes with $X_{-i,l}$ for $i \neq 3$, the commutator $[X_\alpha, X_{-2,l}] \in \langle X_\gamma X_{-2,l-1}, X_\eta X_{-2,l-2}\rangle$ (the second product is 0 for $l = 1$), $[X_\gamma, X_{-2,l}] \in \langle X_\eta X_{-2,l-1}\rangle$, and $[X_\gamma, X_{-3,l}] \in \langle X_\delta X_{-3,l-1}\rangle$. Observe also that $X_\gamma X_{-2,l} Y v = X_\eta X_{-3,k_3'} v = 0$. This implies that $X_\alpha m = 0$ for a vector m of the form (5.28). Hence $U^+(A)$ fixes m in all cases.

D). So the vector m with the required properties has been constructed for all types under consideration. If λ is not one of the weights of Items 1)–10) of Lemma 5.1 or $e = 1$ and λ_1 does not satisfy (5.2), that lemma yields that $d_{M_0}(x_\Gamma) = p^s$ as desired. Now we identify the situations where the exceptional weights of Lemma 5.1 occur. Analyzing the construction of m in Items A)–C), we deduce from Formulae (5.8)–(5.27) that these situations are exhausted by Cases 1)–7) indicated below. It is essential here that φ is not p-large. Observe also that $N(\varphi) < p$ if $\omega = \omega_i$ and $N(\varphi) \neq p$ if $\omega = a_i\omega_i$ or $G = A_r(K)$ and $\omega = a_k\omega_k + a_{p^s}\omega_{p^s}$ with $b \leq k < p^s$; $d - a_{b-1} > 0$ in situations described by Formulae (5.13) and (5.14); we have $d_{\psi_1}(g_1) = p = p^s$ and hence $d_{M'}(x_\Gamma) = p^s$ if $G = B_{p-2}(K)$, $p \geq 7$, $a_{p-4} = a_{p-2} = 0$ and $d = 2a_{p-3} > 0$ or $G = C_{p-1}(K)$, $p \geq 5$, $a_{p-3} = a_{p-1} = 0$ and $d = 2a_{p-2} > 0$. (To check the latter assertion for $G = C_{p-1}(K)$, apply Formula (5.20) and observe that $a_1 + d \neq p$ if $N_1(\varphi) < p$ and $b > 3$.) Now we list the exceptional cases that require special arguments.

1). $G = A_r(K)$, $a_i = 0$ for $b \leq i < p^s - 1$, $d = a_{p^s-1}$, $\lambda_2 = 0$.

2). $G = B_r(K)$ or $C_r(K)$, $a_{b-1} = 0$ and $a_i = 0$ for $i > b$, $d = 2a_b$.

3) $G = C_r(K)$, $\omega = \omega(1, b-2) + \omega_r$, $N(\varphi) = p$.

4). $G = C_r(K)$, $b = 3$, $\omega = (p-2)\omega_1 + \omega_4$.

5). $G = C_r(K)$, $b = 3$, $\omega = (p-3)\omega_1 + \omega_2 + \omega_3$.

6). $G = C_r(K)$, $b = 3$, $\omega = (p-5)\omega_1 + 3\omega_3$.

7). $G = C_4(K)$, $p = 5$, $\omega = 3\omega_1 + \omega_4$.

Case 1). Obviously, $d_\varphi(x) = d_{\varphi^*}(x)$ and $N(\varphi) = N(\varphi^*)$. Recall that $\omega(\varphi^*) = \sum_{i=1}^r a_{n+1-i}\omega_i$. If $b+1 < p^s - 1$ or $a_{p^s} \neq 0$, it follows from the facts proved above that $d_{\varphi^*}(x) = p^{s+1}$ as desired. So we assume that $b+1 = p^s - 1$ and $a_{p^s} = 0$. As $2 \leq b < p$, this forces $s = 1$, $b = p-2$, and $p > 3$. Since $2p-5 \geq p$, Algorithm 1.6 and Theorem 1.14 yield that $d_\mu(g_1) = p$ for $\mu \in \operatorname{Irr}_p \Gamma_1$ unless $\omega(\mu) \in \{0, \omega_1, \omega_{p-3}\}$. The construction of m in Item A) shows that $m = v(b, b+1, a_{b+1})$. Hence Formula (5.10) implies that either $d_{\psi_1}(g_1) = d_N(x_\Gamma) = p$ as required, or $d = a_{b+1} = 1$ and $a_i = 0$ for $i \neq b+1$. But the latter possibility yields a contradiction as $N(\varphi) \geq p$. This completes the analysis of Case 1).

Case 2). We have $N_1(\varphi) = p-1$. Consider the group G_1 and the representation $\varphi_1 \in \operatorname{Irr} G_1$ constructed in Part I of the proof. Set $\omega^1 = \omega(\varphi_1)$. For the group G_1 construct the group $\Gamma^1 = \Gamma_1^1\Gamma_2^1$ and the mapping $\rho_1 : \mathbf{X}(G_1) \to \mathbf{X}(\Gamma^1)$ by the same way as Γ and ρ were constructed for G in Proposition 2.43. Here Γ_j^1 are the simple components of Γ^1 and commute. Put $\lambda^1 = \rho_1(\omega^1)$ and write $\lambda^1 = (\lambda_1^1, \lambda_2^1)$ where $\lambda_i^1 \in \mathbf{X}(\Gamma_i^1)$, $i = 1, 2$. Set $\tau = \varphi(\lambda^1) \in \operatorname{Irr}\Gamma^1$. We have $\tau = \tau_1 \otimes \tau_2$ with $\tau_i \in \operatorname{Irr}\Gamma_i^1$. Let $u_i \in \Gamma_i^1$ be a regular unipotent element. Then $u = u_1u_2$ is a regular unipotent element of Γ^1. The arguments of Lemma 2.42 and Proposition 2.5 yield that $d_{\varphi_1}(g) = p^{s+1}$ for a regular unipotent element $g \in G_1$ if $d_\tau(u) = p^s$. In this situation by Lemma 2.19, $d_\varphi(x) = d_\varphi(g) = p^{s+1}$ as desired.

Observe that $\lambda_2^1 = a_b\omega_2$. Our assumptions imply that $b < r$. Hence $e \neq 1$. As $b \neq 2$, we get $e \neq 3$ if $p = 5$. Since Conjecture (r, s) holds, one can apply Algorithm 1.6 and Theorem 1.7 to the group Γ_2^1 and conclude that $d_\tau(u) = d_{\tau_2}(u_2) = p^s$ if $s > 1$ or $s = 1$, $e \geq (p-1)/2$, and $e \neq 3$.

Now assume that $s = 1$ and $e < (p-1)/2$. Then $b > (p+1)/2$ and so $b > 3$ since $p > 3$. Return to the analysis of the Γ-module M_0 defined at the beginning of Part II and the degree $d_{M_0}(x_\Gamma)$. The construction of m shows that λ_1 satisfies (5.13) for $G = B_r(K)$ and (5.20) for $G = C_r(K)$. Then $\lambda_1 = (a_1 + 2a_b)\omega_1 + \sum_{i=2}^{r_1} c_i\omega_i$. As $N_1(\varphi) < p$, we get $a_1 + 2a_b < p$. Conjecture (r,s), Proposition 1.5, Algorithm 1.6, and Theorem 1.9 imply that $d_{\psi_1}(g_1) = d_{M_0}(x_\Gamma) = p$ as required. Finally, let $(p-1)/2 \leq e = 3$. Then $p = 7$ and $b = 4$. The arguments of the previous paragraph yield that $d_{\psi_1}(g_1) = d_{M_0}(x_\Gamma) = p$ in this case as well. This completes the analysis of Case 2).

Case 3). We construct another vector m' that can play the role of m. The weight $\omega_\Gamma(m')$ will permit us to use the general scheme described at the beginning of Item II. Since b is odd, $b \leq p-2$. Then $\omega(1, r-2) \notin \{0, \omega_1\}$ as $N(\varphi) \geq p$. Fix minimal i with $a_i \neq 0$. We have $i \leq b$. Set $m' = v(b, i, 1)$. By Lemma 2.46, $m' \neq 0$ and $\mathcal{X}_j$ fixes m' for $i \neq b$. Hence $U^+(\Gamma)$ fixes m'. As we have seen at the end of Item C), each element of $U^+(A)$ is a product of root elements associated with roots of the form $\alpha_r + \sum_{t=1}^{r-1} b_t\alpha_t$. This yields that $U^+(A)$ fixes m'. It is clear that $\omega_A(m') = p$. Put $\lambda' = \omega_\Gamma(m')$ and $M' = M(\lambda')$. Arguing as for M_0 at the beginning of Item II, we can show that it suffices to prove that $d_{M'}(x_\Gamma) = p^s$. Write $\lambda' = (\lambda_1', \lambda_2')$ where $\lambda_j' \in \mathbf{X}_j$, $j = 1, 2$. One easily concludes that $\lambda_1' \neq 0$ and $\lambda_2' = \omega_1 + \omega_{q_2}$. Using Lemma 5.1, we deduce that $d_{M'}(x_\Gamma) = p^s$ and complete the proof for Case 3).

Now let one of Cases 4), 5), or 7) occur. All of these cases are dealt with the help of similar arguments. We keep notation G_1, φ_1, ω^1, and g introduced before. Applying Conjecture (r,s) and Theorem 1.7 to the group G_1, we show that in each of these cases $d_{\varphi_1}(g) = p^{s+1}$. Then Lemma 2.19 yields that $d_\varphi(x) = p^{s+1}$ as φ_1 is a factor of $\varphi|G_1$. Set $\xi = (\varphi_1)_{\mathbb{C}}$. Observe that $n(G_1) = p^s + 1$. By Proposition 2.5 b), for $1 \leq j \leq s$ the element g^{p^j} has one Jordan block of size $p^{s-j}+1$ and $p^j - 1$ blocks of size p^{s-j}. We have

$$\omega^1 = \begin{cases} (p-2)\omega_1 + \omega_3 & \text{in Case 4),} \\ (p-3)\omega_1 + \omega_2 + \omega_3 & \text{in Case 5),} \\ 3\omega_1 + \omega_3 & \text{in Case 7).} \end{cases}$$

Let $g_j^0 \in (G_1)_{\mathbb{C}}$ be a unipotent element with the same canonical Jordan form as g^{p^j}, $0 \leq j \leq s$. Proposition 1.5 and Algorithm 1.6 imply that

$$d_\xi(g_0^0) = \begin{cases} p^{s+1} + p^s - 5 & \text{in Case 4),} \\ p^{s+1} + 2p^s - 7 & \text{in Case 5),} \\ 25 & \text{in Case 7)} \end{cases}$$

and

$$d_\xi(g_j^0) = \begin{cases} p^{s-j+1} + p^{s-j} - 1 & \text{in Case 4),} \\ p^{s-j+1} + 2p^{s-j} - 2 & \text{in Case 5),} \\ 5 & \text{in Case 7)} \end{cases}$$

for $1 \leq j \leq s$. Notice that in Case 7) $p = 5$ and $j = 1$. As $p \geq 5$, in all situations being considered we get $d_\xi(g_0^0) \geq p^{s+1}$ and $p^j d_\xi(g_j^0) \geq p^{s+1}$ for $1 \leq j \leq s$. Hence $d_{\varphi_1}(g) = p^{s+1}$ by Theorem 1.7. This completes the analysis of Cases 4), 5), and 7).

Finally, let Case 6) hold. If $r > 4$, set $u_1 = v(r,3,1)$ and $u = X_{-3,3}X_{-4}\dots X_{-(r-1)}u_1$. For $r = 4$ put $u = X_{-3,3}X_{-4}X_{-3}v$. Arguing as for the vector m in Item C), we deduce that $u \neq 0$ and is fixed by $\mathcal{X}_i$ for $i \neq 3$ and by $U^+(A)$. Hence u is fixed by $U^+(\Gamma)$ as well. We have $\omega_1(u) = p-5$, $\omega_2(u) = 4$, $\omega_3(u) = -3$, $\omega_4(u) = 2$, $\omega_i(u) = 0$ for $4 < i \leq r$, and $\omega_A(u) = p$. Set $M(u) = KAu$ and denote by α the root in $R^-(A)$. We claim that $M(u) \cong V(p)$. By Lemma 2.55, it suffices to check that $u' = X_\alpha u \neq 0$. In our assumptions $X_\alpha = aX_{-(\varepsilon_1+\varepsilon_3)} + bX_{-2\varepsilon_2}$ with $a, b \neq 0$. Hence it suffices to show that $X_{-(\varepsilon_1+\varepsilon_3)}u \neq 0$. Set $\varepsilon_1 + \varepsilon_3 = \varepsilon$. It is clear that $\mathcal{X}_\varepsilon$ fixes u. We have $\langle \omega(u), \varepsilon \rangle = p - 3 < p$. Hence by Lemma 2.1, $X_{-\varepsilon}u$ and $u' \neq 0$. Thus $M(u) \cong V(p)$. By Lemma 2.56, $u' \in M_y$. As $u' \in M(u)$ and $\Gamma \subset C_G(A)$, we conclude that $U^+(\Gamma)$ fixes u' and u' generates an indecomposable $K\Gamma$-module N_Γ with highest weight $\omega_\Gamma(u') = \omega_\Gamma(u)$. It is clear that $N_\Gamma \subset M_y$. Set $\omega_\Gamma(u') = (\lambda_1', \lambda_2')$ with $\lambda_i' \in \mathbf{X}_i$ and $\kappa = \varphi(\omega_\Gamma(u'))$. We have $\lambda_1' = 2(p-1)$ and $\lambda_2' = 2\omega_1$. Now Lemma 5.1 forces that $d_\kappa(x_\Gamma) = p^s$ and hence $d_{M_y}(x) = d_{M_y}(x_\Gamma) = p^s$ as desired. All the possibilities have been considered. The proposition is proved. □

COROLLARY 5.5. *If $G \neq D_r(K)$ and $n = p^s + b$ with $0 < b < p$, Theorem 1.7 holds for p-restricted representations provided Conjecture (r, s) holds.*

PROOF. This follows from Corollary 4.11, Lemmas 5.2 and 5.3, and Proposition 5.4. □

6. A special case for $G = B_r(K)$

In this section we assume that $G = B_r(K)$, $n = p^s + b$ with $p < b < 2p$, φ is not p-large, and $\omega = (\sum_{i=1}^{p-1} a_i\omega_i) + a_r\omega_r$ with $a_r \neq 0$. Our goal is to prove Theorem 1.7 for this case provided Conjecture (r, s) holds. The scheme of the proof is close to that of Section 5. We distinguish the following two cases: $s = 1$ and $s > 1$. Recall that $b = 2b_1$ with $b_1 \in \mathbb{Z}^+$. Set $e = p^s - b$. As before, put $\omega(1, p-1) = \sum_{i=1}^{p-1} a_i\omega_i$.

6.1. $s = 1$. First assume that $s = 1$. Then $n = 2p + c$ with $0 < c < p$ and $c = 2c_1 + 1$, $c_1 \in \mathbb{Z}^+$. By Proposition 2.5 b), y has c Jordan blocks of size 3 and e blocks of size 2. Hence $\Gamma_1 = B_{c_1}(K)$ if $c_1 > 0$ and $\Gamma_1 = 1$ for $c_1 = 0$; $\Gamma_2 = C_{e/2}(K)$. We have

$$\zeta(\varepsilon_i) = \begin{cases} 2 & \text{for } 1 \leq i \leq c, \\ 1 & \text{for } c < i \leq p, \\ 0 & \text{for } p < i \leq r. \end{cases} \tag{6.1}$$

LEMMA 6.1. *Let $\mu \in \operatorname{Irr}\Gamma$ with $\omega(\mu) = (\nu_1, \nu_2)$, $\nu_j \in \mathbf{X}_j$. Assume that $d_\mu(x_\Gamma) < p$. Then one of the following holds:*

1) ν_1 or $\nu_2 = 0$;

2) $\nu_j = p^{l_j}\omega_1$ for $j = 1$ and 2;

3) $c = 3$, $\nu_1 = 2p^l\omega_1$, $\nu_2 = p^t\omega_1$;

4) $c = 5$, $\nu_1 = p^l\omega_2$, $\nu_2 = p^t\omega_1$;

5) $c = 7$, $\nu_1 = p^l\omega_3$, $\nu_2 = p^t\omega_1$;

6) $p = 5$, $G = B_6(K)$, $\nu_1 = p^l\omega_1$, $\nu_2 = 2p^t\omega_1$;

7) $p = 7$, $G = B_9(K)$, $\nu_1 = p^l\omega_2$, $\nu_2 = 2p^t\omega_1$;

8) $p = 7$, $G = B_8(K)$, $\nu_1 = p^l\omega_1$, $\nu_2 = p^t\omega_2$.

If $c = 1$, then $\nu_2 = 0$ or $p^t\omega_1$.

PROOF. The lemma follows immediately from Proposition 1.5, Algorithm 1.6, and Theorem 1.14. □

Set

$$N(\varphi) = N_G(\varphi) = \big(2\sum_{i=1}^{c} ia_i + \sum_{i=c+1}^{p-1} a_i(i+c)\big) + a_r(p+c)/2. \tag{6.2}$$

The construction of the group A in Proposition 2.43, Proposition 1.5, Algorithm 1.6, and Theorem 1.14 imply that $d_\varphi(y) = \min\{p, N(\varphi)+1\}$. Observe that $a_r = 1$ if $N(\varphi) < p$.

For $\Delta = B_t(K)$ with $p < t$ and $2t+1 = 2p+f < 3p$ and a representation $\mu \in \operatorname{Irr}\Delta$ with $\omega(\mu) = (\sum_{i=1}^{p-1} b_i\omega_i) + b_t\omega_t$ define $N_\Delta(\mu)$ replacing in Formula (6.2) a_i with b_i, r with t, and c with f, respectively.

LEMMA 6.2. *Assume that $N(\varphi) < p$. Then $d_\varphi(x) = p(N(\varphi)+1)$ unless one of the following holds:*

1) $\omega = \omega(1,c) + \omega_r$;

2) $c \leq 7$, $\omega = \omega_{c+1} + \omega_r$.

PROOF. The proof is quite similar to that of Lemma 5.2, but is easier as $s = 1$. Here we apply Lemma 6.1 instead of Lemma 5.1. Notice that some cases in Lemma 6.1 cannot yield exceptional weights with $N(\varphi) < p$. □

Recall that $z_0 \in G_{\mathbb{C}}$ is a regular unipotent element and $z_1 \in G_{\mathbb{C}}$ is a unipotent element with the same canonical Jordan form as x^p.

LEMMA 6.3. *Let $N(\varphi) < p$. Assume that one of the conditions* 1) *or* 2) *of Lemma* 6.2 *holds. Then $d_\varphi(x) = \min\{d_{\varphi_{\mathbb{C}}}(z_0),\ p(N(\varphi)+1)\}$ in Case 1 and $d_\varphi(x) = d_{\varphi_{\mathbb{C}}}(z_0)$ in Case 2.*

PROOF. The scheme of the proof is the same as in Lemma 5.3. The proof is based on Corollaries 2.44 and 2.41 and the arguments in the proof of Lemma 5.2. As in the proof of Lemma 5.2, set $\lambda = \rho(\omega)$ and $M_0 = M(\lambda)$. Recall that we denote $d_{\varphi_{\mathbb{C}}}(z_i)$ by d_i, $i = 0,1$ (see Section 3). Then $d_1 = N(\varphi)+1$. Keep the following notation introduced in the proof of Lemma 5.3: $d_\varphi(x) = d$ and $d_{M_0}(x_\Gamma) = d'$. As in the proof of that lemma, we deduce that $d \geq pN(\varphi) + d'$, estimate d' using Conjecture (r,s), Formula (1.5), Proposition 1.5, and Algorithm 1.6, and then conclude that either $d \geq d_0$, or $d \geq pd_1$. Then the assertion of the lemma follows from Corollary 2.41. Consider Cases 1) and 2) of Lemma 6.2 separately.

Case 1). Obviously, $\lambda_2 = 0$. If $c = 1$, we have $\lambda_1 = 0$ and $d_0 = 1 + p(p+1)/2 = pN(\varphi) + d'$. Hence $d = d_0$.

Now assume that $c > 1$. The construction of the morphism ρ shows that

$$\lambda_1 = \big(\sum_{i=1}^{c_1-1}(a_i + a_{c-i})\omega_i\big) + (2a_{c_1} + 2a_{c_1+1} + 1)\omega_{c_1}.$$

Observe that λ_1 is p-restricted as ω is such. We have

$$d_0 = 1 + \big(\sum_{i=1}^{c} a_i i(2p+c-i)\big) + (p+c_1)(p+c_1+1)/2,\ d_1 = 1 + (p+c)/2) + 2\sum_{i=1}^{c} a_i i.$$

Let $z_\Gamma \in (\Gamma_1)_\mathbb{C}$ be a regular unipotent element and $\sigma = \varphi(\lambda_1) \in \mathrm{Irr}(\Gamma_1)_\mathbb{C}$. By Proposition 1.5 and Algorithm 1.6,

$$d_\sigma(z_\Gamma) = 1 + \big(\sum_{i=1}^{c} a_i i(c-i)\big) + c_1(c_1+1)/2.$$

Theorem 1.14 yields that $d' = \min\{p, d_\sigma(z_\Gamma)\}$. Observe that $d_0 = pN(\varphi) + d_\sigma(z_\Gamma)$. Hence $d \geq pN(\varphi) + d' = \min\{d_0, pd_1\}$ as required.

Case 2). As $N(\varphi) < p$, one has $p > 7$ for $c = 1$, $p > 11$ for $c = 3$, $p > 23$ for $c = 5$, and $p > 37$ for $c = 7$. Recall that $\Gamma_1 = B_{c_1}(K)$ for $c > 1$ and 1 for $c = 1$. Now $\lambda_1 = \omega_{c_1}$ in the first case and 0 in the second one, $\lambda_2 = \omega_1$, and

$$d_0 = 1 + (c+1)(2p-1) + (p+c_1)(p+c_1+1)/2 = pN(\varphi) + p - c + c_1(c_1+1)/2.$$

Set $c_+ = p - c + c_1(c_1+1)/2$. Observe that $c_+ < p$ for our values of c. If $c = 1$, it is clear that $d' = p-1 = c_+$. Otherwise Proposition 1.5, Algorithm 1.6, Theorem 1.14, and Lemma 2.8 imply that $d' = c_+$. This yields that $d_0 = pN(\varphi) + d' = d$ and completes the proof. □

PROPOSITION 6.4. *Let $N(\varphi) \geq p$. Then $d_\varphi(x) = p^2 = |x|$.*

PROOF. The proof is similar to that of Proposition 5.4. For $r > p$ we apply induction on r assuming that our proposition holds for groups $B_{r_1}(K)$ with $p \leq r_1 < r$. However, now we have to consider the case $r = p$ separately to create the induction base.

Set

$$N_1(\varphi) = \begin{cases} (\sum_{i=1}^{p-1} a_i i) + a_r(p-1)/2 & \text{for } c = 1, \\ (2\sum_{i=1}^{c-2} i a_i + \sum_{i=c-1}^{p-1} a_i(i+c-2)) + a_r(p+c-2)/2 & \text{for } c > 1. \end{cases}$$

Put $G_1 = G(2, \ldots, r)$. We claim that $\varphi|G_1$ has a composition factor $\varphi_1 \in \mathrm{Irr}_p G_1$ with $N_{G_1}(\varphi_1) = N_1(\varphi)$ (here for $c = 1$ the function $N_{G_1}(\varphi_1)$ is defined as in Section 5). For $c = 1$ set $u = v(1, r, a_r)$. If $c > 1$ and $a_i \neq 0$ for some $i \leq c$, choose maximal such i and put $u = v(1, i, a_i)$; otherwise take $u = v$. By Lemma 2.46, in all cases $u \neq 0$ and is fixed by $U^+(G_1)$. Hence u generates an indecomposable G_1-module with highest weight $\omega_u = \omega_{G_1}(u)$. So $\varphi_1 = \varphi(\omega_u)$ is a composition factor of $\varphi|G_1$. One can deduce that $\omega_u = \omega(1, p-2) + (2a_{p-1} + a_p)\omega_{p-1}$ for $c = 1$ and $\omega_u = \omega(1, c-1) + (a_c + a_{c+1})\omega_c + \omega_+(c+1, r-1, 1)$ for $c > 1$. It is clear that φ_1 is p-restricted and is not p-large as φ is such. Now Formulae (6.2) and (5.1) yield our claim on $N_{G_1}(\varphi_1)$.

I. First assume that $N_1(\varphi) \geq p$. Let $g \in G_1$ be a regular unipotent element. By Proposition 5.4 for $c = 1$ and the induction assumption otherwise, we have $d_{\varphi_1}(g) = p^2$. Then our proposition follows from Lemma 2.19.

II. Now let $N_1(\varphi) \leq p - 1$. This yields that $a_{p-1} = 0$, $a_{p-2} = 0$ for $c > 1$, and either $a_r = 1$, or $c = 1$ and $\omega = 2\omega_p$. As in Item II of the proof of Proposition 5.4, we deal with the maximal tori $T \subset G$, $T_\Gamma \subset \Gamma$, and $T_1 \subset A$, respectively, considered in Proposition 2.43. In the majority of cases we construct a nonzero vector $m \in M$ such that m is a weight vector for T, T_Γ and T_1, $U^+(A)$ and $U^+(\Gamma)$ fix m, and $m \in M_y$, and set $M_\Gamma = K\Gamma m$ and $\lambda = \omega_\Gamma(m)$. The construction of m is based on Lemmas 2.46, 2.55, and 2.56. As in Section 5, we write $\lambda = (\lambda_1, \lambda_2)$ where $\lambda_i \in \mathbf{X}_i$, $i = 1, 2$. Recall that $\lambda_1 = 0$ if $\Gamma_1 = 1$. Arguing as in the proof of

Proposition 5.4, we deduce that it suffices to show that $d_{M_\Gamma}(x_\Gamma) = p$. Naturally, the module $M_1 = M(\lambda)$ is a composition factor of M_Γ. Hence we are done if

$$d_{M_1}(x_\Gamma) = p. \tag{6.3}$$

Now we shall verify that in the majority of cases (6.3) holds. For the remaining cases special arguments will be used. As in the proof of Proposition 5.4, $N(\varphi) \neq p$ if $\omega = a_i\omega_i$.

Set $d = N(\varphi) - p + 1$. One easily observes that $d \leq \sum_1^p a_i$ for $c = 1$ and $d \leq 1 + a_{c-1} + 2\sum_{i=c}^{p-1} a_i$ for $c > 1$. Notice that $\zeta(\alpha_c) = \zeta(\alpha_p) = 1$ and $\zeta(\alpha_i) = 0$ for $i \neq c$ or p.

To handle the action of A on M, we conclude that $U^+(A)$ lies in the subgroup generated by certain fixed root subgroups of G and represent the root element $X_\alpha \in L(A)$ associated with the negative root $\alpha \in R(A)$ as a linear combination of certain root elements in L. Put $\delta = \varepsilon_{c_1+1}$ and $\kappa_l = \varepsilon_{c+l} + \varepsilon_{p+1-l}$ for $1 \leq l \leq (p-c)/2$. For $c > 1$ set also $\beta_i = \varepsilon_i - \varepsilon_{p+i}$ and $\gamma_i = \varepsilon_{c+1-i} + \varepsilon_{p+i}$ for $1 \leq i \leq c_1$. Considering the action of A on the standard G-module and arguing as in the proof of Proposition 2.43, one may assume that

(6.4)
$$U^+(A) \subset \begin{cases} \langle \mathcal{X}_\gamma \mid \gamma = \delta,\ \kappa_l,\ 1 \leq l \leq (p-1)/2 \rangle & \text{for } c = 1, \\ \langle \mathcal{X}_\gamma \mid \gamma = \beta_i,\ \gamma_i,\ \delta,\ \kappa_l,\ 1 \leq i \leq c_1,\ 1 \leq l \leq (p-c)/2 \rangle & \text{for } c > 1 \end{cases}$$

and

$$X_\alpha = \begin{cases} gX_{-\delta} + \sum_{l=1}^{(p-1)/2} h_l X_{-\kappa_l} & \text{for } c = 1, \\ \left(\sum_{i=1}^{c_1}(b_i X_{-\beta_i} + f_i X_{-\gamma_i})\right) + gX_{-\delta} + \sum_{l=1}^{(p-c)/2} h_l X_{-\kappa_l} & \text{for } c > 1 \end{cases} \tag{6.5}$$

with $b_i, f_i, g, h_l \in K^*$.

First let $c = 1$. If $d \leq a_1$, set $m = X_{-1,d}v$. Now assume that $d > a_1$. Then there exist j and f such that $1 < j \leq p$, $0 < f \leq a_j$, and $d = f + \sum_{i=1}^{j-1}$. If $j < p$, put $m = v(1, j, f)$. If $\omega = 2\omega_r$, set $m = X_{-r,2}v$. If $p = 3$ and $\omega = \omega_1 + \omega_3$, put $m = X_{-1}X_{-3}v$. Other situations where $j = p$ will be handled with the help of special arguments. Here $a_r = 1$ and $d = \sum_1^p a_i$. By Lemma 2.46, in all cases $m \neq 0$ and is fixed by $\mathcal{X}_i$ for $i > 1$. Hence m is fixed by $U^+(\Gamma)$. Formula (6.4) yields that m is fixed by $U^+(A)$ as well. Set $q = (p-1)/2$, $a_i' = a_i + a_{p-i}$, $a_i^+ = a_{i+1} + a_{p-i}$, $\Delta_{ab} = \sum_{i=a}^b a_i'\omega_i$, $\Delta_{ab}^+ = \sum_{i=a}^b a_i^+\omega_i$, and $\Omega = \sum_{i=p-j+1}^{q-1} a_{i+1}'\omega_i$. Formulae (2.10) imply the following equalities for λ_2. If $d \leq a_1$, we get

$$\lambda_2 = \begin{cases} (a_1^+ + d)\omega_1 + \Delta_{2,q-1}^+ + a_{q+1}\omega_{q+1} & \text{for } p > 3, \\ (a_2 + d)\omega_1 & \text{if } p = 3. \end{cases} \tag{6.6}$$

If $d > a_1$ and $j < p$, one has

$$(6.7)\qquad \lambda_2 = \begin{cases} \Delta_{1,j-2} + (a'_{j-1} + a_j - f)\omega_{j-1} & \\ +(a^+_j + f)\omega_j + \Delta^+_{j+1,q-1} + a_{q+1}\omega_q & \text{if } j < q, \\ \Delta_{1,q-2} + (a_{q-1} + a_q - f + a_{q+2})\omega_{q-1} & \\ +(a_{q+1} + f)\omega_q & \text{for } j = q,\ p > 3, \\ \Delta_{1,q-2} + (a_{q-1} + a_{q+2} + f)\omega_{q-1} & \\ +(a_q + a_{q+1} - f)\omega_q & \text{for } j = q+1,\ p > 3, \\ \Delta_{1,p-j-2} + (a'_{j+1} + f)\omega_{p-j-1} & \\ +(a'_j + a_{j-1} - f)\omega_{p-j} + \Omega + a_q\omega_q & \text{if } q+1 < j < p-1. \end{cases}$$

Now let $c > 1$. Here the construction of m is based on Lemmas 2.55 and 2.56. We have $N(\varphi) = p + l$ where $l = d - 1 \le a_{c-1} + 2\sum_{i=c}^{p-1} a_i$. Observe that $l < N_1(\varphi) - 2$ as $p \ge 5$. Hence $u < p - 2$. For $i \le j \le c$ set $a^+_i = a_i + a_{c-i}$ and $\Lambda^j_i = \sum_{k=i}^{j} a^+_k\omega_k$.

a) First assume that $a_f \ne 0$ for some f with $c < f < r$. As $N_1(\varphi) < p$, one has $f < p - 1$. If $a_f > 1$, or $f > c + 1$, or $c > 7$, or $a_t \ne 0$ for some t with $t < c$, set $M_A = KAv$. We claim that $M_A \cong V(p + l)$ as an A-module. By Lemma 2.55, it suffices to check that $X_{\alpha,l+1}v \ne 0$. One gets $l < \omega_\delta(v) = 1 + 2\sum_{i=c_1+1}^{p-1} a_i < p$ as φ is not p-large. Hence $X^{l+1}_{-\delta}v \ne 0$ by Lemma 2.1 ii). Now Formula (6.5) implies that $X_{\alpha,l+1}v \ne 0$ as required (consider the weight components of the latter vector). So $M_A \cong V(p + l)$. Let $m \in (M_A)_{p+l-2}$ and $m \ne 0$. By Lemma 2.56, $m \in M_y$. As $\Gamma \subset C_G(A)$, the group $U^+(\Gamma)$ fixes m and $\lambda = \omega_\Gamma(v)$. One easily concludes that

$$(6.8)\qquad \lambda_1 = \Lambda^{c_1-1}_1 + (1 + 2a_{c_1} + 2a_{c_1+1})\omega_{c_1},$$

$$(6.9)\qquad \lambda_2 = \Big(\sum_{i=1}^{(p-c-2)/2} (a_{c+i} + a_{p-i})\omega_i\Big) + a_{(p+c)/2}\omega_{(p-c)/2}.$$

b) Next, let $c = 3$, 5, or 7 and $\omega = a_c\omega_c + \omega_{c+1} + \omega_r$. First assume that $a_c \ne 0$. Set $m_1 = X_{-c}v$. Formula (6.4) and the construction of Γ imply that m_1 is fixed by $U^+(A)$ and $U^+(\Gamma)$. If $d = 1$, we have $\omega_A(m_1) = p - 1$ and hence $m_1 \in M_y$ by Lemma 2.54 ii). In this case put $m = m_1$. Now let $d > 1$. Then $\omega_A(m_1) = p + l$ with $l = d - 2 < p - 2$ as $d = 2a_c + 3$ and φ is not p-large. It is clear that $\mathcal{X}_\delta$ fixes m_1 and $\omega_\delta(m_1) = 2a_c + 3 < p$. Hence $X^{l+1}_{-\delta}m_1 \ne 0$ by Lemma 2.1 ii). Applying Lemmas 2.55 and 2.56 and arguing as in the previous paragraph, we deduce that $KAm_1 \cong V(p + l)$ as an A-module, take for m a nonzero vector in KAm_1 with $\omega_A(m) = p - l - 2$ and conclude that $m \in M_y$ and $\lambda = \omega_\Gamma(m_1)$. One has

$$(6.10)\qquad \lambda_1 = \begin{cases} 3\omega_1 & \text{for } c = 3, \\ \omega_1 + \omega_{c_1} & \text{for } c > 3; \end{cases}$$

$$(6.11)\qquad \lambda_2 = 2\omega_1.$$

Now let $\omega = \omega_{c+1} + \omega_r$. Since $N_1(\varphi) < p \le N(\varphi)$, we have $p = 37$ for $c = 7$, $p = 23$ for $c = 5$, and $p = 13$ or 17 for $c = 3$. In all these cases set $m_1 = X_{-c}X_{-(c+1)}v$. By Lemma 2.46, $m_1 \ne 0$ and is fixed by $\mathcal{X}_i$ for $i \ne c$. This implies that m_1 is fixed by $U^+(A)$ and $U^+(\Gamma)$(see Formula (6.4)). If $c = 3$

and $p = 17$ or $c = 7$ and $p = 37$, we have $\omega_A(m_1) = p - 1$ and so $m_1 \in M_y$ by Lemma 2.54 ii). In these cases put $m = m_1$. Otherwise we argue as in the previous paragraph for $a_c \neq 0$ and $d > 1$. One has $\omega_A(m_1) = p + 1$ for $c = 5$ and $p = 23$ or $c = 3$ and $p = 13$. It is clear that $\mathcal{X}_\delta$ fixes m_1 and $\omega_\delta(m_1) = 3 < p$. Hence $X_{-\delta}^2 m_1$ and $X_\alpha^2 m_1 \neq 0$. Applying Lemmas 2.55 and 2.56, we conclude that $KAm_1 \cong V(p+1)$ as an A-module, choose for m a nonzero vector in KAm_1 with $\omega_A(m) = p-3$ and deduce that $m \in M_y$, $U^+(\Gamma)$ fixes m, and $\lambda = \omega_\Gamma(m_1)$. Observe that Formula (6.10) holds for λ_1. One has

$$\lambda_2 = \omega_2. \tag{6.12}$$

c) Finally, assume that $\omega = \omega(1, c) + \omega_r$. First let $a_c \neq 0$. Set $m_1 = X_{-c}v$ and argue as in Item b) for $a_c \neq 0$. If $d = 1$, set $m = m_1$. Then $\omega_A(m) = p - 1$ and $m \in M_y$ by Lemma 2.54 ii). Suppose that $d > 1$. Then $\omega_A(m_1) = p + t$ with $t = d - 2$. We have $t \leq a_{c-1} + 2a_c - 1 \leq \omega_\delta(m_1) - 2 < p - 2$. Hence $X_{-\delta}^{t+1} m_1$ and $X_\alpha^{t+1} m_1 \neq 0$. Using Lemmas 2.55 and 2.56, we choose for m a nonzero vector in KAm_1 such that $\omega_A(m) = p - t - 2$, $m \in M_y$ and $\lambda = \omega_\Gamma(m_1)$. One gets

$$\lambda_1 = \begin{cases} (2a_1^+ + 3)\omega_1 & \text{for } c = 3, \\ (a_1^+ + 1)\omega_1 + \Lambda_2^{c_1 - 1} + (2a_{c_1}^+ + 1)\omega_{c_1} & \text{for } c > 3; \end{cases} \tag{6.13}$$

$$\lambda_2 = \omega_1. \tag{6.14}$$

Now let $a_c = 0$. If $a_{c-1} \neq 0$, set $m_1 = X_{-c}X_{-(c-1)}v$ and construct m as for $a_c \neq 0$ applying Lemmas 2.55 and 2.56. Namely, we have $m = m_1$ and $\omega_A(m) = p - 1$ if $d = 1$; if $d > 1$, one gets $\omega_A(m_1) = p + t$ with $t = d - 2 \leq a_{c-1} - 1 < p - 2$ as $N_1(\varphi) < p$, in this case $\omega_A(m) = p - t - 2$; in all situations $\lambda = \omega_\Gamma(m_1)$. As before, we need to consider the action of $X_{-\delta}$. We have $\omega_\delta(m_1) = 1 + 2\sum_{i=c_1+1}^{p-1} a_i < p$ for $c > 3$ and $2a_2 - 1$ for $c = 3$; hence in both cases $\omega_\delta(m_1) > t$. The following formulae hold:

$$\lambda_1 = \begin{cases} (2a_1^+ + 1)\omega_1 & \text{for } c = 3, \\ (a_1^+ - 1)\omega_1 + (2a_2^+ + 3)\omega_2 & \text{for } c = 5, \\ (a_1^+ - 1)\omega_1 + (a_2^+ + 1)\omega_2 + \Lambda_3^{c_1 - 1} + (2a_{c_1}^+ + 1)\omega_{c_1} & \text{for } c > 5. \end{cases} \tag{6.15}$$

As in the previous paragraph, $\lambda_2 = \omega_1$, i.e (6.14) holds.

Now let $a_{c-1} = 0$. Then $d = 1$. Set $m = v(p, r, 1)$. By Lemma 2.46, $m \neq 0$ and is fixed by $\mathcal{X}_i$ for $i \neq p$. Now Formula (6.4) and the construction of Γ imply that m is fixed by $U^+(A)$ and $U^+(\Gamma)$. We have $\omega_A(m) = (p - 1)\omega_1$ and hence $m \in M_y$ by Lemma 2.54 ii). Since $N(\varphi) \geq p$, one gets that $\omega \neq \omega_r$. It is clear that λ_1 is determined by Formula (6.8) and Formula (6.14) holds for λ_2.

So we have completed the construction of m. Recall that we are done if (6.3) holds, hence it remains to consider the cases where (6.3) fails and the cases where m was not constructed. Lemma 6.1 and Formulae (6.6–6.15) yield that now actually it suffices to handle the following two cases:

1) $c = 1$, $N_1(\varphi) = p - 1$, and $\omega \neq \omega_1 + \omega_3$ for $p = 3$ (in this case m was not constructed);

2) $c = 1$, $\omega = \frac{p-1}{4}\omega_1 + \omega_p$.

Here it is crucial that φ is not p-large, $2a_1 + 2a_2 + 1 < p$ if $c \geq 3$, and that $2a_1 + 2a_2 + 3 < p$ if in the latter case $a_3 \neq 0$ (recall that $N_1(\varphi) < p$).

Let Case 1) hold. Recall that $G_1 = G(2, \ldots, r)$ and g is a regular unipotent element in G_1. As we have shown at the beginning of the proof, $\varphi|G_1$ has a p-restricted composition factor φ_1 with highest weight $\omega_u = \omega(1, p-2) + (2a_{p-1} + a_p)\omega_{p-1}$. We have $N_{G_1}(\varphi_1) = N_1(\varphi) = p - 1$. Let $g' = g^p$, and let g_0 and $g_0' \in (G_1)_{\mathbb{C}}$ be unipotent elements with $J(g_0) = J(g)$ and $J(g_0') = J(g_1)$. By Lemma 5.3, $d_{\varphi_1}(g) = \min\{d_{(\varphi_1)_{\mathbb{C}}}(g_0), pd_{(\varphi_1)_{\mathbb{C}}}(g_0')\}$. As we have mentioned in Section 5, $d_{(\varphi_1)_{\mathbb{C}}}(g_0') = N_{G_1}(\varphi_1) + 1 = p$. As $a_{p-1} = 0$, Proposition 1.5 and Algorithm 1.6 imply that

$$d_{(\varphi_1)_{\mathbb{C}}}(g_0) = 1 + a_p(p-1)p/2 + \sum_{i=1}^{p-2} a_i i(2p-1-i) = pN_1(\varphi) + 1 + \sum_{i=1}^{p-2} a_i i(p-1-i).$$

Since $N_1(\varphi) = p - 1$, we have $\omega \neq \omega_p$ and $\omega \neq \omega_1 + \omega_p$ for $p > 3$. Recall that the weights $\omega_1 + \omega_3$ for $p = 3$ and $2\omega_p$ for all p are excluded. Since φ is not p-large, we get $p > 3$. Now it is clear that $\sum_{i=1}^{p-2} a_i i(p-1-i) \geq 2(p-3) \geq p-1$. Hence $d_{(\varphi_1)_{\mathbb{C}}}(g_0) \geq p^2$ and $d_{\varphi_1}(g) = p^2$. So by Lemma 2.19, $d_\varphi(x) = p^2$.

Now consider Case 2). Naturally, it can occur only for $p \equiv 1(\mod 4)$. The arguments are based on Proposition 2.52. Let $x^- \in U^-(G)$ be a regular unipotent element. In the notation of that proposition take $\lambda_1 = \frac{p-1}{4}\omega_1$, $\lambda_2 = \omega_p$, $f_1 = p(p-1)/2$, and $f_2 = p(p+1)/2 - 1$. Then $f = p^2 - 1$ and $\binom{f}{f_1} \not\equiv 0(\mod p)$ by Lemma 2.7. Hence it remains to show that $(x^- - 1)^{f_i} m_i \neq 0$ for nonzero highest weight vectors $m_i \in M_i = M(\lambda_i)$, $i = 1, 2$, and to verify that $\dim M_\mu = \dim V(\omega)_\mu$ for $\mu \in \mathbf{X}_f(M)$. Set $\nu_i = \varphi(\lambda_i)$. One easily checks that $N(\nu_1) = (p-1)/2$ and $N(\nu_2) = (p+1)/2$. Put $d_0^i = d_{(\nu_i)_{\mathbb{C}}}(z_0)$ and $d_1^i = d_{(\nu_i)_{\mathbb{C}}}(z_1)$. As we have mentioned at the beginning of this subsection, $d_1^i = N(\nu_i) + 1$. By Lemma 6.3, $d_{\nu_i}(x) = \min\{d_0^i, pd_1^i\}$. Using Proposition 1.5 and Algorithm 1.6, one can verify that $d_0^1 = 1 + p(p-1)/2$ and $d_0^2 = 1 + p(p+1)/2$. Hence $d_{\nu_i}(x) = d_0^i$. Now Corollary 2.22 implies that $(x^- - 1)^{f_i} m_i \neq 0$ as desired. Let ω' be the lowest weight of φ. Then $\omega' = -\omega$ and it is clear that $\omega' + \alpha_i \in \mathbf{X}(\varphi)$ just for $i = 1$ or r. Since $f = d(M) - 1$, we have $\mathbf{X}_f(M) = \{\omega', \omega' + \alpha_1, \omega' + \alpha_r\}$. It follows from Theorem 2.33 and Lemma 2.54 that $\dim M_\mu = \dim V(\omega)_\mu = 1$ for $\mu \in \mathbf{X}_f(M)$. Now Proposition 2.52 implies that $(x^- - 1)^f v \neq 0$ and hence $d_\varphi(x) = p^2$ as required. This completes the proof. □

COROLLARY 6.5. *Let $G = B_r(K)$, $n = 2p + c$ with $0 < c < p$, $\varphi \in \mathrm{Irr}_p$, and $\omega = \omega(1, p-1) + a_r\omega_r$. Then Theorem 1.7 holds for φ.*

PROOF. This follows from Corollary 4.11, Lemmas 6.2 and 6.3, and Proposition 6.4. □

6.2. $s > 1$. Now assume that $s > 1$. Then $n = p^s + p + c$ with $c = 2c_1 + 1 < p$ and $c_1 \in \mathbb{Z}^+$; $e = (p-1)p^{s-1} + \ldots + (p-2)p + p - c$ if $s > 2$ and $e = (p-2)p + p - c$ for $s = 2$. By Proposition 2.5 b), y has $p + c$ Jordan blocks of size 2 and $p^s - p - c$ blocks of size 1. One gets $q_1 = (p+c)/2$, $q_2 = (e-1)/2$, $\Gamma_1 \cong C_{q_1}(K)$, and $\Gamma_2 \cong B_{q_2}(K)$. Observe that $\zeta(\varepsilon_i) = 1$ for $1 \leq i \leq p + c$ and 0 for $p + c < i \leq r$.

LEMMA 6.6. *Let $\chi \in \mathrm{Irr}\,\Gamma_2$ and $\omega(\chi) = \sum_{i=1}^{q_2} b_i\omega_i$ with $b_{q_2} \neq 0$. Assume that Conjecture (r, s) holds and $d_\chi(x_2) < p^s$. Then $p = 3$, $n = 13$, and $\omega(\chi) \in \{3^j\omega_2, 2 \cdot 3^j\omega_2, 3^j\omega_1 + 3^l\omega_2\}$, $j, l \in \mathbb{Z}^+$.*

PROOF. As Conjecture (r, s) holds, one concludes that Theorems 1.1 and 1.9 are valid for $\operatorname{Irr}\Gamma_2$. Observe that $p^{s-1} + 2p < 2p^{s-1} < n(\Gamma_2) < p^s$ if $s > 2$ and $3p < n(\Gamma_2) < p^2$ for $p > 3$, $s = 2$. So Theorem 1.9 implies that $p = 3$ and $s = 2$. Then $n = 13$. By Theorem 1.1, χ is not p-large. All such representations with $b_{q_2} \neq 0$ are listed in the statement of the lemma. $\square$

LEMMA 6.7. *Let $\xi \in \operatorname{Irr}\Gamma$ and $\omega(\xi) = (\nu_1, \nu_2)$ with $\nu_i \in \mathbf{X}_i$, $\nu_2 = \sum_{i=1}^{q_2} b_i\omega_i$ and $b_{q_2} \neq 0$. Assume that Conjecture (r, s) holds and $d_\xi(x_\Gamma) < p^s$. Then $p = 3$, $n = 13$, and one of the following holds:*

1) *$\nu_1 = 0$, ν_2 is such as $\omega(\chi)$ in Lemma 6.6;*
2) *$\nu_1 = 3^j\omega_i$, $i = 1, 2$, $\nu_2 = 3^l\omega_2$, $j, l \in \mathbb{Z}^+$.*

PROOF. Set $\xi_i = \varphi(\nu_i)$, $i = 1, 2$. By Theorem 2.9, $d_{\xi_1}(x_1) + d_{\xi_2}(x_2) < p^s$ and hence $d_{\xi_2}(x_2) < p^s$. Lemma 6.6 forces that $p = 3$, $n = 13$, and $\nu_2 \in \{3^l\omega_2, 2 \cdot 3^l\omega_2, 3^j\omega_1 + 3^l\omega_2\}$. Observe that in our case $\Gamma_1 \cong \Gamma_2$, though it is convenient to consider Γ_1 as $C_2(K)$ and Γ_2 as $B_2(K)$. By Proposition 2.5 a), $d_\sigma(g_i) > 6$ for $\sigma \in \operatorname{Irr}\Gamma_i$ if $\omega(\sigma) \notin \{0, 3^j\omega_1, 3^j\omega_2\}$ since in this case $d_\sigma(g_i^3) = 3$ by Theorem 1.14 and Proposition 1.5. We have $d_\sigma(g_2) = 5$ for $\omega(\sigma) = 3^j\omega_1$ and 4 for $\omega(\sigma) = 3^j\omega_2$ since these representations are the standard realizations of $SO_5(K)$ and $Sp_4(K)$, respectively, twisted by field morphisms. Now apply Theorem 2.9 to complete the proof. $\square$

For a group $\Delta = B_l(K)$ with $2l+1 = p^s+p+u$ and $u < p$ and a representation $\mu \in \operatorname{Irr}_p\Delta$ with $\omega(\mu) = (\sum_{i=1}^{p-1} b_i\omega_i) + b_l\omega_l$ set

$$N_\Delta(\psi) = \Big(\sum_{i=1}^{p-1} b_i i\Big) + b_l(p+u)/2. \tag{6.16}$$

Put $N(\varphi) = N_G(\varphi)$. Proposition 1.5, Algorithm 1.6, and Theorem 1.14 imply that

$$d_\varphi(y) = \min\{p, N(\varphi) + 1\}. \tag{6.17}$$

The construction of the homomorphism ζ yields that $\zeta(\omega) = N(\varphi)$ and $\omega_A(v) = N(\varphi)$.

REMARK 6.8. If $n = 3$, $p = 13$, and $N(\varphi) < p$, then $\omega = \omega_6$.

This follows immediately from (6.16).

LEMMA 6.9. *Assume that $N(\varphi) < p$. Then $d_\varphi(x) = p^s(N(\varphi)+1)$ unless $p = 3$, $n = 13$, and $\omega = \omega_6$.*

PROOF. The proof is similar to those of Lemmas 6.2 and 5.2 and is based on Lemma 6.7 and Remark 6.8. $\square$

LEMMA 6.10. *Let $p = 3$, $n = 13$, and $\omega = \omega_6$. Then $d_\varphi(x) = d_{\varphi_C}(z_0) = 22$.*

PROOF. The scheme of the proof is the same as in Lemmas 5.3 and 6.3. The proof is based on Corollaries 2.44 and 2.41. Algorithm 1.6 forces $d_{\varphi_C}(z_0) = 22$. By Corollary 2.41, it suffices to prove that $d_\varphi(x) \geq d_{\varphi_C}(z_0)$. Set $\lambda = \rho(\omega)$. We have $\lambda = (0, \omega_2)$ with $0 \in \mathbf{X}_1$ and $\omega_2 \in \mathbf{X}_2$. By Lemma 2.54, $v \in M_y$. This yields that the $K\Gamma$-module M_y has a composition factor $M_\Gamma = M(\lambda)$ and the $K\Gamma_2$-module M_y has a composition factor $M'_\Gamma = M(\lambda_2)$. It is clear that $d_{M_\Gamma}(x_\Gamma) = d_{M'_\Gamma}(g_2) = 4$. Formula (6.17) implies that $d_\varphi(y) = 3$. Hence $d_{M_y}(x_\Gamma) \geq 4$ and by Corollary 2.44, $d_M(x) \geq 22$ as required. $\square$

PROPOSITION 6.11. *Let $N(\varphi) \geq p$. Then $d_\varphi(x) = p^{s+1} = |x|$.*

PROOF. The proof is similar to those of Propositions 5.4 and 6.4, but is substantially easier. Set $G_1 = G(2, \dots, r)$. If $\omega \neq a_r\omega_r$, fix maximal $j < p$ with $a_j \neq 0$ and put $m = v(1, j, a_j)$; otherwise take $m = v$. By Lemma 2.46, $m \neq 0$ and is fixed by $\mathcal{X}_i$ for $i > 1$. Set $\mu = \omega_{G_1}(m)$ and $\varphi_1 = \varphi(\mu) \in \operatorname{Irr} G_1$. We have $\mu = \omega(1, p-1) + a_r\omega_{r-1}$. Now it is clear that $\varphi|G_1$ has a composition factor φ_1. If $c = 1$, define $N_{G_1}(\varphi_1)$ as in Section 5, for $c > 1$ this parameter is already defined. One can observe that in all cases $N_{G_1}(\varphi_1) = N(\varphi) - a_r$. Let $g \in G_1$ be a regular unipotent element. We shall show that $d_{\varphi_1}(g) = p^{s+1}$ except one special case which will be handled separately. In the general situation Lemma 2.19 yields the proposition.

First let $c = 1$. If $N_{G_1}(\varphi_1) < p$, Formula (5.1) yields that $a_r = 1$ or $\omega = 2\omega_r$. In both cases $N_{G_1}(\varphi_1) = p - 1$. Lemma 5.2 implies that $d_{\varphi_1}(g) = p^{s+1}$ unless $p = 3$, $G_1 = B_5(K)$, and $\mu = \omega_1 + \omega_5$. Then $\omega = \omega_1 + \omega_6$. Set $m_1 = v(4, 6, 1)$. By Lemma 2.46, $m_1 \neq 0$ and is fixed by $\mathcal{X}_i$ for $i \neq 4$. This implies that m_1 is fixed by $U^+(\Gamma)$. Our relations for ζ imply that $\omega_A(m_1) = p - 1$. Put $\beta_1 = \varepsilon_1 + \varepsilon_4$ and $\beta_2 = \varepsilon_2 + \varepsilon_3$. One can assume that $U^+(A) \subset \langle \mathcal{X}_{\beta_1}, \mathcal{X}_{\beta_2} \rangle$. One easily observes that the groups $\mathcal{X}_{\beta_j}$, $j = 1, 2$, and hence $U^+(A)$ fix m_1. So $m_2 \in M_y$ by Lemma 2.54 ii). Put $\lambda = \omega_\Gamma(m_1)$. Recall that $\Gamma_1 = C_2(K)$ and $\Gamma_2 = B_2(K)$. We have $\lambda = (\lambda_1, \lambda_2)$ with $\lambda_1 = 2\omega_1 \in \mathbf{X}_1$ and $\lambda_2 = \omega_2 \in \mathbf{X}_2$. It is clear that $M_1 = K\Gamma m_1$ is an indecomposable Γ-module with highest weight λ and hence has a composition factor $F \cong M(\lambda)$. As $\Gamma \subset C_G(A)$, one gets $M_1 \subset M_y$. Conjecture (r, s) and Lemma 6.7 yield that $d_F(x_\Gamma) = 9$. So $d_{M_y}(x_\Gamma) = d_{M_1}(x_\Gamma) = 9$ and $d_\varphi(x) = 27 = |x|$ by Corollary 2.44.

Now assume that $N_{G_1}(\varphi_1) \geq p$. Then $d_{\varphi_1}(g) = p^{s+1}$ by Proposition 5.4. This completes the proof for $c = 1$.

Suppose that $c > 1$. Then $p > 3$. Now we can apply induction on r assuming that our proposition holds for groups $B_l(K)$ with $2l + 1 = p^s + p + c_1$, $0 < c_1 < c$. If $N_{G_1}(\varphi_1) \geq p$, the inductive assumption forces that $d_{\varphi_1}(g) = p^{s+1}$ as desired. Let $N_{G_1}(\varphi_1) < p$. Then $a_r = 1$ and $N_{G_1}(\varphi_1) = p - 1$. Since $p > 3$, Lemma 6.9 yields that $d_{\varphi_1}(g) = p^{s+1}$ and completes the proof. □

COROLLARY 6.12. *Let $G = B_r(K)$ and $n = p^s + p + c$ with $0 < c < p$ and $s > 1$. Assume that Conjecture (r, s) holds, $\varphi \in \operatorname{Irr}_p$, and $\omega = \omega(1, p-1) + a_r\omega_r$. Then $d_\varphi(x) = \min\{p^s d_{\varphi_C}(z_s), p^{s+1}\}$, unless $p = 3$, $G = B_6(K)$, and $\omega = \omega_6$. In the exceptional case $d_\varphi(x) = d_{\varphi_C}(z_0) = 22$. In particular, Theorem 1.7 holds for the situation considered.*

PROOF. This follows from Corollary 4.11, Lemma 6.9, and Propositions 6.10 and 6.11. □

7. The exceptional cases in Theorem 1.7

In this section $G = A_r(K)$ or $C_r(K)$, $n = p^s + p$, $\omega = p^j\omega_p$ or $p^j\omega_{r+1-p}$ for $G = A_r(K)$ and $\omega = p^j\omega_p$ for $G = C_r(K)$. We prove Theorem 1.7 for this case assuming that Conjecture (r, s) holds. As Frobenius morphisms and passing to the dual representation do not change the minimal polynomials, we may and shall assume that $\omega = \omega_p$ for both groups. Our goal is to prove that $d_\varphi(x) = p^{s+1} - p + 2$. Set $d = p^{s+1} - p + 2$ and $y_1 = x^p$. By Proposition 2.5 b), y_1 has p Jordan blocks of size $p^{s-1} + 1$ on V.

PROPOSITION 7.1. *Let $G = A_r(K)$. Then y_1 has 2 blocks of size p^s on M.*

PROOF. For $1 \leq i \leq p$ and $1 \leq j \leq p^{s-1}$ put $\gamma_{ij} = \varepsilon_{i+(j-1)p} - \varepsilon_{i+jp}$ and $G_i = G(\gamma_{ij} \mid 1 \leq j \leq p^{s-1})$. It is clear that the subgroups G_i and G_t commute for $i \neq t$. Set $\Delta = G_1 G_2 \dots G_p$. One easily observes that Δ is a direct product of p copies of $A_{p^{s-1}}(K)$. We may assume that $xv_i = v_i + v_{i-1}$. Then y_1 is a regular unipotent element in Δ.

By Lemma 2.35, φ can be realized in $\Lambda^p V$. For $0 \leq j \leq p$ let $\varphi_{ij} = \varphi(\omega_j) \in \operatorname{Irr} G_i$ with $\varphi_{i0} = \varphi(0)$. Denote by $\varphi(l_1, l_2, \dots, l_p)$ the representation in $\operatorname{Irr}\Delta$ equivalent to $\otimes_{i=1}^{p} \varphi_{il_i}$. By Proposition 2.36, $\varphi|\Delta = \oplus\varphi(l_1, l_2, \dots, l_p)$ where the sum is taken over all p-tuples $(l_1, l_2, \dots, l_p)$ with $0 \leq l_i \leq p$ and $l_1 + l_2 + \dots + l_p = p$ if $s > 1$ and over all such tuples with $0 \leq l_i \leq 2$ and $l_1 + l_2 + \dots + l_p = p$ for $s = 1$. Let $\xi = \varphi(l_1, l_2, \dots, l_p)$ with some $l_i > 1$. We claim that $d_\xi(y_1) < p^s$. Indeed, obviously, one can assume that $l_1 > 1$. Put $l = p - l_1$, $G^2 = G_2 \dots G_p$ and $\xi_2 = \otimes_{i=2}^{p} \varphi_{il_i} \in \operatorname{Irr} G^2$. Let $e_1 \in G_1$, $e_2 \in G^2$, and $e' \in (G_1)_{\mathbb{C}}$ be regular unipotent elements, $f_1 = d_{\varphi_{1l_1}}(e_1)$, and $f_2 = d_{\xi_2}(e_2)$. Then $d_\xi(y_1)$ is equal to the maximal size d' of the Jordan blocks of the tensor product $J_{f_1} \otimes J_{f_2}$. By Corollary 2.13, $d' < f_1 + f_2$. Set $r' = (p-1)p^{s-1} + p - 2$ and $\Delta_2 = A_{r'}(K)$. Then G^2 can be naturally embedded into Δ_2. Let $\tau = \varphi(\omega_l) \in \operatorname{Irr}\Delta_2$ and $e'_2 \in (\Delta_2)_{\mathbb{C}}$ be a unipotent element with $J(e'_2) = J(e_2)$. Observe that ξ_2 is a component of $\tau|G^2$. Hence $f_2 \leq d_\tau(e_2)$. Put $f'_1 = d_{(\varphi_{1l_1})_{\mathbb{C}}}(e')$ and $f'_2 = d_{\tau_{\mathbb{C}}}(e'_2)$. Corollary 2.40 forces that $f_j \leq f'_j$, $j = 1, 2$. Recall that e'_2 has $p-1$ Jordan blocks of size $p^{s-1}+1$ in the standard realization of $(\Delta_2)_{\mathbb{C}}$ and apply Algorithm 1.6 to compute f'_j. We get $f'_1 = l_1(p^{s-1} + 1 - l_1) + 1$ and $f'_2 = lp^{s-1} + 1$. As $l_1 > 1$, this yields that $f' \leq f'_1 + f'_2 - 1 \leq p^s - l_1(l_1 - 1) + 1 \leq p^s - 1 < p^s$. Hence $d_\xi(y_1) < p^s$ as required.

Now it suffices to consider the representation $\nu = \varphi(1, \dots, 1) \in \operatorname{Irr}\Delta$. As ν is realized in the tensor product of the standard realizations of G_i, $1 \leq i \leq p$, actually our goal is to prove that the tensor product of p blocks $J_{p^{s-1}+1}$ has just 2 blocks of size p^s. Let J^i be the tensor product of i such blocks, $2 \leq i \leq p$, and t_i be the degree of the minimal polynomial of J^i. First we shall show that $t_i = ip^{s-1} + 1$ for $i < p$ and that J^{p-1} has just one Jordan block of size t_{p-1} and other blocks of sizes $\leq t_{p-1} - 2$. Observe that $\binom{ip^{s-1}}{p^{s-1}} \not\equiv 0(\mod p)$ for $2 \leq i \leq p-1$. Using induction on i and Lemma 2.11, we deduce our claim for t_i. Let $J_{\mathbb{C}}$ be the tensor product of $p-1$ blocks $J_{p^{s-1}+1}$ over $\mathbb{C}$. Set $A_{\mathbb{C}} = A_1(\mathbb{C})$. Well-known facts of the representation theory of the group $A_{\mathbb{C}}$ (see, for instance, [**Car85**, ch.5]) imply that one can consider $J_{\mathbb{C}}$ as the image of a unipotent element of $A_{\mathbb{C}}$ in the tensor product $\mathcal{T}$ of $p-1$ copies of the representation $\varphi(p^{s-1}) \in \operatorname{Irr} A_{\mathbb{C}}$. Now the set $\mathbf{X}(\mathcal{T})$ ($\mathcal{T}$ is a representation of $A_{\mathbb{C}}$) and the complete reducibility of representations of semisimple algebraic groups over $\mathbb{C}$ force that $\mathcal{T}$ is a direct sum of $\varphi((p-1)p^{s-1})$ and representations of the form $\varphi(a)$ with $a < (p-1)p^{s-1} - 1$. This implies that $J_{\mathbb{C}}$ has one block of size t_{p-1} and the sizes of other its blocks are less than $t_{p-1} - 1$. Since J^{p-1} has a block of size t_{p-1}, Lemma 2.4 yields the desired facts on the Jordan block structure of J^{p-1}. Hence J^p has just two blocks of size p^s by Theorem 2.9 and Corollary 2.13. This completes the proof. □

PROPOSITION 7.2. *Let $N = (y_1 - 1)^{p^s - 1} M$. Then $d_N(x) = 2$.*

PROOF. First suppose that $G = A_r(K)$. As we have mentioned above, M can be identified with $\bigwedge^p V$. We can assume that $(x-1)v_i = v_{i-1}$ for the vectors v_i of the basis (2.2) with $i > 1$ and $xv_1 = v_1$. Hence $(y_1 - 1)v_i = v_{p+i}$ for $i > p$. For $u_1, u_2, \dots, u_p \in V$ set $\wedge(u_1, u_2, \dots, u_p) = u_1 \wedge u_2 \wedge \dots \wedge u_p$. Put

$$\wedge(i_1, i_2, \dots, i_p) = \wedge(v_{i_1}, v_{i_2}, \dots, v_{i_p}),$$
$$\wedge(i_1, j_1; i_2, j_2; \dots, i_p, j_p) = \wedge((y_1 - 1)^{j_1} v_{i_1}, (y_1 - 1)^{j_2} v_{i_2}, \dots, (y_1 - 1)^{j_p} v_{i_p}).$$

Recall that for $g \in G$

$$\begin{aligned}(g-1)(\wedge(i_1, i_2, \dots, i_p)) = &\wedge((g-1)v_{i_1}, v_{i_2}, \dots, v_{i_p}) + \wedge(v_{i_1}, (g-1)v_{i_2}, \dots, v_{i_p}) \\ &+ \dots + \wedge(v_{i_1}, v_{i_2}, \dots, (g-1)v_{i_p}) \\ &+ \sum_{k_1, \dots, k_p} \wedge(i_1, k_1; i_2, k_2; \dots; i_p, k_p)\end{aligned}$$

where the latter sum is taken over all tuples $(k_1, \dots, k_p)$ such that $k_j = 0$ or 1 and $1 < k_1 + \dots + k_p \le p$. Denote by $M_b(v_{i_1}, v_{i_2}, \dots, v_{i_p})$ the linear span of all vectors of the form $\wedge(i_1, l_1; i_2, l_2; \dots i_p, l_p)$ with $l_1 + l_2 + \dots + l_p > b$. For $1 \le t \le p$ set $j_t = p^s + t$. If $b_1, b_2, \dots, b_l \in \mathbb{Z}^+$ and $b_1 + b_2 + \dots + b_l = b$, put $C(b_1, b_2, \dots, b_l) = b!/(b_1!b_2! \dots b_l!)$. Observe that $C(b_1, b_2, \dots, b_l) = \binom{b}{b_1}\binom{b-b_1}{b_2} \dots \binom{b-b_1-b_2-\dots-b_{l-2}}{b_{l-1}}$. Let $1 \le i_1 < i_2 \dots < i_k \le l$ be all indices for which $b_{i_j} \ne 0$. One easily concludes that

$$\begin{aligned}C(b_1, b_2, \dots, b_l) = &C(b_{i_1} - 1, b_{i_2}, \dots, b_{i_k}) + C(b_{i_1}, b_{i_2} - 1, \dots, b_{i_k}) + \dots \\ &+ C(b_{i_1}, b_{i_2}, \dots, b_{i_k} - 1).\end{aligned}$$

Using the latter equality and induction on b, one can deduce that

$$(y_1 - 1)^b \wedge(j_1, j_2, \dots, j_p) = u + \sum_{b_1 + b_2 + \dots + b_p = b} C(b_1, b_2, \dots, b_p) \wedge(j_1, b_1; j_2, b_2; \dots; j_p, b_p)$$

where the sum is taken over all relevant p-tuples and $u \in M_b(v_{j_1}, v_{j_2}, \dots, v_{j_p})$. Take $b = p^s - 1$ and assume that one of b_t is equal to $p^{s-1} - 1$ and the others are equal to p^{s-1}. Set $C = C(b_1, b_2, \dots, b_p)$ for such p-tuple $(b_1, b_2, \dots, b_p)$. Several applications of Lemma 2.7 yield that $C \not\equiv 0(\mod p)$. Recall that $(y_1 - 1)^{p^{s-1}+1} v_{j_t} = 0$ and $(y_1 - 1)^{p^{s-1}} v_{j_t} = v_t$. Put $m = \wedge(p+1, 2, \dots, p) + \dots + \wedge(1, \dots, t-1, t+p, t+1, p) + \dots + \wedge(1, 2, \dots, p-1, 2p)$. The arguments above imply that

$$(y_1 - 1)^{p^s - 1} \wedge (j_1, j_2, \dots, j_p) = Cm + d \wedge (1, 2, \dots, p-1, p) \tag{7.1}$$

with $d \in K$. Since $\dim N \le 2$ by Proposition 7.1 and x preserves N, it suffices to prove that x acts nontrivially on N. One easily observes that $(x - 1)m = \wedge(p, 1, \dots, p-1)$. As x fixes a highest weight vector $\wedge(1, 2, \dots, p)$, now our claim follows from Formula (7.1). This completes the proof for $G = A_r(K)$.

Now assume that $G = C_r(K)$. The group G can be naturally embedded into $G^+ = A_{2r-1}(K) \cong SL_{2r}(K)$. Let M^+ be the irreducible G^+-module $M(\omega_p)$. Then $V = V(G^+)|G$ and $M^+ \cong \bigwedge^p V(G^+)$. It is well known that M is isomorphic to the head of the G-submodule M_0 of M^+ generated by $\wedge(1, 2, \dots, p)$. (We need no other details of this construction for our arguments.) One can assume that $(x - 1)v_i = \pm v_{i-1}$ for $i > 1$ and $(x - 1)v_1 = 0$. Then $(y_1 - 1)v_i = \pm v_{i-p}$ for $i > p$. As before, y_1 fixes the vectors v_i for $i \le p$. Observe that x is a regular unipotent element in H as well. Set $N_0 = (y_1 - 1)^{p^s - 1} M_0$ and denote by $\overline{\Sigma}$ and $\overline{m}$ the images of a subspace $\Sigma \subset M_0$ and a vector $m \in M_0$ under the canonical mapping of M_0 onto M. Then $N = \overline{N_0}$ and Proposition 7.1 implies that $\dim N \le 2$. One can assume that $v = \wedge(1, 2, \dots, p)$. Put $v' = \wedge(j_1, j_2, \dots, j_p)$ (in the notation introduced above in this proof). As $\langle v'\rangle$ is the weight subspace of weight $-\omega$ in the G-module M^+ and $-\omega \in W\omega$, we have $v' \in M_0$. Now the arguments above used for the case $G = A_r(K)$ imply that $\dim N = d_N(x) = 2$ as required. □

COROLLARY 7.3. *Let $n = p^s + p$ and $G = A_r(K)$ or $C_r(K)$. Assume that $\omega = p^j\omega_p$ or $p^j\omega_{n-p}$ for $G = A_r(K)$ and $\omega = p^j\omega_p$ for $G = C_r(K)$ and that Conjecture (r,s) holds. Then $d_\varphi(x) = p^{s+1} - p + 2$. Hence Theorem 1.7 holds.*

PROOF. This follows from Propositions 2.5 and 7.2. □

8. Theorem 1.9 for regular unipotent elements and groups of types A, B, and C

In this section we assume that Conjecture (r, s) holds and $G = A_r(K)$, $B_r(K)$, or $C_r(K)$ and prove Theorem 1.9 for regular unipotent elements of these groups. Until the end of the section $p^s + p \le n < p^{s+1}$, some $a_i \ne 0$ with $p \le i \le r + 1 - p$ for $G = A_r(K)$ and $p \le i \le r$ for $G = B_r(K)$ or $C_r(K)$; we also assume that $i < r$ if $G = B_r(K)$ and $p^s + p < n < p^s + 2p$.

LEMMA 8.1. *Assume that Conjecture (r,s) holds. Let $\Delta = A_l(K)$ or $C_l(K)$ with $n(\Delta) = p^s - p$ and $s > 1$. Assume that $\chi \in \mathrm{Irr}_p(\Delta)$ with $\omega(\chi) \notin \{0, \omega_1, \omega_l\}$ for $\Delta = A_l(K)$ and $\omega(\chi) \notin \{0, \omega_1\}$ for $\Delta = C_l(K)$. Then $d_\chi(h) > p^s - p$ for a regular unipotent element $h \in \Delta$.*

PROOF. Observe that $n(\Delta) = (p-1)(p^{s-1} + \ldots + p)$. So $n(\Delta) = 2p$ for $p = 3$ and $s = 2$ and $n(\Delta) > p^{s-1} + p$ otherwise. Let $h_j \in \Delta_{\mathbb{C}}$, $0 \le j \le s-1$, be unipotent elements with $J(h_j) = J(h^{p^j})$. By Proposition 2.5 b), for $j > 1$ the element h_j has $(p-1)(p^{j-1} + \ldots + p)$ Jordan blocks of size p^{s-j} and p blocks of size $p^{s-j} - 1$, h_1 has p blocks of size $(p-1)(p^{s-2} + \ldots + 1)$ (of size $p - 1$ for $s = 2$). Now Proposition 1.5 and Algorithm 1.6 imply that $p^j d_{\chi_{\mathbb{C}}}(h_j) > p^s$ for all j. By Conjecture (r, s), Theorem 1.7 holds for Δ. This forces that $d_\chi(h) = 7 > 6$ if $n = 6$, $p = 3$ and $\omega(\chi) = \omega_3$ and $d_\chi(h) = p^s$ otherwise. We are done. □

LEMMA 8.2. *Let $G = A_r(K)$ or $C_r(K)$ and $n = p^s + p$. Assume that $\omega \ne \omega_p$ or ω_{r+1-p} for $G = A_r(K)$ and $\omega \ne \omega_p$ for $G = C_r(K)$. Then $d_\varphi(x) = p^{s+1} = |x|$.*

PROOF. Set $G_1 = G(2, 3, \ldots, r)$. Then $G_1 \cong A_{r-1}(K)$ or $C_{r-1}(K)$ for $G = A_r(K)$ or $C_r(K)$, respectively. Passing to φ^* if necessary, we can and shall assume that for $G = A_r(K)$ the coefficient $a_i \ne 0$ for some i with $p \le i \le (r+1)/2$.

First let $\omega = \omega_i$. Then $i > p$ and $s > 1$. By Proposition 2.5 b), y has p Jordan blocks of size 2 and $p^s - p$ blocks of size 1. Applying Proposition 2.43, we conclude that $e = p^s - p$, $\zeta(\varepsilon_j) = 1$ for $1 \le j \le p$, 0 for $p < j < r + 2 - p$ if $G = A_r(K)$ and for $p < j \le r$ if $G = C_r(K)$, and -1 for $G = A_r(K)$ with $r + 2 - p \le j \le r + 1$; $\Gamma_1 = A_{p-1}(K)$ for $G = A_r(K)$ and $B_{(p-1)/2}(K)$ for $G = C_r(K)$; $\Gamma_2 = A_{e-1}(K)$ for $G = A_r(K)$ and $C_{e/2}(K)$ for $G = C_r(K)$. If $G = A_r(K)$, set $\gamma_j = \varepsilon_j - \varepsilon_{p^s+j}$, $1 \le j \le p$. For $G = C_r(K)$ put $\delta_j = \varepsilon_j + \varepsilon_{p+1-j}$ with $1 \le j \le (p-1)/2$ and $\kappa = 2\varepsilon_{(p+1)/2}$. Next, set $U = \langle \mathcal{X}_{\gamma_j} \mid 1 \le j \le p\rangle$ if $G = A_r(K)$ and $U = \langle \mathcal{X}_\kappa, \mathcal{X}_{\delta_j} \mid 1 \le j \le (p-1)/2\rangle$ for $G = C_r(K)$. One can assume that $U^+(A) \subset U$. Put $m = v(p, i, 1)$. By Lemma 2.46, $m \ne 0$ and is fixed by $\mathcal{X}_l$ for $l \ne p$. This implies that m is fixed by $U^+(\Gamma)$. The construction of m shows that U and hence $U^+(A)$ fix m as well and that $\omega_A(m) = (p-1)$. Set $\lambda = \omega_\Gamma(m)$. As before, we write $\lambda = (\lambda_1, \lambda_2)$ with $\lambda_t \in \mathbf{X}_t$, $t = 1, 2$. One easily concludes that $\lambda_1 = \omega_{p-1}$ for $G = A_r(K)$, $\lambda_1 = \omega_1$ for $G = C_r(K)$, $p > 3$, and $\lambda_1 = 2\omega_1$ for $G = C_r(K)$ with $p = 3$; $\lambda_2 = \omega_{i-p+1}$ in all cases. By Lemma 2.54, $m \in M_y$. Hence $N = K\Gamma m \subset M_y$ since $\Gamma \subset C_G(y)$. By Corollary 2.44, it suffices to prove that $d_N(x_\Gamma) = p^s$. It follows from the arguments above that N has a quotient $N' \cong M(\lambda)$. By Lemma 8.1, $d_{\varphi(\lambda_2)}(g_2) > p^s - p$ as $i > p$. Since $\lambda_1 \neq 0$, Proposition 2.5 a) and Theorem 2.9 yield that $d_{N'}(x_\Gamma) = p^s$ and hence $d_N(x_\Gamma) = p^s$ as desired.

Now assume that $\omega \neq \omega_i$. Set $m = X_{-1,a_1}v$ ($m = v$ if $a_1 = 0$). By Lemma 2.46, $m \neq 0$ and is fixed by $\mathcal{X}_i$ for $i > 1$ and hence by $U^+(G_1)$. Put $\omega^1 = \omega_{G_1}(m)$ and $\varphi_1 = \varphi(\omega^1) \in \operatorname{Irr} G_1$. Then m generates an indecomposable KG_1-module with highest weight ω^1 and $\varphi|G_1$ has a composition factor isomorphic to φ_1. Define $N_{G_1}(\varphi_1)$ as in Section 5. One easily concludes that $N_{G_1}(\varphi_1) \geq p$ unless $G = C_r(K)$ and either $\omega = \omega_1 + \omega_i$, or $p = 3$ and $\sum_{i=1}^r a_i = 2$. If $N_{G_1}(\varphi_1) \geq p$, by Proposition 5.4, $d_{\varphi_1}(g) = p^{s+1}$ for a regular unipotent element $g \in G_1$. Then $d_\varphi(x) = p^{s+1}$ by Lemma 2.19.

Now consider the exceptional cases. Here $G = C_r(K)$. First suppose that $r = p$ and $\omega = \omega_1 + \omega_p$. Then $\Gamma_2 = 1$. As earlier in this proof, we can assume that $U^+(A) \subset U$. Set $l = X_{-p}v$, $\mu = \omega_{\Gamma_1}(l)$, and $M_\mu = M(\mu)$. By Lemma 2.1 ii), $l \neq 0$. Arguing as for m in the case $\omega = \omega_i$, we show that l is fixed by $U^+(\Gamma)$ and $U^+(A)$, $\omega_A(l) = p - 1$, and the $K\Gamma$-module M_y has a composition factor isomorphic to M_μ. We have $\mu = 3\omega_1$ for $p > 3$ and $6\omega_1$ for $p = 3$. Algorithm 1.6 and Theorem 1.14 yield that $d_{M_\mu(g_1)} = p$. Hence $d_{M_y}(g_1) = p$ and Corollary 2.44 completes the proof in this case.

Now let $r > p$ and $\omega = \omega_1 + \omega_i$. If $i > p$, Lemma 5.2 implies that $d_{\varphi_1}(g) = p^{s+1}$. So we can complete the proof as in the general case. Suppose that $\omega = \omega_1 + \omega_p$. Set $m_+ = v(1, p, 1)$, $\omega_+ = \omega_{G_1}(m_+)$, and $\varphi_+ = \varphi(\omega_+)$. Then $\omega_+ = \omega_1 + \omega_p$. By Lemma 2.46, $m_+ \neq 0$ and is fixed by $\mathcal{X}_j$ for $j > 1$. By Lemma 5.2, $d_{\varphi_+}(g) = p^{s+1}$. Arguing as for m in the general case, one can conclude that $d_\varphi(x) = p^{s+1}$ as well.

Finally, let $p = 3$ and $\Sigma_{i=1}^r a_i = 2$. It remains to consider the cases where $\omega = \omega_i + \omega_j$ with $j \geq i \geq 3$ or $\omega = \omega_2 + \omega_i$ with $i \geq 3$.

If $\omega = \omega_2 + \omega_3$, Proposition 2.34 and Conjecture (r, s) yield that $d_\varphi(x) = p^{s+1}$ since Theorem 1.9 holds for the group $G^+ = A_{2r-1}(K)$.

Now let $G = C_3(K)$ and $\omega = 2\omega_3$. Then $G_1 \cong C_2(K)$ and $\omega^1 = 2\omega_2$. Due to Conjecture (r, s), Theorem 1.7 holds for G_1. Let $g_0 \in (G_1)_{\mathbb{C}}$ be a regular unipotent element, $g_0' \in (G_1)_{\mathbb{C}}$, and $J(g_0') = J(g^p)$. Put $\sigma = (\varphi_1)_{\mathbb{C}}$. Propositions 1.5 and 2.5 and Algorithm 1.6 imply that $d_\sigma(g_0) = 3d_\sigma(g_0') = 9$. Hence $d_{\varphi_1}(g) = d_\varphi(x) = 9$ by Theorem 1.7 and Lemma 2.19.

If $\omega = \omega_2 + \omega_i$ with $i > p$, notice that $s > 1$ and argue as for $\omega = \omega_1 + \omega_i$.

Next, suppose that $s > 1$ and $\omega = \omega_i + \omega_j$ with $j \geq i \geq 3$. Then $\omega^1 = \omega_{i-1} + \omega_{j-1}$. Since $i - 1 \geq 2$, Lemma 5.2 yields that $d_{\varphi_1}(g) = p^{s+1}$. Now Lemma 2.19 completes the proof. □

LEMMA 8.3. *Let $G = A_r(K)$ and $n = p^s + p + 1$ or $G = C_r(K)$ and $n = p^s + p + 2$. Assume that $\omega = \omega_p$ or ω_{r+1-p} for $G = A_r(K)$ and $\omega = \omega_p$ for $G = C_r(K)$. Then $d_\varphi(x) = p^{s+1} = |x|$.*

PROOF. Passing to the dual representation if necessary, one can assume that in both cases $\omega = \omega_p$. First let $s = 1$. Set $G_1 = G(2, \ldots, r)$ and $N = KG_1v$. By Theorem 2.33, the KG_1-module N is isomorphic to $M(\omega_{p-1})$. Observe that in both cases $n(G_1) = 2p$. Conjecture (r, s) implies that Theorem 1.7 holds for a regular unipotent element $g \in G_1$. Proposition 2.5 b) forces that g^p has p blocks of size 2 in $V(G_1)$. Let u_0 and u_1 be unipotent elements in $(G_1)_{\mathbb{C}}$ with $J(u_0) = J(g)$ and $J(u_1) = J(g^p)$, and let $\sigma \in \mathrm{Irr}(G_1)_{\mathbb{C}} = \varphi(\omega_{p-1})$. Algorithm 1.6 yields that $d_\sigma(u_0) = p^2$ and $d_\sigma(u_1) = p$. Hence $d_N(g) = p^2$ by Theorem 1.7. Then $d_\varphi(g) = p^2$ and $d_\varphi(x) = p^2$ by Lemma 2.19.

Now assume that $s > 1$. By Proposition 2.5 b), for $G = A_r(K)$ the element y has $p+1$ blocks of size 2 and $p^s - p - 1$ blocks of size 1 on V and for $G = C_r(K)$ it has $p+2$ blocks of size 2 and $p^s - p - 2$ blocks of size 1. We have

$$\begin{array}{lllll} e = p^s - p - 1 & \text{for } G = A_r(K) & \text{and} & p^s - p - 2 & \text{for } G = C_r(K), \\ \Gamma_1 = A_p(K) & \text{for } G = A_r(K) & \text{and} & B_{(p+1)/2}(K) & \text{for } G = C_r(K), \\ \Gamma_2 = A_{e-1}(K) & \text{for } G = A_r(K) & \text{and} & C_{e/2}(K) & \text{for } G = C_r(K). \end{array}$$

If $G = A_r(K)$, we get $\zeta(\varepsilon_i) = 1$ for $1 \leq i \leq p+1$, 0 for $p+1 < i \leq p^s$, and -1 for $p^s < i \leq n$; if $G = C_r(K)$, then $\zeta(\varepsilon_i) = 1$ for $1 \leq i \leq p+2$ and 0 for $p+2 < i \leq r$. Set $\gamma_i = \varepsilon_i - \varepsilon_{p^s+i}$ with $1 \leq i \leq p+1$ and $U_1 = \langle \mathcal{X}_{\gamma_i} \mid 1 \leq i \leq p+1 \rangle$ for $G = A_r(K)$. Put $\delta_i = \varepsilon_i + \varepsilon_{p+3-i}$ with $1 \leq i \leq (p+1)/2$, $\sigma = 2\varepsilon_{(p+3)/2}$, and $U_1 = \langle \mathcal{X}_\sigma, \mathcal{X}_{\gamma_i} \mid 1 \leq i \leq (p+1)/2 \rangle$ for $G = C_r(K)$. One can assume that $U^+(A) \subset U_1$. We have $\zeta(\omega) = p$. Put $m = v(t, p, 1)$ with $t = p+1$ for $G = A_r(K)$ and $t = p+2$ for $G = C_r(K)$. By Lemma 2.46, $m \neq 0$ and is fixed by $\mathcal{X}_i$ for $i \neq t$. Now we can conclude that m is fixed by $U^+(\Gamma)$ and $U^+(A)$ as well. Hence m generates an indecomposable $K\Gamma$-module M_Γ with highest weight $\lambda = \omega_\Gamma(m)$. Writing $\lambda = (\lambda_1, \lambda_2)$ with $\lambda_j \in \mathbf{X}_j$, $j = 1, 2$, one gets that $\lambda_1 = \omega_{p-1}$ for $G = A_r(K)$, ω_3 for $G = C_r(K)$ and $p \neq 3$, and $2\omega_2$ for $G = C_r(K)$ with $p = 3$; $\lambda_2 = \omega_1$ in all cases. Observe that $\omega_A(m) = p - 1$ and so $m \in M_y$ by Lemma 2.54. Hence $M_\Gamma \subset M_y$. Obviously, M_Γ has a composition factor $M' \cong M(\lambda)$. Set $M_j = M(\lambda_j)$. It is clear that $d_{M_2}(g_2) = e$. Put $\Delta = (\Gamma_1)_{\mathbb{C}}$. Let h_0 and $h_1 \in \Delta$ be unipotent elements with $J(h_0) = J(g_1)$ and $J(h_1) = J(g_1^p)$. Proposition 2.5 b) implies that g_1^p and h_1 are transvections for $G = A_r(K)$ and that these elements have 2 blocks of size 2 and $p-2$ blocks of size 1 for $G = C_r(K)$. Let $\mu = \varphi(\lambda_1) \in \mathrm{Irr}\,\Delta$. Conjecture (r, s) yields that $d_{M_1}(g_1) = \min\{d_\mu(h_0), pd_\mu(h_1), p^2\}$. Algorithm 1.6 and Proposition 1.5 force that $d_\mu(h_0) = 2p - 1$ and $d_\mu(h_1) = 2$ for $G = A_r(K)$ and $d_\mu(h_0) = 3p - 2$ and $d_\mu(h_1) = 3$ for $G = C_r(K)$. This implies that $d_{M_1}(g_1) = 2p - 1 > p + 1$ for $G = A_r(K)$ and $d_{M_1}(g_1) = 3p - 2 > p + 2$ for $G = C_r(K)$. Since $M' \cong M_1 \otimes M_2$, Theorem 2.9 implies that $d_{M'}(x_\Gamma) = p^s$ and so $d_{M_y}(x) = p^s$. Now Corollary 2.44 yields that $d_\varphi(x) = p^{s+1}$ and completes the proof. □

LEMMA 8.4. *Let $G = B_r(K)$ and $n = p^s + p + 1$. Then $d_\varphi(x) = p^{s+1}$.*

PROOF. If $\omega \neq \omega_i$, argue just as in the relevant case of Lemma 8.2. However, here the arguments are much easier since in all cases $N_{G_1}(\varphi_1) \geq p$. For $\omega = \omega_i$ the lemma follows from Conjecture (r, s), Theorem 1.9, and Proposition 2.34. □

The following lemma deals with a special case that satisfies the assumptions of Lemma 8.6 below, but cannot be settled with the help of the arguments of that lemma.

LEMMA 8.5. *Let $p = 5$, $G = B_7(K)$, and $\omega = \omega_7$. Then $d_\varphi(x) = 25$.*

PROOF. Recall that G can be naturally embedded into the group $D = D_8(K)$ (see Lemma 2.23). Set $\sigma = \varphi(\omega_7) \in \mathrm{Irr}\,D$. Then $\sigma|G = \varphi$ (see, for instance, Seitz

[**Sei87**, Table 1]). Observe that x is a regular unipotent element of D as well. Now we shall compute $d_\sigma(x)$.

We shall use the embedding of D into $E_8(K)$ and Lawther's results [**Law95**, Table 9] on the Jordan block sizes of unipotent elements in the representation $\delta = \varphi(\omega_8)$ of $E_8(K)$. Put $E = E_8(K)$, and let $\Pi(E) = \{\beta_1, \ldots, \beta_8\}$. Denote by β_m the maximal root in $R(E)$. Set $D_1 = E(\beta_2, \ldots, \beta_8, -\beta_m)$, $\gamma_1 = -\beta_m$, and $\gamma_j = \beta_{8-j+2}$ for $2 \leq j \leq 8$. It is well known that $D_1 \cong D$. The roots γ_j, $1 \leq j \leq 8$, constitute $\Pi(D_1)$ and the labelling is standard. To simplify the notation, we identify D and D_1. Put $\delta_1 = \delta|D$ and $\lambda = \varphi(\omega_2) \in \operatorname{Irr} D$. We claim that $\delta_1 = \lambda \oplus \sigma$. Probably, this is well known, but we failed to find a reference.

For a weight $\mu \in \mathbf{X}(E)$ set $c_j(\mu) = \langle \mu, \gamma_j \rangle$, $1 \leq j \leq 8$, and $\bar{\mu} = \sum_{j=1}^{8} c_j(\mu)\omega_j \in \mathbf{X}(D)$. It is clear that there exists a homomorphism from $\mathbf{X}(E)$ to $\mathbf{X}(D)$ induced by the restriction of weights from some maximal torus $T \subset E$ to the maximal torus $T \cap D$ in D that sends μ to $\bar{\mu}$. We have $\overline{\omega_8} = -\varepsilon_1 + \varepsilon_2$ and $\overline{\beta_1} = -\omega_7$. This implies that $\mathbf{X}(\delta_1)$ contains the orbits of these weights under the action of $W(D)$. In particular, ω_2 and $\omega_7 \in \mathbf{X}(\delta_1)$. The orbits mentioned above are distinct and one easily checks that their union contains 240 weights. It is well known that δ is the adjoint representation of E and $\dim \delta = 248$. Hence the dimension of the zero weight subspace in the E-module $M(\delta)$ is equal to $r(E) = 8$, all other weight subspaces are one-dimensional, and $\mathbf{X}(\delta)$ contains just 240 nonzero weights. This forces that $\mathbf{X}(\delta_1)$ contains no nonzero dominant weights, except ω_2 and ω_7. Hence vectors of these weights in $M(\delta)|D$ are fixed by $U^+(D)$. As $M(\omega_i) \cong V(\omega_i)$ for $i = 2$ or 7, one concludes that $M(\omega_2) \oplus M(\omega_7) \subset M(\delta)|D$. Comparing the dimensions, we get the desired equality.

Now consider x as an element of E. According to [**Law95**, Table 9], $\delta(x)$ has 3 Jordan blocks of size 25 as $p = 5$. Hence it suffices to prove that $\lambda(x)$ has at most two blocks of size 25.

Recall that $J(x) = (15, 1)$ (as an element of D). Set $\Sigma_1 = A_{15}(K)$ and let Σ be a group of type A_{14} embedded into Σ_1 in a standard way. Put $\xi = \varphi(\omega_2) \in \operatorname{Irr} \Sigma_1$. It is well known (see, for instance, [**Sei87**, Table 1]) that λ can be identified with the composition of the standard homomorphism of D onto $SO_{16}(K)$ and the restriction $\xi|SO_{16}(K)$. The image of x under the homomorphism above is conjugate in Σ_1 to a regular unipotent element u of Σ. So consider $\xi(u)$. For $i = 1, 2$ set $\xi_i = \varphi(\omega_i) \in \operatorname{Irr} \Sigma$. By Proposition 2.36, $\xi|\Sigma = \xi_1 \oplus \xi_2$. Let M_2 be the irreducible Σ-module $M(\omega_2)$. Set $u_1 = u^5$. By Proposition 2.5 b), u_1 has 5 Jordan blocks of size 3 in the standard realization of Σ (i.e. in ξ_1). Embed u_1 into a Zariski closed subgroup A^+ of type A_1 and denote by ξ_i^A and M_2^A the restrictions of these objects to A^+. One can assume that $\mathbf{X}(\xi_1^A) = \{-2, 0, 2\}$ and $\dim(\xi_1^A)_\mu = 5$ for each weight $\mu \in \mathbf{X}(\xi_1^A)$. As ξ_2 is the wedge square of ξ_1, one easily concludes that 4 is the maximal weight of ξ_2^A and $\dim(M_2^A)_4 = 10$. By [**Sup98**, Lemma 3], the module M_2^A is completely reducible as all its composition factors are p-restricted. Now it follows from the standard facts on representations of $A_1(K)$ that M_2^A has just 10 summands isomorphic to $M(4)$ and other irreducible components of this module have the form $M(a)$ with $a < 4$. By Lemma 2.54, $\xi_2(u_1)$ has only 10 Jordan blocks of size 5. Now Proposition 2.5 b) yields that $\xi_2(u)$ has at most 2 blocks of size 25. Naturally, this holds for $\xi(u)$ as well. This completes the proof. □

LEMMA 8.6. *Let $G = B_r(K)$ and $n = p^s + 2p$. Assume that $\omega = \omega(1, p-1) + a_r\omega_r$ with $a_r \neq 0$. Then $d_\varphi(x) = p^{s+1}$.*

PROOF. Observe that $p > 3$ or $s > 1$ as $n < p^{s+1}$ by our assumptions. Set $G_1 = G(2, \ldots, r)$, $m = X_{-1,a_1}v$, $\omega^1 = \omega_{G_1}(m)$, and $\varphi_1 = \varphi(\omega^1)$. By Lemma 2.54, $m \neq 0$. Obviously, $U^+(G_1)$ fixes m. Hence m generates an indecomposable KG_1-module with highest weight ω^1 and $\varphi|G_1$ has a composition factor φ_1. Set $N = N(\varphi_1)$ in the sense of Section 6. Let $g \in G_1$ be a regular unipotent element. Proposition 6.11 and Lemma 6.9 imply that $d_{\varphi_1}(g) = p^{s+1}$ if $N \geq p$ or $s > 1$ and $N = p - 1$. By Lemma 2.19, $d_\varphi(x) = p^{s+1}$ in all such cases.

Observe that $N \geq p - 1$. Now it is clear that it remains to handle the case where $\omega = \omega_r$ and $s = 1$. For $p = 5$ this situation was settled in Lemma 8.5. So suppose that $p > 5$. We have $\Gamma_1 \cong B_{(p-1)/2}(K)$, $\Gamma_2 = 1$, $\omega_{\Gamma_1}(v) = \omega_{(p-1)/2}$, and $\zeta(\omega) = p$. Let α be the negative root of A. Set $\beta_i = \varepsilon_{i+p} - \varepsilon_i$, $\gamma_i = -\varepsilon_{p+1-i} - \varepsilon_{i+p}$, $1 \leq i \leq (p-1)/2$, and $\delta = -\varepsilon_{(p+1)/2}$. One can assume that $X_\alpha = cX_\delta + Y$ where Y lies in the Lie algebra generated by X_{β_i} and X_{γ_i}, $1 \leq i \leq (p-1)/2$; $c \neq 0$. Observe that $\langle \omega, -\delta \rangle = 1$ and hence $X_\delta v \neq 0$. This forces that $X_\alpha v \neq 0$. Then Lemmas 2.55 and 2.56 imply that $KAv \cong V(p)$ and the vector $m = X_\alpha v \in M_y$. Put $M_\Gamma = K\Gamma m$. Since $\Gamma \subset C_G(A)$, we conclude that $U^+(\Gamma)$ fixes m and $M(\Gamma) \subset M_y$. As the Weyl module $V(\omega_{(p-1)/2})$ for the group Γ_1 is irreducible, we have $M_\Gamma \cong M(\omega_{(p-1)/2})$. Algorithm 1.6 and Theorem 1.14 imply that $d_{M_\Gamma}(x_\Gamma) = \min\{1+(p-1)(p+1)/8, p\} = p$ (observe that $(p-1)(p+1)/8 \geq p - 1$ if $p \geq 7$). Now our claim follows from Corollary 2.44. □

Proof of Theorem 1.9 *for regular unipotent elements and groups of types A, B, and C.* Using Lemmas 8.2, 8.3, 8.4, and 8.6, one can assume the following: if $G = A_r(K)$, then $n > p^s + p$ and $n > p^s + p + 1$ for $\omega = \omega_p$ or ω_{n-p}; if $G = B_r(K)$, then $n > p^s + p + 1$ if $a_i \neq 0$ for some i with $p \leq i < r$, and $n > p^s + 2p$ if $\omega = \omega(1, p-1) + a_r\omega_r$ with $a_r \neq 0$; if $G = C_r(K)$, then $n > p^s + p$ and $n > p^s + p + 2$ for $\omega = \omega_p$. Set $G_1 = G(2, \ldots, r)$. Passing to the dual representation if necessary, one may assume that $a_i \neq 0$ for some i with $p \leq i \leq (r+1)/2$ if $G = A_r(K)$. Put $m = v(1, p, a_p)$ if $a_p \neq 0$, $m = v(1, p+1, 1)$ if $\omega = \omega_{p+1}$ and $G \neq B_r(K)$ or $C_{p+1}(K)$, $m = X_{-1} \ldots X_{-p}X_{-(p+1)}v$ for $G = C_{p+1}(K)$ and $\omega = \omega_{p+1}$, and $m = v$ otherwise. By Lemma 2.46, $m \neq 0$ and is fixed by $U^+(G_1)$. Set $\omega^1 = \omega_{G_1}(m)$ and $\varphi_1 = \varphi(\omega^1) \in \operatorname{Irr} G_1$. It is clear that $\varphi|G_1$ has a composition factor isomorphic to φ_1. Let $g \in G_1$ be a regular unipotent element. Assume that $G \neq B_{p+1}(K)$ and $\omega \neq \omega_{p+1}$ for $G = C_{p+1}(K)$. Then φ_1 satisfies the assumptions of Theorem 1.9. Let $x^1 \in G_1$ be a regular unipotent element. Conjecture (r, s) implies that Theorem 1.9 holds for φ_1 and hence $d_{\varphi_1}(g) = p^{s+1}$.

Now let $G = B_{p+1}(K)$ and $a_p \neq 0$. We have $\omega_r(m) \geq 2a_p$. Define $N_{G_1}(\varphi_1)$ as in Section 6. Then $N_{G_1}(\varphi_1) > p$. Hence $d_{\varphi_1}(g) = p^2$ by Proposition 6.4.

Finally, assume that $G = C_{p+1}(K)$ and $\omega = \omega_{p+1}$. Then $\omega_1 = \omega_{p-1}$ and $G_1 = C_p(K)$. By Conjecture (r, s), Theorem 1.7 holds for G_1. Set $\varphi_0 = (\varphi_1)_{\mathbb{C}}$. Let $u_0, u_1 \in (G_1)_{\mathbb{C}}$ be unipotent elements with $J(u_0) = J(g)$ and $J(u_1) = J(g^p)$. Algorithm 1.6 yields that $d_{\varphi_0}(u_0) = pd_{\varphi_0}(u_1) = p^2$. Hence $d_{\varphi_1}(g) = p^2$ by Theorem 1.7. Thus we have proved that in all cases $d_{\varphi_1}(g) = p^{s+1}$. Now Lemma 2.19 completes the proof. □

9. The general case for regular elements

In this section we complete the proof of Theorem 1.7 for regular unipotent elements of the groups $A_r(K)$, $B_r(K)$, and $C_r(K)$ with $p^s < n < p^{s+1}$ $(s > 0)$

and p-restricted representations. Taking into account the results of Sections 5–8, we can and shall assume the following: $n \geq p^s + p$; $a_i = 0$ for $p \leq i \leq n - p$ if $G = A_r(K)$ and $a_i = 0$ for $i \geq p$ otherwise. Now our goal is to prove

PROPOSITION 9.1. *Let $G \neq D_r(K)$, $s > 0$, and $p^s + p \leq n < p^{s+1}$. Assume that $a_i = 0$ for $p \leq i \leq n - p$ if $G = A_r(K)$ and $a_i = 0$ for $i \geq p$ otherwise. Then Theorem 1.7 holds for G if Conjecture (r, s) holds.*

Throughout this section the coefficients a_i satisfy the assumptions of Proposition 9.1. Recall that $z_j \in G_{\mathbb{C}}$, $0 \leq j \leq s$, are the elements defined in Theorem 1.7 and we denote $d_{\varphi_{\mathbb{C}}}(z_j)$ by d_j. Write the p-adic expansion: $n = b_s p^s + \ldots + b_j p^j + \ldots + b_1 p + b_0$. For $G = A_r(K)$ set $c_i = a_i + a_{n-i}$, otherwise put $c_i = a_i$, $1 \leq i < p$. Observe that our assumptions on n imply that $r \geq p$. The proof of Proposition 9.1 will be subdivided into several subcases.

I. First assume that $b_j \neq 0$ for some j with $0 < j < s$. Then y has at least p blocks of size $b_s + 1$ by Proposition 2.5 b). Put $N = b_s \sum_{i=1}^{p-1} c_i$. Then Proposition 1.5, Algorithm 1.6, and Theorem 1.14 imply that $d_\varphi(y) = \min\{N + 1, p\}$.

LEMMA 9.2. *Assume that $n = b_s p^s + \ldots + b_j p^j + \ldots + b_1 p + b_0$ with $b_s \neq 0$ and $b_j \neq 0$ for some j with $0 < j < s$. Let $N < p$. Then $d_\varphi(x) = \min\{p^{s+1}, p^j d_j \mid 0 \leq j \leq s\}$.*

PROOF. Set $n_1 = n - b_s p^s$ and $e = p^s - n_1$. If $G = A_r(K)$, one gets $q_1 = n_1 - 1$ and $q_2 = e - 1$, otherwise $r_1 = [n_1/2]$ and $r_2 = [e/2]$. We have $\zeta(\varepsilon_i) = b_s$ for $1 \leq i \leq n_1$, $\zeta(\varepsilon_{n_1+1}) = b_s - 1$, and $\zeta(\varepsilon_i) \geq \zeta(\varepsilon_j)$ for $i < j$. Then $\zeta(\omega) = N$ and by Lemma 2.54, $v \in M_y = (y-1)^N M$. For $G = A_r(K)$ we have $\Gamma_i = A_{q_i}(K)$, otherwise $\Gamma_1 = B_{q_1}(K)$ if n_1 is odd and $C_{q_1}(K)$ if n_1 is even, $\Gamma_2 = B_{q_2}(K)$ for odd and $C_{q_2}(K)$ for even e. (Naturally, $\Gamma_2 = 1$ if $e = 1$.) Set $\lambda = \omega_\Gamma(v)$. As before, write $\lambda = (\lambda_1, \lambda_2)$ with $\lambda_j \in \mathbf{X}_j$, $j = 1, 2$. The construction of the homomorphism ρ in Proposition 2.43 implies that $\lambda_1 = \omega(1, p-1) + \omega_+(q_1 - p + 2, q_1, r - q_1)$ for $G = A_r(K)$ and $\lambda_1 = \omega(1, p-1)$ otherwise; $\lambda_2 = 0$ in all cases. Hence $K\Gamma_1 v$ has a composition factor M' isomorphic to $M(\lambda_1)$.

Fix maximal $j < s$ with $b_j \neq 0$. By our assumptions, $j > 0$. We have $|g_1| = p^j$ if $n_1 = p^j$ and p^{j+1} otherwise. Proposition 2.5 b) implies the following. For $0 < k \leq s$ set $n(k) = \sum_{l=0}^{k-1} b_l p^l$ and $n'(k) = \sum_{l=k}^{s} b_l p^{l-k}$. Then x^{p^k} has $n(k)$ Jordan blocks of size $n'(k) + 1$ and $p^k - n(k)$ blocks of size $n'(k)$. For $k < j + 1$ the element $g_1^{p^k}$ has $n(k)$ blocks of size $n'(k) - b_s p^{s-k} + 1$ and $p^k - n(k)$ blocks of size $n'(k) - b_s p^{s-k}$, for $k \geq j + 1$ one has $n(k) = n_1 \geq p$. Let $z'_t \in (\Gamma_1)_{\mathbb{C}}$, $0 \leq t \leq j$, be a unipotent element with $J(z'_t) = J(g_1^{p^t})$. Set $\mu = \varphi(\lambda_1)$, $\delta_k = p^k d_{\mu_{\mathbb{C}}}(z'_k)$ for $0 \leq k \leq j$, and $\delta_k = p^k$ for $j + 1 \leq k \leq s$. Proposition 1.5, Algorithm 1.6, the assumptions on a_i in Proposition 9.1, and the arguments above imply that $p^k d_k = N p^s + \delta_k$ for $0 \leq k \leq s$. Put $j' = j$ if $n_1 = p^j$ and $j' = j + 1$ otherwise. Conjecture (r, s) and Theorem 1.7 force that $d_\mu(g_1) = \min\{p^{j'}, \delta_k \mid 0 \leq k < j'\}$. It is clear that $d_{M_y}(g_1) \geq d_\mu(g_1)$. By Corollary 2.44, $d_\varphi(x) \geq N p^s + d_\mu(g_1)$. If $d_\mu(g_1)$ is equal to some δ_k, we get that $d_\varphi(x) \geq p^k d_k$. Now Corollary 2.41 shows that $d_\varphi(x) = p^k d_k$ as desired. If $d_\mu(g_1) = p^{j'}$, similar arguments yield that $d_\varphi(x) = p^{j'} d_{j'}$ (observe that $\delta_j = p^j$ if $n_1 = p^j$). This completes the proof. □

LEMMA 9.3. *Let n be as in Lemma 9.2. Assume that $N \geq p$. Then $d_\varphi(x) = |x| = p^{s+1}$.*

PROOF. Put $G_1 = G(2,\dots,r)$. Let $g \in G_1$ be a regular unipotent element. Passing to the dual representation if necessary, one can suppose that for $G = A_r(K)$ some $a_i \neq 0$ with $i < p$, for other groups this is obvious as φ is nontrivial. Choose maximal $j < p$ with $a_j \neq 0$ and set $m = v(1, j, a_j)$. By Lemma 2.46, $m \neq 0$ and is fixed by $U^+(G_1)$. Set $r' = r(G_1)$, $\omega' = \omega_{G_1}(m)$, and $\varphi' = \varphi(\omega')$. Obviously, φ' is a composition factor of $\varphi|G_1$. Observe that if $G \neq B_p(K)$ or $j < p-1$, then $\omega' = \omega(1, p-1) + \omega(r'-p+2, r')$ for $G = A_r(K)$ and $\omega' = \omega(1, p-1)$ otherwise. In the exceptional case $\omega' = \omega(1, p-2) + 2a_{p-1}\omega_{p-1}$.

The proof is based on computing $d_{\varphi'}(g)$ with the use of Conjecture (r,s) and induction by r. If $d_{\varphi'}(g) = p^{s+1}$, Lemma 2.19 yields that $d_\varphi(x) = p^{s+1}$ as required.

Set $y_1 = g^{p^s}$. Let $u \in (G_1)_{\mathbb{C}}$ be a unipotent element with $J(u) = J(y_1)$. If

$$d_{\varphi'_{\mathbb{C}}}(u) > p, \tag{9.1}$$

Conjecture (r,s) and Corollary 3.3 imply that $d_{\varphi'}(g) = p^{s+1}$ and hence $d_\varphi(x) = p^{s+1}$. If $n > b_s p^s + p$ or $G = A_r(K)$, we have $n(G_1) \geq b_s p^s + p - 1$. So, by Proposition 2.5 b), y_1 has at least $p-1$ Jordan blocks of size $b_s + 1$ on $V(G_1)$. This implies that $d_{\varphi'_{\mathbb{C}}}(u) = N + 1 > p$.

Now let $n = b_s p^s + p$ and $G \neq A_r(K)$. Then $n(G_1) = b_s p^s + p - 2$. Proposition 2.5 b) implies that y_1 has $p-2$ blocks of size $b_s + 1$ and $p^s - p + 2$ blocks of size b_s. Proposition 1.5 and Algorithm 1.6 yield that (9.1) holds unless $G = C_r(K)$, $n = p^s + p$ and either $\omega = \omega_1 + \omega_{p-1}$, or $p = 3$ and $\omega = 2\omega_{p-1}$. If (9.1) holds, we are done. So it remains to consider the exceptional cases.

Let $G = C_r(K)$ with $n = p^s + p$. Set $G^+ = A_{n-1}(K)$. The facts on the standard realization of G mentioned in the Introduction show that G can be embedded into G^+ in the standard way such that $U^\pm \subset U^\pm(G^+)$ and the maximal torus $T \subset G$ normalizing both U^+ and U^- lies in the maximal torus $T^+ \subset G^+$ that normalizes $U^+(G^+)$ and $U^-(G^+)$. Let $\omega^+ \in \mathbf{X}(G^+)$ be the weight defined by the same formula as ω and $\varphi^+ = \varphi(\omega^+) \in \operatorname{Irr} G^+$. Recall that x is a regular unipotent element in G^+ as well. Conjecture (r,s) and the arguments above yield that $d_{\varphi^+}(x) = p^{s+1}$.

First suppose that $p = 3$. Then Proposition 2.34 implies that $d_\varphi(x) = d_{\varphi^+}(x) = p^{s+1}$ as desired.

Next, assume that $p > 3$ and $\omega = \omega_1 + \omega_{p-1}$. Then $N = p$. Our assumptions on s and j yield that $s > 1$. Let $\Pi(G^+) = \{\beta_1, \beta_2, \dots, \beta_{n-1}\}$ and let $Y_{\pm i}$, $1 \leq i \leq n-1$, be the root elements of $L(G^+)$ associated with the roots $\pm\beta_i$. One may assume that

$$X_i = Y_i - Y_{n-i},\ X_{-i} = Y_{-i} - Y_{-(n-i)},\ 1 \leq i < r,\ X_{\pm r} = Y_{\pm r} \tag{9.2}$$

(see [**Bou75**, ch. VIII, §13]). Set $M^+ = M(\omega^+)$. Let $m \in M^+$ be a nonzero highest weight vector. Put $V_G = KGm$. Then V_G is an indecomposable G-module with highest weight ω. Denote by M_G the quotient module of V_G by the maximal G-submodule and by $\bar{l}$ the image of a vector $l \in V_G$ under the canonical homomorphism $V_G \to M_G$. It is clear that $M_G \cong M$. Let $x_- \in U^-$ be a regular unipotent element. Now it suffices to find a vector $u \in V_G$ such that

$$\overline{(x_- - 1)^{p^{s+1}-1} u} \neq 0. \tag{9.3}$$

Let ω' be the lowest weight of φ^+. Set

$$\Lambda_1 = \{\mu \in \mathbf{X}(M^+) \mid \omega^+ - \mu = \sum_{i=1}^{n-1} c_i\beta_i, \sum_{i=1}^{n-1} c_i \leq 2p-1\},$$

$$\Lambda_2 = \{\nu \in \mathbf{X}(M^+) \mid \nu - \omega' = \sum_{i=1}^{n-1} d_i\beta_i, \sum_{i=1}^{n-1} d_i \leq 2p-1\}.$$

Put $V_j = \sum_{\gamma\in\Lambda_j} M^+_\gamma$, $j = 1, 2$. Now our goal is to prove the following

Claim. i) $V_j \subset V_G$ for $j = 1, 2$;

ii) $\bar{w} \neq 0$ for $w \in V_2 \setminus \{0\}$;

iii) there exists $u \in V_1$ with $(x_- - 1)^{p^{s+1}-1}u \neq 0$;

iv) $(x_- - 1)^{p^{s+1}-1}M^+ \subset V_2$.

It is clear that this claim implies the existence of a vector $u \in V_G$ that satisfies (9.3). Let $m' \in M^+$ be a nonzero vector of weight ω'. Observe that $m' \in V_G$ and $\overline{m'} \neq 0$. This follows immediately from the fact that $-\omega \in \mathbf{X}(M)$ and that $\langle m'\rangle$ is the weight subspace of weight $-\omega$ in the G-module M^+. Denote by Ω_1 the set of vectors of the form $Y_{-j_k} \dots Y_{-j_2}Y_{-j_1}m$ with $1 \leq j_1, j_2, \dots, j_k < 3p-2$ and by Ω_2 the set of vectors of the form $Y_{l_t} \dots Y_{l_2}Y_{l_1}m'$ with $n-3p+2 < l_1, l_2, \dots, l_t \leq n-1$. Let V^1 be the linear span of m and Ω_1 and V^2 be the linear span of m' and Ω_2. We claim that $V_j \subset V^j$ for $j = 1, 2$. Indeed, as φ^+ is p-restricted, it follows from [**Bor70**, Lemma 6.2 and Theorem 6.4] and the commutator relations in $L(G^+)$ that V_1 is the linear span of m and all vectors of the form $Y_{-j_f} \dots Y_{-j_2}Y_{-j_1}m$ with $f \leq 2p-1$. Using induction on f and the commutator relations between Y_{-i} and analyzing the weight structure of M^+, one can deduce that $j_1, j_2, \dots, j_f \leq p-2+f$ if such vector is nonzero. This forces that $V_1 \subset V^1$. For V_2 and V^2 argue similarly proving that V_2 is the linear span of m' and all vectors of the form $Y_{j_f} \dots Y_{j_2}Y_{j_1}m'$ with $f \leq 2p-1$ and $j_1, j_2, j_f \geq n-p+2-f$. Observe that $n \geq 6p$ as $p > 3$ and $s > 1$. Hence $r \geq 3p$. It is clear that Y_{-t} annihilates m and all vectors in Ω_1 if $t \geq r$ since in this case Y_{-t} commutes with Y_{-j} for $j < 3p-2$. Now formula (9.2) forces that V^1 and hence $V_1 \subset U^-m \subset V_G$. Similar arguments with m' and Y_j yield that Y_t annihilates m' and all vectors in Ω_2 for $t \leq r$ and $V_2 \subset Um' \subset V_G$. Hence i) holds. Next, let $w \in V_2 \setminus \{0\}$. Then [**Bor70**, Lemma 6.2 and Theorem 6.4] imply that $Y_{-f_l} \dots Y_{-f_2}Y_{-f_1}w = cm'$ with $c \in K^*$ for a suitable sequence $f_1, \dots, f_l$. Since $V_2 \subset V^2$, we conclude that $f_1, \dots, f_l > n-3p+2$. A glance at the weight structure of M^+ yields that Y_{-h} annihilates V^2 for $h \leq r$. Now Formula (9.2) implies that $cm' \in U^-w$. Hence $\bar{w} \neq 0$ since $\overline{cm'} \neq 0$. The assertion ii) is proved. Using [**Ste68**, Lemma 72], we conclude that iv) holds and that $(x_- - 1)^{p^{s+1}-1}M^+_\mu = 0$ if $\mu \notin \Lambda_1$. As $d_{\varphi_+}(x_-) = p^{s+1}$, this yields iii) and completes the proof of the claim and of the lemma. □

II. Now assume that $n = b_sp^s + b_0$ with $b_s > 1$. The element y has b_0 blocks of size $b_s + 1$ and $p^s - b_0$ blocks of size b_s in the standard realization of G. Put

$$N' = b_s\sum_{i=1}^{b_0} ic_i + \sum_{i=b_0+1}^{p-1} c_i(ib_s + b_0 - i). \tag{9.4}$$

Proposition 1.5, Algorithm 1.6, and Theorem 1.14 imply that $d_\varphi(y) = \min\{N' + 1, p\}$. Set $n_1 = b_0$ and $n_2 = p^s - b_0$. Observe that $\Gamma_1 = 1$ for $n_1 = 0$ or 1 and $\Gamma_2 = 1$ if $n_2 = 1$. If $n_1 > 1$, we have $q_1 = n_1 - 1$ if $G = A_r(K)$ and $q_1 = [n_1/2]$

otherwise. If $e > 1$, then $q_2 = n_2 - 1$ if $G = A_r(K)$ and $q_2 = [n_2/2]$ otherwise. If $\Gamma_i \neq 1$, one gets $\Gamma_i = A_{q_i}(K)$ for $G = A_r(K)$, otherwise $\Gamma_i = B_{q_i}(K)$ for odd and $C_{q_i}(K)$ for even n_i. In the arguments below in this section we assume that $b_0 > 0$ whenever a_{b_0} or a_{n-b_0} occurs.

The construction of ζ in Proposition 2.43 implies that

$$\begin{aligned} &\zeta(\omega) = N', \\ &\zeta(\alpha_i) = 0 \quad \text{if} \quad i < p^s \quad \text{and} \quad i \neq b_0, \\ &\zeta(\alpha_{b_0}) = 1 \quad \text{if} \quad b_0 > 0, \\ &\zeta(\alpha_{p^s}) = 1, \end{aligned} \tag{9.5}$$

except the case where $G = C_{p^s}(K)$. In this exceptional case we have

$$\zeta(\alpha_{p^s}) = 2. \tag{9.6}$$

Set $\lambda = \omega_\Gamma(v)$ and $M_\Gamma = M(\lambda)$. Write $\lambda = (\lambda_1, \lambda_2)$ with $\lambda_i \in \mathbf{X}_i$, $i = 1, 2$. Put $d_\Gamma = d_{M_\Gamma}(x_\Gamma)$.

LEMMA 9.4. *Assume that* $n = b_s p^s + b_0$ *with* $b_s > 1$ *and* $N' < p$. *Then* $d_\varphi(x) = \min\{d_0, pd_1, p^s d_s\}$.

PROOF. Observe that $d_s = d_\varphi(y)$ by Theorem 1.14. Arguing as in the proof of Lemma 9.2, we can conclude that $v \in M_y$. Then the $K\Gamma$-module $M' = K\Gamma v \subset M_y$ and Corollary 2.44 implies that $d_\varphi(x) \geq N'p^s + d_\Gamma$ as $d_\Gamma \leq d_{M'}(x_\Gamma)$. In particular, if $d_\Gamma = p^s$, one gets $d_\varphi(x) = (N'+1)p^s = p^s d_s$ since $d_\varphi(x) \leq p^s d_\varphi(y)$ by Proposition 2.5 a).

Now assume that $d_\Gamma < p^s$. As $N' < p$, one can deduce from (9.4) that λ_1 and λ_2 are p-restricted. Since $r \geq p$, our assumptions on ω imply that $a_r = 0$ and hence $\lambda_i \neq \omega_{q_i}$ for $\Gamma_i = B_{q_i}(K)$. Therefore Lemma 3.1, Theorem 2.9, and Lemma 2.8 yield that one of the following holds:

1) $\lambda_2 = 0$;

2) $\lambda_1 = 0$, $s = 1$;

3) $\lambda_i \in \{0, \omega_1, \omega_{q_i}\}$ for $G = A_r(K)$, $\lambda_i \in \{0, 2\omega_1\}$ if $\Gamma_i = B_1(K)$, and $\lambda_i \in \{0, \omega_1\}$ otherwise, $i = 1, 2$.

Now consider these special cases. Set $M_i = M(\lambda_i)$ and $d_{\Gamma,i} = d_{M_i}(g_i)$, $i = 1, 2$. We assume that $\lambda_1 \neq 0$ in Case 1) and $\lambda_2 \neq 0$ in Case 2) since the situation where $\lambda_1 = \lambda_2 = 0$ is covered by 3).

Let 1) hold. Then our assumptions on ω yield that $b_0 > 1$, $\omega = \omega(1, b_0) + \omega(r - b_0 + 1, r)$ for $G = A_r(K)$ and $\omega = \omega(1, b_0)$ otherwise. For $1 \leq i \leq q_1$ set $f_i = a_i + a_{n-b_0+i}$ for $G = A_r(K)$ and put $f_i = a_i + a_{b_0-i}$ otherwise. By (2.10), $\lambda_1 = \sum_{i=1}^{q_1} f_i\omega_i$ for $G = A_r(K)$, for other groups $\lambda_1 = \sum_{i=1}^{q_1-1} f_i\omega_i + a_{q_1}\omega_{q_1}$ for even and $\sum_{i=1}^{q_1-1} f_i\omega_i + 2f_{q_1}\omega_{q_1}$ for odd b_0. Since $\lambda_2 = 0$, it is clear that $d_\Gamma = d_{\Gamma,1}$. Set

$$N_1 = \begin{cases} \sum_{i=1}^{q_1} f_i i(b_0 - i) & \text{for} \quad G = A_r(K) \quad \text{or} \quad \Gamma_1 = B_{q_1}(K), \\ a_{q_1} q_1^2 + \sum_{i=1}^{q_1-1} f_i i(b_0 - i) & \text{for} \quad \Gamma_1 = C_{q_1}(K). \end{cases}$$

By Theorem 1.14, $d_{\Gamma,1} = \min\{p, N_1 + 1\}$. Using Proposition 1.5 and Algorithm 1.6, observe that $d_0 = N'p^s + N_1 + 1$ and $pd_1 = p^s N' + p$. Now Corollary 2.44 forces that $d_\varphi(x) \leq \min\{d_0, pd_1\}$. Hence $d_\varphi(x) = \min\{d_0, pd_1\}$ by Corollary 2.41 and so $d_\varphi(x) = d(x, \varphi)$.

Next, assume that 2) holds. Then $d_\Gamma = d_{\Gamma,2}$. We have $\omega = \omega(b_0, p-1) + \omega(n - p + 1, n - b_0)$ for $G = A_r(K)$ and $\omega = \omega(b_0, p-1)$ otherwise. Define f_i, $1 \leq i \leq q_2$, as follows: $f_i = a_{i+b_0} + a_{n-p+i}$ for $G = A_r(K)$, $f_{q2} = a_{q2+b_0}$ for $G \neq A_r(K)$ and even n_2, and $f_i = a_{i+b_0} + a_{p-i}$ in all other cases where $G \neq A_r(K)$. Then $\lambda_2 = \sum_{i=1}^{q_2} f_i\omega_i$ for $\Gamma_2 \neq B_{q_2}(K)$ and $\lambda_2 = (\sum_{i=1}^{q_2-1} f_i\omega_i) + 2f_{q_2}\omega_{q_2}$ if $\Gamma_2 = B_{q_2}(K)$. Set $N_2 = \sum_{i=1}^{q_2} f_i i(n_2-i)$. By Theorem 1.14, $d_{\Gamma,2} = \min\{p, N_2+1\}$. Proposition 1.5 and Algorithm 1.6 imply that now $d_0 = N'p + N_2 + 1$. By definition, $d_1 = N' + 1$. Arguing as at the end of the analysis of Case 1), we deduce that $d_\varphi(x) = d(x, \varphi) = \min\{d_0, pd_1\}$.

Finally, let 3) hold. Then $d_{\Gamma,i} = 1$ or n_i. Since $b_0 < p$, by Corollary 2.15, $d_\Gamma = p^s - 1$ if both $d_{\Gamma,1}$ and $d_{\Gamma,2} \neq 1$. If $b_0 = 0$ and $G = A_r(K)$, then $\omega \in \{\omega_1, \omega_{p-1}, \omega_{r-p+2}, \omega_r\}$ for $s = 1$ and $\omega \in \{\omega_1, \omega_r\}$ otherwise. For $G \neq A_r(K)$ and $b_0 = 0$ we get $\omega \in \{\omega_1, \omega_{p-1}\}$ for $s = 1$ and $\omega = \omega_1$ otherwise. If $b_0 = 1$, set

$$\Omega = \begin{cases} \{\omega_2, \omega_{p-1}, \omega_{r-p+2}, \omega_{r-1}\} & \text{for } G = A_r(K) \text{ and } s = 1, \\ \{\omega_2, \omega_{r-1}\} & \text{for } G = A_r(K) \text{ and } s > 1, \\ \{\omega_2, \omega_{p-1}\} & \text{for } G \neq A_r(K) \text{ and } s = 1, \\ \{\omega_2\} & \text{for } G \neq A_r(K) \text{ and } s > 1. \end{cases}$$

If $1 < b_0 < p - 1$, put

$$\Omega = \{\omega_1, \omega_{b_0-1}, \omega_{b_0+1}, \omega_{p-1}, \omega_{r-p+2}, \omega_{r-b_0}, \omega_{r+2-b_0}, \omega_r, \omega_1 + \omega_{b_0+1}, \omega_1 + \omega_{p-1},$$
$$\omega_1 + \omega_{r-p+2}, \omega_1 + \omega_{r-b_0}, \omega_{b_0-1} + \omega_{b_0+1}, \omega_{b_0-1} + \omega_{p-1}, \omega_{b_0-1} + \omega_{r-p+2},$$
$$\omega_{b_0-1} + \omega_{r-b_0}, \omega_{b_0+1} + \omega_{r+2-b_0}, \omega_{b_0+1} + \omega_r, \omega_{p-1} + \omega_{r+2-b_0}, \omega_{p-1} + \omega_r,$$
$$\omega_{r-p+2} + \omega_{r+2-b_0}, \omega_{r-p+2} + \omega_r, \omega_{r-b_0} + \omega_{r+2-b_0}, \omega_{r-b_0} + \omega_r\}$$
$$\text{for } G = A_r(K) \text{ and } s = 1;$$

$$\Omega = \{\omega_1, \omega_{b_0-1}, \omega_{b_0+1}, \omega_{r-b_0}, \omega_{r+2-b_0}, \omega_r, \omega_1 + \omega_{b_0+1}, \omega_1 + \omega_{r-b_0}, \omega_{b_0-1} + \omega_{b_0+1},$$
$$\omega_{b_0-1} + \omega_{r-b_0}, \omega_{b_0+1} + \omega_{r+2-b_0}, \omega_{b_0+1} + \omega_r, \omega_{r-b_0} + \omega_{r+2-b_0}, \omega_{r-b_0} + \omega_r\}$$
$$\text{for } G = A_r(K) \text{ and } s > 1;$$

$$\Omega = \{\omega_1, \omega_{b_0-1}, \omega_{b_0+1}, \omega_{p-1}, \omega_1 + \omega_{b_0+1}, \omega_1 + \omega_{p-1}, \omega_{b_0-1} + \omega_{b_0+1}, \omega_{b_0-1} + \omega_{p-1}\}$$
$$\text{for } G \neq A_r(K) \text{ and } s = 1;$$

$$\Omega = \{\omega_1, \omega_{b_0-1}, \omega_{b_0+1}, \omega_1 + \omega_{b_0+1}, \omega_{b_0-1} + \omega_{b_0+1}\} \text{ for } G \neq A_r(K) \text{ and } s > 1.$$

Finally, if $b_0 = p - 1$, set

$$\Omega = \{\omega_1, \omega_{p-2}, \omega_{r-p+3}, \omega_r\} \text{ for } G = A_r(K) \text{ and } \Omega = \{\omega_1, \omega_{p-2}\} \text{ otherwise.}$$

Then $\omega = a_{b_0}\omega_{b_0} + a_{n-b_0}\omega_{n-b_0} + \mu$ for $G = A_r(K)$ and $\omega = a_{b_0}\omega_{b_0} + \mu$ otherwise with $\mu \in \Omega \cup \{0\}$ in both cases. Using Proposition 1.5 and Algorithm 1.6, one concludes that $d_0 = N'p^s + d_\Gamma$ for all these ω. As in Items 1) and 2), Corollaries 2.41 and 2.44 force that $d_\varphi(x) = d_0 = d(x, \varphi)$ as required. This completes the proof. □

PROPOSITION 9.5. *Let n be as in Lemma 9.4 and $N' \geq p$. Then $d_\varphi(x) = |x| = p^{s+1}$.*

PROOF. Set $G_1 = G(2, \dots, r)$ and $n^- = n(G_1)$. Then $n^- = n - 1$ for $G = A_r(K)$ and $n - 2$ otherwise. Observe that $n^- \geq p^s + p$, except the case where $s = 1$ and $n = 2p$ or $2p + 1$, and in all situations $n^- > p^s$. If $b_0 = 0$, set

$$N^- = \begin{cases} (b_s - 1)\sum_{i=1}^{p-1} c_i i & \text{for } G = A_r(K) \text{ or } s > 1, \\ a_{p-1}((b_1 - 1)(p-2) + b_1 - 2) & \\ +(b_1 - 1)\sum_{i=1}^{p-2} a_i i & \text{if } G \neq A_r(K) \text{ and } s = 1. \end{cases}$$

If $b_0 = 1$, for $G = A_r(K)$ or $s > 1$ define N^- as in the case $b_0 = 0$ and for other groups put $N^- = (b_1 - 1)\sum_{i=1}^{p-1} a_i i$. Finally, if $b_0 > 1$, set

$$N^- = b_s\big(\sum_{i=1}^{b_0-1} c_i i + (b_0 - 1)\sum_{i=b_0}^{p-1} c_i\big) + (b_s - 1)\sum_{i=b_0}^{p-1} c_i(i - b_0 + 1)$$

for $G = A_r(K)$ and

$$N^- = b_s\big(\sum_{i=1}^{b_0-2} a_i i + (b_0 - 2)\sum_{i=b_0-1}^{p-1} a_i\big) + (b_s - 1)\sum_{i=b_0-1}^{p-1} a_i(i - b_0 + 2)$$

otherwise. Passing to the dual representation if necessary, one can assume that $a_i \neq 0$ for some $i < p$ if $G = A_r(K)$. Now fix maximal $i < p$ with $a_i \neq 0$ and set $m = v(1, i, a_i)$. By Lemma 2.46, $m \neq 0$ and generates an indecomposable G_1-module with highest weight $\omega^1 = \omega_{G_1}(m)$. Hence $\varphi|G_1$ has a composition factor $\varphi_1 \cong \varphi(\omega^1)$. One easily concludes that $\omega^1 = \omega(1, i) + \omega(r - p + 1, r - 1)$ for $G = A_r(K)$ and $\omega^1 = \omega(1, i)$ otherwise. Let $g \in G_1$ be a regular unipotent element and $y_1 = g^{p^s}$. Since $n^- > p^s$, one has $|g| = |x|$ and $y_1 \neq 1$. By Lemma 2.19, $d_\varphi(x) \geq d_\varphi(g) \geq d_{\varphi_1}(g)$. Hence $d_\varphi(x) = |x|$ and we are done if $d_{\varphi_1}(g) = |z| = p^{s+1}$. Proposition 1.5 and Algorithm 1.6 imply that $N^- = d_{(\varphi_1)_{\mathbb{C}}}(y_0)$ for an element $y_0 \in (G_1)_{\mathbb{C}}$ with $J(y_0) = J(y_1)$. Hence N^- plays for g the same role as N' for x. Conjecture (r, s) and Corollary 3.3 imply that $d_{\varphi_1}(g) = |g|$ if $N^- \geq p$. Notice that due to our assumptions on the coefficients a_i the representation φ_1 cannot be one of the exceptional representations in Theorems 1.7 and 1.10 for which Formula (1.4) does not hold. Observe that $N' = N^-$ if $b_0 = 0$ and $G = A_r(K)$, or $s > 1$, or $a_{p-1} = 0$.

Now assume that $N^- < p$. The arguments above yield that $b_0 > 0$ for $G = A_r(K)$, for $s > 1$ and for $a_{p-1} = 0$. Let $G \neq A_r(K)$, $n = b_1 p$, and $a_{p-1} > 0$. Then $G = C_p(K)$, $b_1 = 2$ and either $\omega = \omega_1 + \omega_{p-1}$, or $p = 3$ and $\omega = 2\omega_2$. One gets $N' - N = 1$ in the first case and 2 in the second one.

Since $N^- \leq p - 1$, we conclude that $a_{p-1} + a_{n-p+1} = 0$ for $G = A_r(K)$ and $b_0 > 1$ and $a_{p-1} = 0$ for $G \neq A_r(K)$ and $b_0 > 2$; $b_s = 2$ if $a_{p-1} \neq 0$; if $p > 3$, then $a_{p-2} = 0$ if $b_0 > 2$ for $G = A_r(K)$ and $b_0 > 3$ otherwise; for such p one has $b_s = 2$ if $a_{p-2} \neq 0$. Furthermore, if $G \neq A_r(K)$, $b_0 = b_s = 2$ and $a_{p-1} \neq 0$, then $\omega = \omega_{p-1}$. Observe also that if $b_0 = p - 1$, one gets $N^- = N'$ if $G = A_r(K)$ and $a_{p-1} + a_{n-p+1} = 0$ or $G \neq A_r(K)$ and $a_{p-2} + a_{p-1} = 0$.

We have $N' - N^- = \sum_{i=b_0}^{n-b_0} a_i$ for $G = A_r(K)$, $N' - N^- = a_{b_0-1} + 2\sum_{i=b_0}^{p-1} a_i$ for $G \neq A_r(K)$ and $b_0 > 1$, and $N' - N^- = \sum_{i=1}^{p-1} a_i$ for $G \neq A_r(K)$ and $b_0 = 1$. In the majority of cases we construct a nonzero weight vector $m \in M_y$ such that $U^+(A)$ and $U^+(\Gamma)$ fix m. Set $\mu = \omega_\Gamma(m)$. Obviously, the module $M' = K\Gamma m \subset M_y$ and has a composition factor $M_\Gamma \cong M(\mu)$. Therefore $d_{M_\Gamma}(x_\Gamma) \leq d_{M'}(x_\Gamma)$. Now Corollary 2.44 implies that $d_\varphi(x) = |x|$ if $d_{M_\Gamma}(x_\Gamma) = p^s$. This yields our claim for such cases.

The construction of m depends upon G and the difference $N' - (p - 1)$. We have to distinguish the following cases.

1. First let $G = A_r(K)$ or $b_0 = 1$. Then $N' = p - 1 + u$ with $u \leq \sum_{i=b_0}^{p-1}(a_i + a_{n-i})$ for $G = A_r(K)$ and $u \leq \sum_{i=1}^{p-1} a_i$ otherwise. If $u \leq a_{b_0}$, set $m = X_{-b_0,u}v$. If $G = A_r(K)$ and $a_{b_0} < u \leq a_{b_0} + a_{n-b_0}$, put $t = u - a_{b_0}$ and $m = X_{-(n-b_0),t}X_{-b_0,u}v$. Next, assume that $a_{b_0} + a_{n-b_0} < u \leq a_{n-b_0} + \sum_{i=b_0}^{p-1} a_i$ for $G = A_r(K)$ and $a_1 < u$

for other groups. Put $a = a_{b_0} + a_{n-b_0}$ for $G = A_r(K)$ and $a = a_1$ otherwise. Then there exist j with $b_0 < j < p$ and $d \leq a_j$ such that $u = a + d + \sum_{i=b_0+1}^{j-1} a_i$. Now put $m = X_{-(n-b_0),a_{n-b_0}} v(b_0, j, d)$ for $G = A_r(K)$ and $m = v(1, j, d)$ otherwise. Finally, if $G = A_r(K)$ and $a_{n-b_0} + \sum_{i=b_0}^{p-1} a_i < u$, then there exist k with $b_0 < k < p$ and $c \leq a_{n-k}$ such that $u = c + (\sum_{i=b_0}^{p-1} a_i) + \sum_{i=n-k+1}^{n-b_0} a_i$. Then fix maximal i with $b_0 \leq i < p$ and $a_i \neq 0$ and put $m_1 = v(b_0, i, a_i)$ and $m = m_1(n - b_0, n - k, c)$.

2. Now assume that $G \neq A_r(K)$ and $b_0 > 1$. We have $N' = p - 1 + w$ with $w \leq a_{b_0-1} + 2\sum_{i=b_0}^{p-1} a_i$.

2.1. If $w \leq \sum_{i=b_0}^{p-1} a_i$, construct m as in Case 1 putting $a = a_{b_0}$.

2.2. Let $\sum_{i=b_0}^{p-1} a_i < w \leq \sum_{i=b_0-1}^{p-1} a_i$. Set $t = w - \sum_{i=b_0}^{p-1} a_i$. One has $t \leq a_{b_0-1}$. If $a_j > 0$ for some $j > b_0$, fix maximal such j, put $m_1 = v(b_0 + 1, j, a_j)$, and $m = m_1(b_0, b_0 - 1, t)$; otherwise set $m = v(b_0, b_0 - 1, t)$.

2.3. Finally, let $w > \sum_{i=b_0-1}^{p-1} a_i$. If $a_{p-1} \neq 0$, we have already shown that $b_0 = b_s = 2$ and $\omega = \omega_{p-1}$. Assume that $a_{p-1} = 0$. There exist j and t such that $b_0 \leq j < p - 1$, $0 < t \leq a_j$, and $w = t + (\sum_{i=b_0-1}^{p-2} a_i) + \sum_{i=b_0}^{j-1} a_i$. Put $m_1 = v(b_0 - 1, j, t)$; if $a_k > 0$ for some k with $k > j$, fix maximal such k and set $m = m_1(b_0, k, \omega_k(m_1))$, otherwise put $m = m_1(b_0, j + 1, t)$.

3. Let $p = 3$, $G = C_3(K)$, and $\omega = 2\omega_2$. Then $N' = 4 = p + 1$. Set $m = v(3, 2, 1)$.

The cases where $G \neq A_r(K)$, $b_0 = b_s = 2$, and $\omega = \omega_{p-1}$, or $G = C_p(K)$ and $\omega = \omega_1 + \omega_{p-1}$ will be handled later with the use of specific arguments.

Lemmas 2.46 and 2.48 imply that $m \neq 0$ and is fixed by all $\mathcal{X}_i$ with $i \neq b_0$ or $n - b_0$ in Case 1 for $G = A_r(K)$, $i \neq b_0$ in Case 1 for other groups and in Subcases 2.1 and 2.2, and $i \neq 3$ in Case 3. In Case 1 if $G = A_r(K)$ and we construct m via m_1, first we apply Lemma 2.46 to show that $m_1 \neq 0$ and is fixed by all $\mathcal{X}_i$ with $i \neq b_0$; since $\mathcal{X}_i$ commutes with $X_{-l,f}$ for $i < p$ and $l > n - p$, this forces that $\mathcal{X}_i$ fixes m if $i < p$ and $i \neq b_0$; then consider the indecomposable $KG(n-p+1, n-p+2, \dots, r)$-module generated by m_1 and apply Lemma 2.46 to this module to get the required assertions for m. In Case 2.3 similar arguments enable one to conclude that $m \neq 0$ and is fixed by $\mathcal{X}_i$ for $i \neq b_0 - 1$ or b_0. It remains to check that $\mathcal{X}_{b_0-1}$ fixes m. Since φ is p-restricted, we have $\omega(m) = \omega - \sum_{i=b_0-1}^{p-1} c_i\alpha_i$ with $c_{b_0-1} = t + \sum_{i=b_0-1}^{j-1} a_i < p$. Hence it suffices to prove that $X_{b_0-1}m = 0$. Set $\xi = \omega(m) + \alpha_{b_0-1}$. We claim that $\xi \notin \mathbf{X}(M)$ which yields that $X_{b_0-1}m = 0$. Put $w = w_j \dots w_{b_0+1}w_{b_0}$ (for $j = b_0$ take $w = w_{b_0}$) and set $\nu = w\xi$. One can directly check that $\nu = \omega - \sum_{i=b_0-1}^{j} e_i\alpha_i - \sum_{i=j+1}^{p-1} c_i\alpha_i$ and $\langle \nu, \alpha_{j+1}\rangle = -(c_{j+1} + 1)$. Now it is clear that $w_{j+1}\nu \notin \mathbf{X}(M)$. As W preserves $\mathbf{X}(M)$, this forces that $\xi \notin \mathbf{X}(M)$ as desired.

Now one can conclude that in all the cases considered m is fixed by $U^+(\Gamma)$ and hence the Γ-module $F = K\Gamma m$ has a composition factor isomorphic to $M(\mu)$.

Next, we claim that $U^+(A)$ fixes m. Set $V_1 = \langle v_1, v_2, \dots, v_{p^s}\rangle$. It is clear that y acts trivially on V_1 and $(y - 1)v_{p^s+1} \in \langle v_1\rangle$. First consider Cases 1 and 2. We have $r \geq p^s$ as $b_s > 1$. Furthermore, our assumptions on b_0 yield that $G = B_{p^s}(K)$ if $r = p^s$. Denote by R_1 the set of all roots in R^+ that are linear combinations of the roots α_i with $i < p^s$, by R_2 the similar set for the roots α_i with $i \leq p^s$, and set $\alpha = \sum_{i=1}^{p^s} \alpha_i$. Put $R' = R^+ \setminus R_2$ if $G \neq B_{p^s}(K)$ and $R' = \{\varepsilon_i + \varepsilon_j \mid 1 \leq i < j \leq p^s\}$ for $G = B_{p^s}(K)$. Now we shall show that

$y = x_\alpha(t_\alpha)\prod_{\gamma\in R'} x_\gamma(t_\gamma)$. Set $R'_2 = R_2 \setminus (R_1 \cup \{\alpha\})$ for $G \neq B_{p^s}(K)$ and $R'_2 = R_2 \setminus (R_1 \cup R' \cup \{\alpha\})$ if $G = B_{p^s}(K)$. Using [**Ste68**, Lemma 17], we can write $y = \prod_{\delta\in R_1} x_\delta(t_\delta)\prod_{\mu\in R'_2} x_\mu(t_\mu)x_\alpha(t_\alpha)\prod_{\gamma\in R'} x_\gamma(t_\gamma)$ for every fixed orderings of roots in R_1, R'_2, and R'. Analyzing the action of y on V_1 and $\langle V_1, v_{p^s+1}\rangle$ and using induction on the sum of the coefficients in the expansion of a root from R_1 in a linear combination of the simple ones, we conclude that $t_\delta = 0$ for all $\delta \in R_1$ and $t_\mu = 0$ for all $\mu \in R'_2$. This yields the required form for y. As each non-unity element y' of $U^+(A)$ is conjugate to y under the action of T, we obtain that

$$y' = x_\alpha(t'_\alpha)\prod_{\gamma\in R'} x_\gamma(t'_\gamma). \tag{9.7}$$

Furthermore, for $G = A_r(K)$ the action of y on $V/\langle v_1,\dots,v_{n-p^s}\rangle$ and arguments similar to those above show that $t'_\gamma = 0$ for all roots γ of the form $\sum_j^k \alpha_i$ with $n - p^s + 1 \leq j \leq k \leq r$. Now it is clear that $U^+(A)$ fixes m in Cases 1 and 2. In Case 3 one can assume that $y' = x_{\alpha_1+\alpha_2+\alpha_3}(t)x_{2\alpha_2+\alpha_3}(u)$ for $y' \in U^+(A)$. This implies immediately that $U^+(A)$ fixes m in this case as well.

Formulae (9.5) and (9.6) yield that $\omega_A(m) = \zeta(\omega(m)) = p-1$. Hence $m \in M_y$.

Now set $d_\Gamma = d_{M_\Gamma}(x_\Gamma)$ and consider the value of d_Γ for Cases 1–3. Write $\mu = (\mu_1, \mu_2)$ with $\mu_i \in \mathbf{X}_i$. Applying Lemma 3.1, Theorem 2.9, and Lemma 2.8 as in the proof of Lemma 9.3, one can deduce that $d_\Gamma = p^s$ unless one of the following holds:

I) $\mu_2 = 0$;

II) $\mu_1 = 0$, $s = 1$;

III) $\mu_i \in \{0, \omega_1, \omega_{q_i}\}$ for $G = A_r(K)$, $\mu_i \in \{0, (p^j + p^k)\omega_1\}$ if $\Gamma_i = B_1(K)$, and $\mu_i \in \{0, \omega_1\}$ otherwise, $i = 1, 2$.

Here μ_i can be not p-restricted, but since φ is not p-large, this can occur only for $\Gamma_i = B_{q_i}(K)$ and only certain coefficients in the canonical expansion of μ_i can exceed $p - 1$.

Observe that $N' \neq p$ if $\omega = a_i\omega_i$ with $a_i > 1$ or $i \leq b_0$ and if $G = A_r(K)$ and $\omega = a_i\omega_i + a_{n-i}\omega_{n-i}$ with $a_i + a_{n-i} > 1$ or $i \leq b_0$. Now the construction of m in Cases 1–3 and the arguments on the weights with a_{p-2} or $a_{p-1} \neq 0$ imply that it remains to consider the following possibilities:

i) $G = C_p(K)$, $\omega = \omega_1 + \omega_{p-1}$;

ii) Case 3) above;

iii) $b_0 = 1$, $s = 1$;

iv) $b_0 = b_s = 2$, $G = C_r(K)$, $\omega = \omega_{p-1}$;

v) $s = 1$, $p > 3$, $b_0 = b_s = 2$, $\omega = \omega_{p-2}$;

vi) $p > 3$, $G \neq A_r(K)$, $b_0 = 2$, $s = 1$, $a_2 = 0$, $N' = a_1 + p - 1$;

vii) $p > 3$, $G \neq A_r(K)$, $b_0 = 2$, $s = 1$, $a_{p-1} = 0$, $N^- = p - 1$, $\sum_{i=2}^{p-2} a_i > 0$;

viii) $p = 3$, $G = C_r(K)$, $s = 1$, $b_0 = 2$, $\omega = 2\omega_1$;

ix) $G = B_r(K)$, $s = 1$, $b_0 = 3$, $b_1 = 2$, $\omega = \omega_{p-2}$.

Observe that by Corollary 2.44, $d_\varphi(x) = |x|$ if M_y contains a module $F = K\Gamma l$ for a weight vector l and $d_F(x_\Gamma) = p^s$. So we are done when the latter equality is valid.

Assume that i) holds. Then $N' = p$. Set $M_A = KAv$, $\lambda = \omega_\Gamma(v)$, and $M_0 = M(\lambda)$. We claim that $M_A \cong V(p)$. Let $Y \in L(A)$ be the root element associated with the negative root in $R(A)$. By Lemma 2.55, it suffices to show that $Yv \neq 0$. Set $\gamma_i = \varepsilon_i + \varepsilon_{p+1-i}$ for $1 \leq i \leq (p-1)/2$ and $\gamma_{(p+1)/2} = 2\varepsilon_{(p+1)/2}$. One can see that

$y = \prod_{i=1}^{(p+1)/2} x_{\gamma_i}(t_i)$ with all $t_i \neq 0$. Hence $U^+(A)$ consists of elements y' of the form $\prod_{i=1}^{(p+1)/2} x_{\gamma_i}(u_i)$ with all $u_i \neq 0$ for non-unity y'. Analyzing the construction of the group A and the bases (2.11) in Proposition 2.43, one can assume that $U^-(A)$ consists of the elements of the form $\prod_{i=1}^{(p+1)/2} x_{-\gamma_i}(w_i)$ with all $w_i \neq 0$ for non-unity elements. Hence $Y = \sum_{i=1}^{(p+1)/2} l_i X_{-\gamma_i}$ with all $l_i \neq 0$. We have $\gamma_1 = \sum_{i=1}^{p} \alpha_i$. So $\langle \omega, \gamma_1 \rangle = 2$. Now Lemma 2.1 i) implies that $X_{-\gamma_1} v \neq 0$. Since $X_{-\gamma_1} v$ is a weight component of Yv, this forces $Yv \neq 0$ and yields our claim on M_A. By Lemma 2.56, $Yv \in M_y$. As $\Gamma \subset C_G(Y)$, the vector Yv generates an indecomposable $K\Gamma$-module M' with highest weight λ that has a composition factor isomorphic to M_0. Hence $d_{M'}(x_\Gamma) \geq d_{M_0}(x_\Gamma)$. Observe that $\Gamma_1 = 1$, $\Gamma_2 \cong B_{(p-1)/2}(K)$, and $\lambda_2 = 2\omega_1$. By Theorem 1.14 and Algorithm 1.6, $d_{M_0}(x_\Gamma) = p$. Hence there exists a vector $w \in M'$ with $(x_\Gamma - 1)^{p-1} w \neq 0$. Proposition 2.43 3) implies that now the set J defined in that proposition is equal to $\{1, \ldots, r-1\}$. Denote by Σ_J the set of all integer linear combinations of the simple roots lying in J. One easily observes that all weight components of the vector Yv have weights of the form $\omega - \alpha_r - \sigma$ with $\sigma \in \Sigma_J$. Since $\Gamma \subset G_J$, this holds for the vector w as well. Now apply Lemma 2.42 with $I = J$ and conclude that $d_{M_y}(x) = d_{M_0}(x_\Gamma) = p$. Then $d_\varphi(x) = p^2$ by Proposition 2.5 b). This completes the analysis of Case i).

Proposition 2.34 and Conjecture (r, s) imply that $d_\varphi(x) = p^{s+1}$ in Cases ii), viii), and ix).

Next, assume that iii) holds. Then $\Gamma_1 = 1$ and $\Gamma_2 \cong A_{p-2}(K)$ or $C_{(p-1)/2}(K)$. First let $N' - p + 1 \leq a_1 + a_r$ for $G = A_r(K)$ and $N' - p + 1 \leq a_1$ otherwise. Recall that $N' \neq p$ if $\omega = a_1\omega_1 + a_r\omega_r$ for $G = A_r(K)$ and $\omega = a_1\omega_1$ for other groups. Hence in this case $\sum_{i=2}^{r-1} \omega_i(m) > 1$ for $G = A_r(K)$ and $\sum_{i=2}^{r} \omega_i(m) > 1$ otherwise. Now suppose that $N' - p + 1 > a_1 + a_r$ for $G = A_r(K)$ and $N' - p + 1 > a_1$ for other groups. Passing to the dual representation if necessary, one can assume that $a_i \neq 0$ for some i with $1 < i < p$. Fix maximal such i. Since $N^- < p \leq N'$, we get that either $i < p - 2$ or $b_1 = 2$ and $\omega = \omega_1 + \omega_{p-2}$ or ω_{p-1}. First let $i < p - 2$. The construction of m yields that $\omega_j(m) \neq 0$ for some j with $2 < j < p - 1$. In all situations that occurred in Case iii) until now, we can see that $\mu_2 \neq \omega_1$ and also $\neq \omega_{p-2}$ if $\Gamma_2 \cong A_{p-2}(K)$. Then Algorithm 1.6 and Theorem 1.14 imply that $d_{M(\mu_2)}(g_2) = p$ and hence $d_\Gamma = p$ as desired.

Now let $b_1 = 2$ and $\omega = \omega_1 + \omega_{p-2}$ or ω_{p-1}. Then $N^- = p - 1$. Set $\varphi_0 = (\varphi_1)_{\mathbb{C}}$. Let u_0 and $u_1 \in (G_1)_{\mathbb{C}}$ be unipotent elements with $J(u_0) = J(g)$ and $J(u_1) = J(g^p)$. Proposition 1.5 and Algorithm 1.6 imply that $d_{\varphi_0}(u_0)$ and $pd_{\varphi_0}(u_1) \geq p^2$ unless $G = B_p(K)$ and ω_{p-1}. Then Conjecture (r, s) and Lemma 2.19 yield that in other cases considered in this paragraph $d_\varphi(x) = d_{\varphi_1}(g) = p^2$.

Finally, assume that $G = B_p(K)$ and ω_{p-1}. Apply Conjecture (r, s) and Proposition 2.34 to complete the proof in Case iii).

Now let iv) hold or v) hold with $G = A_r(K)$. Then $N^- = p - 1$. Applying the arguments of Lemma 9.4 to φ_1, one can deduce from Conjecture (r, s) that $d_{\varphi_1}(g) = p^{s+1}$. Hence $d_\varphi(x) = p^{s+1}$ by Lemma 2.19.

Next, assume that v) holds with $G \neq A_r(K)$. Then $G = C_r(K)$ and $N' = p$. We construct a vector $m' \in M$ that will play the role of m. Set $m' = v(p, p-2, 1)$. By Lemma 2.46, $m' \neq 0$ and is fixed by $\mathcal{X}_i$ for $i \neq 2$. Hence $U^+(\Gamma)$ fixes m'. By Formulae (9.5), $\omega_A(m') = p - 1$. This implies that $U^+(A)$ fixes m' since $\omega_A(u) \leq p$ for all weight vectors $u \in M$. Then $m \in M_y$ by Lemma 2.54. Set $\mu' = \omega_\Gamma(m')$ and write $\mu' = (\mu'_1, \mu'_2)$ with $\mu'_i \in \mathbf{X}_i$, $i = 1, 2$. Put $M'_\Gamma = M(\mu')$. Applying

Corollary 2.44 as earlier for m, we can conclude that $d_\varphi(x) = |x|$ if $d_{M'_\Gamma}(x_\Gamma) = p$. Assume that $p > 5$. We have $\Gamma_1 = C_1(K)$, $\Gamma_2 = B_{(p-3)/2}(K)$, $\lambda_1 = 1$, $\lambda_2 = \omega_3$ for $p > 7$ and $2\omega_2$ for $p = 7$. Then Algorithm 1.6 and Theorem 1.14 yield that $d_{M'_\Gamma}(x_\Gamma) = d_{M'_\Gamma}(g_2) = p$ as desired.

For $p = 5$ the weight $\lambda_2 = 0$ and so a specific approach is needed. Here we argue as in Lemma 9.3 for $G = C_r(K)$ and $\omega = \omega_1 + \omega_{p-1}$ with $p > 3$. The notation G^+, M^+, ω^+, x_-, β_i, $Y_{\pm i}$, ω', V_G, M_G, and $\bar{l}$ from that proof are used. Now $G^+ = A_{11}(K)$. Conjecture (r, s) and the arguments in Case iv) imply that $d_{\varphi^+}(x) = 25$. Set

$$\Lambda_1 = \{\mu \in \mathbf{X}(M^+) \mid \omega^+ - \mu = \sum_{i=1}^{11} c_i\beta_i,\ \sum_{i=1}^{11} c_i \leq 3\},$$

$$\Lambda_2 = \{\nu \in \mathbf{X}(M^+) \mid \nu - \omega' = \sum_{i=1}^{11} d_i\beta_i,\ \sum_{i=1}^{11} d_i \leq 3\}$$

and define V_j, $j = 1, 2$, as in Lemma 9.3. One gets

$$\Lambda_1 = \{\omega^+, \omega^+ - \beta_3, \omega^+ - \beta_3 - \beta_4, \omega^+ - \beta_2 - \beta_3, \omega^+ - \sum_{i=1}^{3} \beta_i, \omega^+ - \sum_{i=3}^{5} \beta_i\},$$

and Λ_2 can be obtained from Λ_1 by replacing ω^+ by ω', β_i by β_{12-i}, and all "-" signs by "+" in the latter formula. Using (9.2) and arguing as in Lemma 9.3, we easily deduce that $V_j \subset V_G$, $(x_- - 1)^{24}M^+ \subset V_2$, $\bar{w} \neq 0$ for $w \in \overline{V_2} \setminus 0$, and there exists $u \in V_1$ with $(x_- - 1)^{24}u \neq 0$. This implies that $(x_- - 1)^{24}M \neq 0$ and hence $d_\varphi(x) = 25$ as desired.

Now let vi) hold. Recall that $\Gamma_2 \cong B_{(p-3)/2}(K)$. One easily observes that $\mu_2 \neq p^j\omega_{(p-3)/2}$. As $N' \geq p$, the weight $\omega \neq \omega_1$. Hence $\mu_2 \neq 2\omega_1$ for $p = 5$ and $\mu_2 \neq \omega_1$ otherwise. If $p = 5$, we claim that $\mu_2 = 4$ or 8. Indeed, since $N^- < p$ and $N' = a_1 + p - 1 \geq p$, for $p = 5$ one can deduce that either $b_1 = 3$, $G = B_r(K)$, and $\omega = 2\omega_1$, or $b_1 = 2$, $G = C_r(K)$, and $\omega = 4\omega_1$. The construction of m yields that $\mu_2 = 4$ in the first case and 8 in the second one. Now Theorem 2.2, Lemma 2.8, Proposition 1.5, Algorithm 1.6, and Theorem 1.14 imply that $d_{M_\Gamma}(g_2) = p$ both for $p > 5$ and $p = 5$. This completes the proof in Case vi).

Next, consider Case vii). First assume that $a_{p-2} = 0$. Then $a_j > 0$ for some j with $2 \leq j < p - 2$. The construction of m implies that $\omega_{j+2}(m) \neq 0$. For $p = 5$ we have $\omega = \omega(1, 2)$ and $\mu_2 = 2(a_1 + a_2)$. For $p > 5$ the weight μ_2 is p-restricted since φ is not p-large. Using Algorithm 1.6 and Theorem 1.14, we conclude that $d_{M(\mu_2)}(g_2) = p$ if $\mu_2 \neq \omega_1$ for $p > 5$ and 2 or 6 for $p = 5$. Now Corollary 2.44 and the construction of m show that it remains to consider the situations where $\omega = \omega_{p-3}$ or $p = 5$ and $a_1 + a_2 = 3$. First suppose that $\omega = \omega_{p-3}$. Then $b_1 > 2$ since otherwise $N' = p - 1$. But then $N^- \geq 2p - 6$. Since $N^- = p - 1$, we get $p = 5$ and $b_1 = 3$. So $G = B_r(K)$. To complete the proof for this subcase, argue as in Case iii) for $G = B_r(K)$ and $\omega = \omega_{p-1}$ using Proposition 2.34.

If $p = 5$ and $a_1 + a_2 = 3$, one can deduce that $b_1 = 2$ and $\omega = 2\omega_1 + \omega_2$ since $N^- = 4$. Let $u \in (G_1)_\mathbb{C}$ be a regular unipotent element. Proposition 1.5 and Algorithm 1.6 imply that $d_{(\varphi_1)_\mathbb{C}}(u) > 25$. Now it follows from Conjecture (r, s) and Lemma 2.19 that $d_\varphi(x) = d_{\varphi_1}(g) = 25$.

Finally, assume that $a_{p-2} \neq 0$. As $N^- = p - 1$, we have $b_1 = 2$ and $\omega = \omega_1 + \omega_{p-2}$. Complete the analysis of Case vii) arguing as in the previous paragraph. Now all the possibilities have been considered. The proposition is proved. □

Proposition 9.1 follows directly from Results 9.2–9.5.

10. Theorem 1.3 for groups of types A_r and B_r and regular elements

In this section we prove Theorem 1.3 for regular unipotent elements of the groups $A_r(K)$ and $B_r(K)$ and p-restricted representations under the assumption that Conjecture (r, s) holds. Namely, we suppose that $G = A_r(K)$ or $B_r(K)$, $n = p^{s+1}$, $s > 0$, φ is p-restricted, and $\omega \neq 0$ and prove that $d_\varphi(x) = p^{s+1}$ for a regular unipotent element $x \in G$. We need some special arguments in the case where $G = B_4(K)$, $p = 3$, and $\omega = \omega_4$.

LEMMA 10.1. *Let $p = 3$, $G = B_4(K)$, and $\omega = \omega_4$. Then $d_\varphi(x) = 9$.*

PROOF. Taking into account that ω is a miniscule weight, one can determine explicitly the action of the elements $x_i(1)$, $1 \leq i \leq 4$, in a fixed base of M consisting of weight vectors. Then take $x = x_1(1) \dots x_4(1)$ and check directly that $(x-1)^8 u \neq 0$ for a nonzero lowest weight vector $u \in M$. This forces $d_\varphi(x) = 9$ as $|x| = 9$. □

Unfortunately, at present we cannot propose a proof of this lemma without direct calculations.

All other cases are handled by general arguments.

PROPOSITION 10.2. *Theorem* 1.3 *holds for regular unipotent elements and p-restricted representations provided Conjecture* (r, s) *holds.*

PROOF. Due to Lemma 10.1, we can exclude the case settled there. Obviously, $d_\varphi(x) = p^{s+1}$ if $\omega = \omega_1$ or $G = A_r(K)$ and $\omega = \omega_r$. So we eliminate these cases as well.

By Theorem 1.1, Theorem 1.3 is valid for p-large representations. So suppose that φ is not p-large. Put $G_1 = G(2, \dots, r)$. For $\omega = \omega_2$ set $m = X_{-1}X_{-2}v$, otherwise put $m = X_{-1,a_1}v$. By Lemmas 2.46 and 2.1, $m \neq 0$. It is clear that $U^+(G_1)$ fixes m. Set $\mu = \omega_{G_1}(m)$ and observe that $\varphi|G_1$ has a composition factor $\tau \cong \varphi(\mu)$. Let $g \in G_1$ be a regular unipotent element. We claim that $d_\tau(g) = p^{s+1}$. Then Lemma 2.19 yields the assertion of the theorem.

One has $G_1 \cong A_{r-1}(K)$ or $B_{r-1}(K)$ for $G = A_r(K)$ or $B_r(K)$, respectively, and

$$n(G_1) \geq n - 2 > p^{s+1} - p \geq p^s + p.$$

Since φ is not p-large, τ is p-restricted. As $p > 2$ and n is a power of p, the group $G \neq A_3(K)$. Now one easily sees that $\mu \notin \{0, \omega_1\}$ and that $\mu \neq \omega_r$ as well if $G = A_r(K)$. Observe also that $\mu \neq \omega_3$ for $G = B_4(K)$ as by our assumptions, $\omega \neq \omega_4$ in this case. (That is why we need Lemma 10.1!) Then Conjecture (r, s) and Lemma 3.1 imply that $d_\tau(g) = p^{s+1}$ and complete the proof. □

REMARK 10.3. For $G = A_r(K)$ Proposition 10.2 can be deduced from Hartley's results **[Har86]** on restrictions of representations of $GL_n(K)$ to finite subgroups F for which the standard $GL_n(K)$-module contains a direct summand isomorphic to KF, but we wished to present a unified proof.

COROLLARY 10.4. *Assume that $G = A_r(K)$, $B_r(K)$, or $C_r(K)$ and Conjecture (r, s) holds for G. Then Theorems* 1.7 *and* 1.10 *hold for regular unipotent elements and p-restricted representations.*

PROOF. Theorem 1.7 under given assumptions follows from Corollaries 4.11, 5.5, 6.5, 6.12, 7.3, results of Section 8 (see the end of that section), and Propositions 9.1 and 10.2. Then Lemma 2.31 implies Theorem 1.10. □

11. Proofs of the main theorems

In this section the proofs of the main results will be completed. In fact, we have to do the following. Assume that Conjecture (r, s) holds for a fixed rank r and elements of order p^{s+1} and that Theorem 1.1 holds for our group G. Then prove Theorems 1.10, 1.9, and 1.3 for unipotent elements $x \in G$ with at least 2 blocks in $J(x)$ as well as for non p-restricted representations, Theorem 1.7 for $G = D_r(K)$, and Propositions 1.11, 1.12, and 1.13. Notice that for elements with a single Jordan block and p-restricted representations Theorems 1.7 and 1.10 are already proved under our assumptions in Sections 4–10 (see Corollary 10.4). Naturally, we can assume that $|x| > p$. We keep the notation given in the Introduction before the statement of Theorem 1.10. Theorem 1.1 allows us to assume that φ is not p-large.

The following trivial lemma will be used in the proofs of all theorems indicated at the beginning of this section.

LEMMA 11.1. *Let $\omega = p^i\mu$ with μ p-restricted. Then $d_\varphi(x) = d_{\varphi(\mu)}(x)$ for each unipotent $x \in G$.*

PROOF. Set $\chi = \varphi(\mu)$. By Theorem 2.2, $\varphi \cong \chi \cdot \mathrm{Fr}$. Hence the minimal polynomials of $\varphi(x)$ and $\chi(x)$ coincide. □

LEMMA 11.2. *If the assertions of Theorems* 1.3 *and* 1.9 *hold for a fixed unipotent element of G and all representations in* Irr_p, *then they hold for the image of this element in each $\varphi \in \mathrm{Irr}$.*

PROOF. This follows immediately from Theorems 2.2 and 2.9 and Lemma 11.1. □

Fix unipotent $x \in G$. Assume that $|x| = p^{s+1}$ with $s > 0$. Let $k_1 \geq k_2 \geq \ldots \geq k_t$ with $k_1 + k_2 + \ldots + k_t = n$ be the block sizes in the canonical Jordan form of x and $z_f \in G_{\mathbb{C}}$ be unipotent elements with the same labelled Dynkin diagram as x^{p^f}, $0 \leq f \leq s$.
The end of the proof of Theorem 1.3. Let $k_1 = p^{s+1}$. Then $k_2 = p^{s+1}$ if $G = C_r(K)$. Proposition 10.2 and Lemma 11.2 enable us to assume that $t > 1$. Set $r' = k_1 - 1$, $(k_1-1)/2$, $k_1 - 1$, or $(k_1+1)/2$ for $G = A_r(K)$, $B_r(K)$, $C_r(K)$, or $D_r(K)$, respectively. Put $G_1 = G(r - r' + 1, r - r' + 2, \ldots, r)$ for $G = A_r(K)$ or $B_r(K)$ and $G_1 = G(1, 2, \ldots, r')$ for $G = C_r(K)$. If $G = D_r(K)$, construct a subgroup $B \cong B_{r-1}(K)$ in G as in Lemma 2.23 taking in that lemma $I_1 = \{1, 2, \ldots, r-1\}$ and $I_2 = \varnothing$. If $T \subset G$ is the maximal torus whose elements have diagonal matrices in the base (2.2), one easily observes that $T_B = T \cap B$ is a maximal torus in B and the restriction of weights from T to T_B determines a map $\sigma : \mathbf{X} \to \mathbf{X}(B)$ such that $\sigma(\varepsilon_i) = \varepsilon_{i,B}$ for $i < r$ and $\sigma(\varepsilon_r) = 0$. Let $\beta_1, \beta_2, \ldots, \beta_{r-1}$ be the simple roots of B with respect to T_B. Set $G_1 = B(\mathcal{X}_{\beta_{r-r'}}, \mathcal{X}_{\beta_{r-r'+1}}, \ldots, \mathcal{X}_{\beta_{r-1}})$. For all types let $g \in G_1$ be a regular unipotent element. One easily observes that $J(g)$ (in V)

is $J(p^{s+1}, 1, \dots, 1)$ for $G \neq C_r(K)$ and $J(p^{s+1}, p^{s+1}, 1, \dots, 1)$ for $G = C_r(K)$ and that g has a single Jordan block of size k_1 in $V(G_1)$. Lemma 1.4 i) implies that $\mathrm{cl}(g)$ is completely determined by $J(g)$. Hence by Corollary 2.28, $g \in \overline{\mathrm{cl}(x)}$. Since G_1 is a simple algebraic group, it is clear that $\varphi|G_1$ has a nontrivial composition factor ξ. Then Conjecture (r, s) yields that $d_\xi(g) = p^{s+1}$ and hence $d_\varphi(x) = p^{s+1}$ by Lemma 2.19. This completes the proof. □

Now assume that $k_1 < |x|$. Construct a collection $u_1, \dots, u_c$, $u_1 \geq \dots \geq u_c > 1$, the groups H_j^p over K and H_j over $\mathbb{C}$ for $1 \leq j \leq c$, $H^p = \prod_{j=1}^c H_j^p$, $H = \prod_{j=1}^c H_j$, unipotent elements $h_f \in H$, $0 \leq f \leq s$, the homomorphism $\theta : \mathbf{X} \to \mathbf{X}(H)$, and the representation $\psi \in \mathrm{Irr}\, H$ as in the Introduction. Recall that actually in the Introduction we had constructed a sequence $u_1, \dots, u_l$ with $u_1 \geq \dots \geq u_l \geq 1$ connected with the sequence (2.5) for x in a certain way and then fixed maximal c with $u_c > 1$. Furthermore, $\omega(\psi) = \theta(\overline{\omega})$ and the homomorphism θ was constructed as a sequence $(\theta_1, \dots, \theta_l)$ with $\theta_j : \mathbf{X} \to \mathbf{X}(H_j)$ and $\theta(\lambda) = (\theta_1(\lambda), \dots, \theta_c(\lambda))$ for $\lambda \in \mathbf{X}$. Set $\overline{\omega} = \omega(\varphi)$ and write $\overline{\omega} = \sum_{i=1}^r \overline{a_i} \omega_i$. For $1 \leq i \leq k \leq r$ set $\overline{\omega}(i, k) = \sum_{t=i}^k \overline{a_t} \omega_t$. We proceed to realize our scheme for the general case described in Section 3. For $\lambda \in \mathbf{X}^+$ denote by ψ_λ the representation in $\mathrm{Irr}\, H$ with highest weight $\theta(\lambda)$. Set $\varphi_i = \varphi(\omega_i)$ and $\sigma_i = \varphi(\theta(\omega_i)) \in \mathrm{Irr}\, H$ for $1 \leq i \leq r$. Each element h_f can be uniquely represented as $\prod_{j=1}^c h_{f,j}$ with $h_{f,j} \in H_j$. The construction of the elements h_f yields that the canonical Jordan form of $h_{f,j}$ on $V(H_j)$ coincides with that of the p^fth power of a regular unipotent element of H_j^p on $V(H_j^p)$. We have $\psi_\lambda \cong \bigotimes_{j=1}^c \psi_{\lambda,j}$ where $\psi_{\lambda,j} \in \mathrm{Irr}\, H_j$ and $\omega(\psi_{\lambda,j}) = \theta_j(\lambda)$. Put $d_{f,j,\lambda} = d_{\psi_{\lambda,j}}(h_{f,j})$, $d_{f,\lambda} = d_{\psi_\lambda}(h_f)$, $\psi_j = \psi_{\overline{\omega},j}$, $d_{f,j} = d_{f,j,\overline{\omega}}$, and $d_f = d_{f,\overline{\omega}}$ for $0 \leq f \leq s$ and $1 \leq j \leq c$. Recall that $r_j = r(H_j)$ for $1 \leq j \leq c$. If $G = D_r(K)$ and all u_j are improper for G, define the conjugacy classes C_1 and C_2 that consist of unipotent elements g with $J(g) = J(x)$ such as in the Introduction.

Proposition 11.3. *Let $G = D_r(K)$. Assume that Theorem* 1.10 *holds for G. Then Theorem* 1.7 *holds for G.*

Proof. Let $x \in G$ be a regular unipotent element. We have $k_1 = 2r - 1$ and $k_2 = 1$. If $k_1 = p^{s+1}$, Theorem 1.3 forces that $d_\varphi(x) = p^{s+1}$ for every nontrivial representation φ and hence Theorem 1.7 holds. So assume that $k_1 < p^{s+1}$. Then $\theta = \theta_1$. Rules 1–9 imply that $\theta(\varepsilon_i) = \varepsilon_i(H)$ for $1 \leq i \leq r - 1$ and $\theta(\varepsilon_r) = 0$. Now it follows from Algorithm 1.6 and Proposition 1.5 that $d_f = d_{\varphi_C}(z_f)$ for $0 \leq f \leq s$. This completes the proof. □

Until the proof of Theorem 1.9 assume that $k_1 < n$. In Proposition 2.27 we have constructed a subgroup $S = S(x) \subset G$ such that S is isomorphic to a quotient of H^p by its central subgroup, $\mathrm{cl}(x)$ contains a regular unipotent element of S, $T \cap S$ is a maximal torus in S, and the homomorphism $\mathbf{X} \to \mathbf{X}(H)$ induced by the restriction of weights from T to $T \cap S$ coincides with θ. We can and shall assume that $x \in S$. Recall that $S = \prod_{j=1}^c S_j$ where S_j is isomorphic to a quotient of H_j^p by its central subgroup and S_i and S_j commute for $i \neq j$. As in Section 2, write $x = \prod_{j=1}^c x_j$ where x_j is a regular unipotent element in S_j. Set $\omega_S = \theta(\omega)$ and $\chi = \varphi(\omega_S) \in \mathrm{Irr}\, S$. Observe that $\chi \cong \bigotimes_{j=1}^c \chi_j$ where $\chi_j \in \mathrm{Irr}\, S_j$ and $\omega(\chi_j) = \theta_j(\omega)$. By Corollary 2.30, $d_\varphi(x) \geq d_\chi(x)$.

Now our goal is to show that $d_\varphi(x) \leq d(x,\varphi)$ and $d_\chi(x) = d(x,\varphi)$ for $\varphi \in \mathrm{Irr}_p$ if the pair (x,φ) does not satisfy the assumptions of Theorem 1.9. This and Corollary 2.30 would imply Theorem 1.10 for such φ.

LEMMA 11.4. *Assume that $\varphi \in \mathrm{Irr}_p$ and either $G \neq C_r(K)$, or $a_1+2\sum_{i=2}^r a_i < p$. Then χ is p-restricted.*

PROOF. Let $j \leq c$ and $\theta_j(\omega_i) \neq 0$. Then the construction of θ implies that $\theta_j(\omega_i)$ is a fundamental weight of H_j for $G = A_r(K)$ and a fundamental weight or a sum of two fundamental weights for other G; furthermore, $\theta_j(\omega_1) = \omega_1$ and $\theta_j(\omega_i)$ are fundamental weights for $G = B_r(K)$, $i = r$ and for $G = D_r(K)$, $i \geq r-1$. To complete the proof, recall that φ is not p-large and apply Lemma 2.3. □

To compare $d_\varphi(x)$ and $d(x,\varphi)$, we need a series of estimates and formulae for $d_{f,\lambda}$ and d_f.

LEMMA 11.5. *One has $d_{f,\lambda} - 1 = \sum_{j=1}^c (d_{f,j,\lambda} - 1)$.*

PROOF. This follows immediately from Lemma 2.11 and the definitions of relevant values given above. □

COROLLARY 11.6. *Assume that $a_i = 0$ for $p \leq i \leq n-p$ if $G = A_r(K)$ and $a_i = 0$ for $i \geq p$ otherwise. Then $p^f(d_f - 1) \leq p^k(d_k - 1)$ for $0 \leq k < f \leq s$.*

PROOF. This follows directly from Lemmas 3.2 and 11.5. □

LEMMA 11.7. *Let $\overline{\omega} = \lambda_1 + \lambda_2$ with $\lambda_t \in \mathbf{X}^+$, $t = 1,2$. Then $d_{f,j} - 1 = (d_{f,j,\lambda_1} - 1) + (d_{f,j,\lambda_2} - 1)$ and $d_f - 1 = (d_{f,\lambda_1} - 1) + (d_{f,\lambda_2} - 1)$ for $0 \leq f \leq s$ and $1 \leq j \leq c$. In particular,*

$$d_{f,j} - 1 = \sum_{i=1}^r \overline{a_i}(d_{f,j,\omega_i} - 1), \quad d_f - 1 = \sum_{i=1}^r \overline{a_i}(d_{f,\omega_i} - 1).$$

PROOF. The first assertion follows immediately from Proposition 1.5. Then apply induction by $\sum_{i=1}^r \overline{a_i}$. □

COROLLARY 11.8. *One has*

$$d_f = 1 + \sum_{i=1}^r \overline{a_i} \sum_{j=1}^c (d_{f,j,\omega_i} - 1) \tag{11.1}$$

for $0 \leq f \leq c$.

PROOF. This follows directly from Lemmas 11.5 and 11.7. □

PROPOSITION 11.9. *Let $G \neq A_r(K)$. Set $G^+ = A_{n-1}(K)$. Assume that $x^+ \in G^+$ and $J(x) = J(x^+)$. Let $\varphi^+ \in \mathrm{Irr}\, G^+$ and $\omega(\varphi^+) = \omega(1,r)$. Suppose that $a_r = 0$ for $G = B_r(K)$ and that $a_{r-1} = a_r = 0$ for $G = D_r(K)$. Denote by d_f^+, $0 \leq f \leq s$, the integers that can be constructed for x^+ and φ^+ in the same manner as d_f for x and φ. Then $d_f^+ = d_f$.*

PROOF. Set $\omega^+ = \omega(\varphi^+)$. Applying Lemma 11.7, one can conclude that it suffices to prove the proposition for the fundamental representations. So assume that $\omega = \omega_i$. Hence $\omega = \overline{\omega}$. In this proof H^+, H_j^+, ψ^+, etc. denote the objects associated with x^+ and φ^+ in the same manner as the relevant objects introduced for the statement of Theorem 1.10 and denoted by the symbols without the "plus" sign, are connected with x and φ. Denote by ε_a^+, $1 \le a \le n$, the elements of $\mathcal{E}(G^+)$. The notation $\mathbf{j}(\varepsilon_a)$, $\mathbf{j}(\varepsilon_a^+)$, $\mathbf{q}(\varepsilon_a)$, and $\mathbf{q}(\varepsilon_a^+)$ mean the same as in the Introduction. For $1 \le j \le c$ set $\Sigma_j = SL_{u_j}(\mathbb{C})$ if u_j is proper for G and $SL_{u_j}(\mathbb{C}) \times SL_{u_j}(\mathbb{C})$ otherwise. Obviously, $H^+ \cong \prod_{j=1}^c \Sigma_j$ and each Σ_j is equal either to H_l^+, or to $H_l^+ H_{l+1}^+$ for some l. In this situation put $l = \mathbf{l}(j)$. Set $d'_{f,j} = d_{f,l}^+$ and $\nu_j = \psi_l^+$ in the first case and $d'_{f,l} = d_{f,l}^+ + d_{f,l+1}^+ - 1$ and $\nu_j = \psi_l^+ \otimes \psi_{l+1}^+$ in the second one. By Lemma 11.5, now it suffices to prove that $d'_{f,j} = d_{f,j}$. Assume that $a < r$ for $G = D_r(K)$ and $a \le n/2$ otherwise. Analyzing Rules 1)—9) that determine the maps θ and θ^+, we can deduce that one of the following holds:

1) $\mathbf{j}(\varepsilon_a) = j$, $\mathbf{j}(\varepsilon_a^+) = \mathbf{l}(j)$, $\mathbf{q}(\varepsilon_a) = \mathbf{q}(\varepsilon_a^+)$;

2) $\theta(\varepsilon_a) = 0$, $\mathbf{j}(\varepsilon_a^+) = l$, $l = \mathbf{l}(j)$, u_j is proper for G;

3) $\mathbf{j}(\varepsilon_a) = j$, $\mathbf{j}(\varepsilon_a^+) = \mathbf{l}(j) + 1$, $\mathbf{q}(\varepsilon_a^+) = q$, $\mathbf{q}(\varepsilon_a) = -(u_j - q + 1) < 0$, u_j is improper for G;

4) $\theta(\varepsilon_a) = 0$, $\theta^+(\varepsilon_a^+) = 0$.

Define the homomorphisms $\sigma_j : \mathbf{X}(G^+) \to \mathbf{X}(\Sigma_j)$ as follows: $\sigma_j = \theta_g^+$ if $\Sigma_j = H_g^+$ and $\sigma_j = (\theta_g^+, \theta_{g+1}^+)$ for $\Sigma_j = H_g^+ H_{g+1}^+$. Here in the second case we mean that σ_j maps a weight $\lambda \in \mathbf{X}(G^+)$ onto the pair $(\theta_g^+(\lambda), \theta_{g+1}^+(\lambda))$ regarded as a weight of Σ_j. Then $\sigma_j(\omega^+) = \omega(\nu_j)$. For proper u_j one has the following possibilities:

a) $\sigma_j(\omega^+) = \omega_m$, $\theta_j(\omega) = \omega_m$;

b) $u_j = 2m + 1$, $\sigma_j(\omega^+) = \omega_m$ or ω_{m+1}, $\theta_j(\omega) = 2\omega_m$;

c) $\sigma_j(\omega^+) = 0$, $\theta_j(\omega) = 0$.

For improper u_j the following possibilities arise:

a) $\sigma_j(\omega^+) = (\omega_m, \omega_m)$, $\theta_j(\omega) = \omega_m + \omega_{u_j - m}$;

b) $\sigma_j(\omega^+) = (\omega_m, \omega_{m-1})$, $\theta_j(\omega) = \omega_m + \omega_{u_j - m + 1}$ (here we assume that $\omega_0 = \omega_{u_j} = 0$);

c) $\sigma_j(\omega^+) = 0$, $\theta_j(\omega) = 0$.

Let $g_j \in \Sigma_j$ be a regular unipotent element. By Lemma 2.11, $d'_{f,j} = d_{\nu_j}(g_j)$. Now Algorithm 1.6 implies that $d'_{f,j} = d_{f,j}$ as required. In fact, it occurs that these values are computed by the same rules. This completes the proof. □

LEMMA 11.10. *i) Let $G = B_r(K)$ and $\omega = \omega_r$. Then $\theta_j(\omega) = \omega_m$ if $u_j = 2m+1$ or $2m$.*

ii) Let $G = D_r(K)$ and $\omega = \omega_{r-1}$ or ω_r. If $J(x)$ contains a block of odd size or $\mathbf{j}(\varepsilon_n) \neq j$, then $\theta_j(\omega)$ is such as in Item i).

Assume that all blocks in $J(x)$ have even sizes, $\mathbf{j}(\varepsilon_n) = j$, and $u_j = 2m$. Then $\theta_j(\omega_{r-1}) = \omega_{m+1}$ and $\theta_j(\omega_r) = \omega_m$ for $x \in C_1$ and $\theta_j(\omega_{r-1}) = \omega_m$ and $\theta_j(\omega_r) = \omega_{m+1}$ for $x \in C_2$.

PROOF. This follows immediately from Rules 1–9 and the definition of the classes C_1 and C_2. Observe that $\theta(\varepsilon_n) = 0$ if $G = D_r(K)$ and $J(x)$ contains a block of odd size. □

Recall some notation of Section 2 connected with certain fundamental representations. If $\Delta = B_g(K)$, then $\xi(\Delta) = \varphi(\omega_g) \in \operatorname{Irr}\Delta$, for $\Delta = D_g(K)$ let the symbols $\xi_-(\Delta)$ and $\xi_+(\Delta)$ denote the representations $\varphi(\omega_{g-1})$ and $\varphi(\omega_g) \in \operatorname{Irr}\Delta$, respectively. If Δ is fixed, the indication of Δ is sometimes omitted. If $\Sigma = A_{t-1}(K)$ with $t = 2l$, set $\varphi_i^t = \varphi(\omega_i) \in \operatorname{Irr}\Sigma$ for $0 \le i \le t$ and $\mu_i^t = \varphi_{l+i}^t$ for $-l \le i \le l$; here $\omega_0 = \omega_t = 0$.

LEMMA 11.11. *Let $G = B_r(K)$ or $D_r(K)$. Denote by J_1 and J_2 the sets of all $j \leq c$ with proper and improper u_j, respectively. If $J_1 \neq \varnothing$, let Λ be the representation of the group $\prod_{j\in J_1} S_j$ isomorphic to $\bigotimes_{j\in J_1} \xi(S_j)$. Otherwise let Λ be the trivial representation of S.*

a) Let $G = B_r(K)$. Then $\xi|S$ is a direct sum of components of the form $\Lambda \bigotimes \bigotimes_{j\in J_2} \varphi_{l_j}^{u_j}$ with $0 \leq l_j \leq u_j$, $\varphi_{l_j}^{u_j} \in \operatorname{Irr} S_j$, and all possible combinations occur. (If $J_2 = \varnothing$, all the components are isomorphic to Λ.)

b) Let $G = D_r(K)$. If some u_j is proper for G, the assertion a) with ξ replaced by ξ_- or ξ_+ holds. Assume that all u_j are improper. Define the sets $I_{1,t}$ and $I_{2,t}$ as in Lemma 2.37b). Set $l_j = u_j/2$. Then for $x \in C_1$ and $\sigma = \xi_-$ or ξ_+ the restriction $\sigma|S$ is a direct sum of components of the form $\bigotimes_{j=1}^{c} \mu_{a_j}^{u_j}$ with $-l_j \leq a_j \leq l_j$, $\mu_{a_j}^{u_j} \in \operatorname{Irr} S_j$, for each component the number of indices j with $a_j \in I_{1,u_j}$ is odd for $\sigma = \xi_-$ and even for $\sigma = \xi_+$, and all possible combinations occur. If $x \in C_2$, one has to interchange ξ_+ and ξ_- in the previous assertion.

PROOF. Recall the construction of S for these groups in the proof of Proposition 2.27 and use the notation of that proof. If $J_1 \neq \varnothing$, set $G_+ = G_{I_{j_1}}$ if $G = B_r(K)$ and $q = 1$, $G_+ = G_{I_{j_q}} \prod_{l=1}^{q'} G_l$ if $G = B_r(K)$ and $q > 1$, and $G_+ = \prod_{l=1}^{q'} G_l$ for $G = D_r(K)$. Write $G_+ = \prod_{a=1}^{d} \tilde{G}_a$ where $d = q' + 1$ for $G = B_r(K)$ and $d = q'$ for $G = D_r(K)$, $\tilde{G}_a = G_m$ or $G_{I_{j_q}}$ (here we want to show explicitly the multipliers whose product forms G_+). If $J_2 \neq \varnothing$, put $G_- = \prod_{j\in J_2} G_j$. Set $G_+ = 1$ and $d = 0$ if $J_1 = \varnothing$ and $G_- = 1$ if $J_2 = \varnothing$. Put $\tilde{r}_a = r(\tilde{G}_a)$ and $r_j^- = r(G_j)$ for $j \in J_2$. The following facts on the restrictions $\xi|G_+G_-$ for $G = B_r(K)$ and $\xi_\pm|G_+G_-$ for $G = D_r(K)$ are well known and can be easily deduced if one considers the restriction of weights of relevant representations from T to $T \cap G_+G_-$. If both G_+ and $G_- \neq 1$, for $G = B_r(K)$ we get that $\xi|G_+G_-$ is a direct sum of components of the form $\bigotimes_{a=1}^{d} \lambda_a \bigotimes \bigotimes_{j\in J_2} \nu_j$ with $\lambda_a = \xi(\tilde{G}_a)$ if $\tilde{G}_a = G_{I_{j_q}}$ and $\lambda_a = \xi_\pm(\tilde{G}_a)$ otherwise, $\nu_j = \xi_\pm(G_j)$; furthermore, all possible combinations occur. The same holds for $G = D_r(K)$ and the representations ξ_- and ξ_+ if $r(G_+G_-) < r$. Now assume that $G = D_r(K)$ and $r(G_+G_-) = r$. For every tensor product TP of the form described above set $A(TP) = \{a \mid \lambda_a = \xi_-(\tilde{G}_a)\}$, $J(TP) = \{j \mid j \in J_2, \nu_j = \xi_-(G_j)\}$, and $N(TP) = |A(TP)| + |J(TP)|$. The restriction $\xi_-|G_+G_-$ is a direct sum of components TP with odd $N(TP)$ and $\xi_+|G_+G_-$ is a direct sum of such components with even $N(TP)$; in both cases all possible combinations satisfying the additional conditions imposed above occur. To describe relevant restrictions for G_- or $G_+ = 1$, we take for TP tensor products of the form $\bigotimes_{a=1}^{d} \lambda_a$ and set $N(TP) = |A(TP)|$ in the first case and consider components TP of the form $\bigotimes_{j\in J_2} \nu_j$ with $N(TP) = |J(TP)|$ in the second one.

By [**Sei87**, Table 1], if $\tilde{G}_a$ is of type d and $B = S_{j_{2l-1}} S_{j_{2l}}$ or $S_j \subset \tilde{G}_a$ is the subgroup constructed in the proof of Proposition 2.27, then $\xi_\pm(\tilde{G}_a)|B \cong \xi(S_{j_{2l-1}}) \otimes \xi(S_{j_{2l}})$ in the first case and $\xi(S_j)$ in the second one. Now the assertion of the lemma follows from Lemma 2.37. □

LEMMA 11.12. *Let x and φ satisfy the assumptions of Theorem 1.9. Then $p^f d_f > p^{s+1}$ for $0 \leq f \leq s$.*

PROOF. Recall that $p^s + p \leq k_1 < p^{s+1}$. Lemma 11.7 allows one to reduce the question to fundamental representations. Hence assume that $\overline{\omega} = \omega_i$ with

$p \le i \le n-p$ for $G = A_r(K)$ and $i \le p$ otherwise. Furthermore, suppose that $i < r$ or $k_1 \ge p^s + 2p$ for $G = B_r(K)$ and $i < r-1$ or $k_1 \ge p^s + 2p$ for $G = D_r(K)$.

First assume that $G = A_r(K)$. Passing to the dual representation if necessary, one can suppose that $i \le n/2$. Rules 1–9 and Algorithm 1.6 imply that $d_0 - 1$ is a sum of i nonnegative integers and the sum of p maximal summands is at least $p(k_1 - p) \ge p^{s+1}$ and that for $f > 0$ the difference $d_f - 1$ is a sum of i nonnegative integers at least p of which are $\ge p^{s-f}$. This yields the lemma in the case being considered.

Now Proposition 11.9 enables one to extend the result to relevant fundamental representations of other groups, except the spinor representation of $B_r(K)$ and the semispinor representations of $D_r(K)$.

So it remains to consider the cases where $k_1 \ge p^s + 2p$, $G = B_r(K)$, and $i = r$ or $G = D_r(K)$ and $i = r-1$ or r. Recall that $k_1 \ne 9$ by our assumptions. Then Lemma 11.10 implies that $\theta_1(\overline{\omega}) = \omega_{r_1}$ for odd k_1 and ω_a with $p < a < k_1 - p$ for even k_1. Algorithm 1.6 and Lemma 11.5 imply that in this situation $p^f d_f \ge p^f d_{f,1} \ge p^{s+1}$. Indeed, if k_1 is odd, one can directly check that $d_{0,1} = r_1(r_1+1)/2 + 1 > p^{s+1}$ and observe that for $f > 0$ the element $h_{f,1}$ has either at least $2p$ blocks of size $> p^{s-f}$, or at least p blocks of size $> p^{s-f} + 1$. For $s = 1$ the difference $d_{1,1} - 1$ is a halfsum of nonnegative integers at least p of which are ≥ 2, for $s > 1$ we have $r_1 > 2p$ and $d_{f,1} - 1$ is a halfsum of more than $2p$ nonnegative summands with either at least $2p$ summands $\ge p^{s-f}$, or at least p pairs of summands where the sum $\ge 2p^{s-f}$. This yields the claim for odd k_1. For even k_1 argue as for $G = A_r(K)$. Now all the possibilities have been considered. □

LEMMA 11.13. *Suppose that ω_i does not satisfy the assumptions of Theorem 1.9. Then*

$$d_{f,j,\omega_i} - 1 \ge p^{s-f}(d_{s,j,\omega_i} - 1) \tag{11.2}$$

for $0 \le f < s$ and $1 \le j \le c$.

PROOF. Obviously, (11.2) holds if $d_{s,j,\omega_i} = 1$. So it is valid if $u_j \le p^s$ or $\theta_j(\omega_i) = 0$. Therefore assume that $u_j > p^s$ and $\theta_j(\omega_i) \ne 0$. By Lemma 3.2, Formula (11.2) holds if $i < p$ or $G = A_r(K)$ and $n - p < i \le r$. Hence we have to consider the following cases:

1) $u_j = p^s + b_{0,j}$ with $0 < b_{0,j} < p$;

2) $p^s + p \le u_j < p^s + 2p$ and either $G = B_r(K)$ and $i = r$, or $G = D_r(K)$ and $i = r-1$ or r.

For fixed i and j set $D_f = d_{f,j,\omega_i} - 1$. The estimates below are based on Rules 1–9, Proposition 1.5, and Algorithm 1.6.

Let 1) hold. Passing to the dual representation if necessary, one can suppose that $i \le n/2$ for $G = A_r(K)$. First assume that u_j is proper for G, $i < r$ for $G = B_r(K)$, and $i < r-1$ for $G = D_r(K)$. Then $\theta_j(\omega_i) = \omega_{i_j}$ with $i_j \le u_j/2$ or $H_j = B_{r_j}(\mathbb{C})$ and $\theta_j(\omega_i) = 2\omega_{r_j}$, in both cases $D_0 = i_j(p^s + b_{0,j} - i_j)$. By Proposition 2.5 b), for $f > 0$ the element $h_{f,j}$ has $b_{0,j}$ Jordan blocks of size $p^{s-f}+1$ and $p^f - b_{0,j}$ blocks of size p^{s-f}. Hence $D_s = \min\{b_{0,j}, i_j\}$. If $i_j \le b_{0,j}$, we get $D_0 > p^s i_j = p^s D_s$ and $D_f = p^{s-f} i_j$ for $f > 0$. For $i_j > b_{0,j}$ one has $D_0 \ge b_{0,j} p^s = p^s D_s$ by (2.8) and $D_f > b_{0,j} p^{s-f}$ if $0 < f < s$. This forces that in all cases (11.2) holds.

Now let u_j be improper for G and i be such as in the previous paragraph. The arguments in the proof of Proposition 11.9 imply that $\theta_j(\omega_i) = \omega_k + \omega_{u_j - k}$

or $\omega_k + \omega_{u_j-k+1}$ with $k \le u_j/2$ in the first case and $k \le (u_j+1)/2$ in the second one. Here we assume that $\omega_{u_j} = 0$. Let the first possibility occur. Then $D_s = 2\min\{k, b_{0,j}\}$. If $k \le b_{0,j}$, we get $D_0 = 2k(p^s + b_{0,j} - k) \ge p^s D_s$ and $D_f = 2kp^{s-f}$ for $f > 0$. If $k > b_{0,j}$, using Formula (2.8) as above, we conclude that $D_0 > 2b_{0,j}p^s$ and $D_f > 2b_{0,j}p^{s-f}$ for $0 < f < s$.

Next, consider the second possibility. First let $k < (u_j+1)/2$. If $k \le b_{0,j}$, one concludes that $D_0 > (2k-1)p^s$ and $D_f = (2k-1)p^{s-f}$ for $f > 0$. If $k > b_{0,j}$, we have $D_s = 2b_{0,j}$, $D_0 > 2b_{0,j}p^s$ by (2.8), and $D_f > 2b_{0,j}p^{s-f}$ for $0 < f < s$.

If $k = (u_j+1)/2$, using Formula (2.8) and taking into account that $b_{0,j} \le (u_j-1)/2$, one can deduce that $D_s = 2b_{0,j}$, $D_0 \ge 2b_{0,j}p^s$, and $D_f > 2b_{0,j}p^{s-f}$ for $0 < f < s$. This implies that in all cases (11.2) holds for improper u_j and the values of i considered above.

Finally, assume that 1) or 2) holds and either $G = B_r(K)$ with $\omega = \omega_r$, or $G = D_r(K)$ with $\omega = \omega_{r-1}$ or ω_r. Apply Lemma 11.10. First let u_j be proper for G. Then $H_j = B_{r_j}(\mathbb{C})$ with $u_j = 2r_j + 1$, $u_j = p^s + 2g$ with $g \in \mathbb{Z}^+$ and $0 < g < p$, and $\theta_j(\omega_i) = \omega_{r_j}$. Suppose that $g \le (p-1)/2$ or $s > 1$. Then $r_j \ge 2g$ and by Proposition 2.5 b), $h_{s,j}$ has $2g$ blocks of size 2 and $p^s - 2g$ blocks of size 1. Hence $D_s = g$. By (2.8), $D_0 \ge gp^s$. If $g > (p-1)/2$, one has $2g = p + k$ with $k < p$. Then Proposition 2.5 b) yields that $h_{1,j}$ has k blocks of size $p^{s-1}+2$ and $p-k$ blocks of size $p^{s-1}+1$. So Algorithm 1.6 implies that $D_1 > (k(p^{s-1}+1) + (p-k)p^{s-1} + k(p^{s-1}-1))/2 = gp^{s-1}$. One easily concludes that $D_f > gp^{s-f}$ for $1 < f < s$. Next, let $s = 1$ and $g > (p-1)/2$. Then $u_j = 2p + 2d + 1$ with $d \in \mathbb{Z}^+$ and $2d+1 < p$. Hence $h_{1,j}$ has $2d+1$ blocks of size 3 and $p-2d-1$ blocks of size 2 by Proposition 2.5 b). Therefore $D_0 = (p+d)(p+d+1)/2$ and $D_1 = (p+2d+1)/2$. Now one can see that (11.2) holds for proper u_j.

Next, assume that u_j is improper for G. Then $u_j = p^s + 2g + 1$ with $g \in \mathbb{Z}^+$ and $0 \le g < p$. By Lemma 11.10, $\theta_j(\omega_i) = \omega_{u_j/2}$ or $\omega_{u_j/2+1}$. First suppose that $2g+1 < p$ or $s > 1$. Then $2g+1 \le u_j/2 - 1$. So Proposition 2.5 and Formula 2.8 imply that $D_s = 2g+1$ and $D_0 \ge (2g+1)p^s$. If $2g+1 \ge p$, one has $u_j = p^s + p + 2k$ with $k \in \mathbb{Z}^+$ and $2k < p$. Arguing as for proper u_j and $2g > p$, we deduce that in this case $D_1 \ge 4kp^{s-1} + (p-2k)p^{s-1} = (2g+1)p^{s-1}$ and that in all situations under consideration $D_f \ge (2g+1)p^{s-f}$ for $f > 0$.

Finally, let $s = 1$ and $2g+1 \ge p$. Then $u_j = 2p + 2d$ with $d \in \mathbb{Z}^+$ and $2d < p$. By Proposition 2.5 b), $h_{1,j}$ has $2d$ blocks of size 3 and $p - 2d$ blocks of size 2. If $d > 0$, one gets $u_j/2 - 1 \ge p > 2d$. Lemma 11.10 implies that in this case $D_0 \ge p(p+2d)$ and $D_1 = p + 2d$. If $d = 0$, the same lemma forces that either $D_0 = p^2$ and $D_1 = p$, or $D_0 = p^2 - 1$ and $D_1 = p - 1$. Hence (11.2) is valid for improper u_j as well. All the possibilities have been considered. This completes the proof. □

COROLLARY 11.14. *Assume that ω does not satisfy the assumptions of Theorem 1.9. We have $p^f(d_{f,j}-1) \ge p^s(d_{s,j}-1)$ and $p^f(d_f-1) \ge p^s(d_s-1)$ for $1 \le j \le c$ and $0 \le f \le s$. If $p^f d_{f,j} > p^s d_{s,j}$ for some f and j, then $p^f d_f > p^s d_s$. If $a_i \ne 0$ and $p^f d_{f,j,\omega_i} > p^s d_{s,j,\omega_i}$ for some f and j, then $p^f d_{f,j} > p^s d_{s,j}$ and hence $p^f d_f > p^s d_s$.*

PROOF. The definition of $\overline{\omega}$ yields that $\overline{a_i} \ne 0$ if and only if $a_i \ne 0$. Our assumptions imply that ω_i does not satisfy the assumptions of Theorem 1.9 whenever

$a_i \neq 0$. So (11.2) holds for ω_i by Lemma 11.13. It remains to apply Lemma 11.5 and Formulae (11.1) and (11.2). □

COROLLARY 11.15. *Assume that* $p^s d_s > p^{s+1}$. *Then* $p^f d_f > p^{s+1}$ *for* $0 \leq f < s$.

PROOF. Lemma 11.12 implies the claim of the proposition for the weights ω satisfying the assumptions of Theorem 1.9. Hence we can exclude these weights and, using Lemma 11.13, suppose that (11.2) holds for ω_i whenever $a_i \neq 0$. Observe that $d_s - 1 \geq p$. Now it follows from (11.1) and (11.2) that $p^f(d_f - 1) \geq p^s(d_s - 1) \geq p^{s+1}$ which yields the corollary. □

Proof of Proposition 1.12. The assertions of the proposition can be deduced from Rules 1–9, Propositions 1.5 and 2.5, Algorithm 1.6, and Lemma 11.10. Set $r' = r$ for $G = B_r(K)$ and $r' = r - 1$ for $G = D_r(K)$. One have to consider the following situations:

i) $u_1 = p^s + b_0$ with $0 < b_0 < p$;

ii) $u_1 \geq p^s + p$, $a_i = 0$ for $p \leq i \leq n - p$ if $G = A_r(K)$ and $a_i = 0$ for $i \geq p$ otherwise;

iii) $G = B_r(K)$ or $D_r(K)$, $p^s + p \leq u_1 < p^s + 2p$, $a_i = 0$ for $p \leq i < r'$, and $a_j \neq 0$ for some $j \geq r'$.

Actually Proposition 1.5 reduces the question to fundamental weights $\overline{\omega}$. Naturally, d_s depends only upon $\theta_j(\overline{\omega})$ for $u_j > p^s$. Using the results mentioned above, one can check directly that the assertions of the proposition always hold in Cases i) and ii) and that they hold in Case iii) if $J(x)$ has a block of odd size. In Case ii) Rules 6 and 7 are especially important. If $G = D_r(K)$ and all u_j are even, in Cases i) and iii) one has to consider the situations where $x \in C_1$ and $x \in C_2$ separately taking into account different possibilities described by Lemma 11.10. Next, in the assumptions of Item iii) suppose that $\overline{\omega} = \omega_i$ with $i \geq r'$ and all blocks in $J(x)$ have even sizes. This situation requires some additional comments. Naturally, here $G = D_r(K)$. Rule 9) forces $\mathbf{j}(r) = c$. If $h_{s,c}$ has a block of odd size, Algorithm 1.6 implies that $d_{s,c}$ is the same for $\theta_c(\overline{\omega}) = \omega_{u_c/2}$ or $\omega_{u_c/2+1}$. Here we can argue as for $G = B_r(K)$. Proposition 2.5 b) implies that such block exists unless $s = 1$ and $u_c = 2p$.

Assume that $s = 1$, $c > 1$, and $u_c = 2p$. Proposition 2.5 b) shows that the element $h_{1,c}$ has p blocks of size 2 in $V(H_c)$ and the elements $h_{1,c-1}$ and z_1 have all blocks of size 2 and, may be, 3, in $V(H_{c-1})$ and $V(G_\mathbb{C})$, respectively. So z_1 has $\geq 2p$ blocks of size ≥ 2 on $V(G_\mathbb{C})$. By our assumptions, $\overline{\omega} = \omega_{r-1}$ or ω_r. Lemma 11.10 implies that $\theta_{c-1}(\overline{\omega}) = \omega_{u_{c-1}/2}$ and $\theta_c(\overline{\omega}) = \omega_p$ or ω_{p+1}. Applying Algorithm 1.6, one can deduce that both d_1 and $d_{\varphi_\mathbb{C}}(z_1) > p$.

Now it remains to consider the situation where $G = D_{2p}(K)$, $J(x) = (2p, 2p)$, and $\overline{\omega} = \omega_{2p-1}$ or ω_{2p}. By Proposition 2.5 b), h_1 has p blocks of size 2 in $V(H)$ and z_1 has $2p$ blocks of size 2 in $V(G_\mathbb{C})$. Lemma 11.10 forces that $\theta_1(\overline{\omega}) = \omega_{p+1}$ if $x \in C_1$ and $\overline{\omega} = \omega_{2p-1}$ or $x \in C_2$ and $\overline{\omega} = \omega_{2p}$, otherwise $\theta_1(\overline{\omega}) = \omega_p$. Hence $d_1 = p$ in the first case and $p + 1$ in the second one.

Below in this proof by abuse of notation we use the same symbols to denote conjugacy classes in G and $G_\mathbb{C}$ with the same labelled Dynkin diagram. Namely, let C_1' and C_2' be such classes containing elements u with $J(u) = J(z_1)$ $(= J(x^p))$. Assume that the $(r-1)$th label on the labelled Dynkin diagram is 0 for C_1' and 2 for C_2'. We claim that x^p and $z_1 \in C_t'$ if $x \in C_t$, $t = 1, 2$. Indeed, the analysis of the action of x^p on the subspace V_1 constructed in the proof of Lemma 2.25 enables one to conclude that there exists a subgroup A_x of type A_1 containing x^p such that $T \cap A_x$ is a maximal torus in A_x and the mapping $\kappa_x : \mathbf{X} \to \mathbb{Z}$ determined by restricting weights from T to $T \cap A_x$ has the following properties: $\kappa_x(\varepsilon_i) = 1$ for $i < r$, $\kappa_x(\varepsilon_r) = 1$ if $x \in C_1$ and -1 for $x \in C_2$. Since $\kappa_x(\alpha_i) \geq 0$ for $1 \leq i \leq r$, the mapping κ_x determines the labelled Dynkin diagram of x^p. This yields that $x^p \in C'_t$ just when $x \in C_t$. By the definition of z_1, the element $z_1 \in C'_t$ if and only if $x^p \in C'_t$. Now Algorithm 1.6 yields that $d_1 = d_{\varphi_{\mathbb{C}}}(z_1)$ and completes the proof. □

LEMMA 11.16. *One has $d_{\varphi_i}(x) \leq d_{\sigma_i}(h_0)$.*

PROOF. In this proof φ^j_a means the representation $\varphi(\omega_a) \in \operatorname{Irr} S_j$. Obviously, it suffices to consider the cases where $d_{\sigma_i}(h_0) < p^{s+1}$. First assume that $G = A_r(K)$. Passing to the dual representation if necessary, suppose that $i \leq n/2$. One has $S \cong \prod_{j=1}^{c} A_{u_j-1}(K)$. Denote by Δ the set of all $(c+1)$-tuples of integers $(i_1, i_2, \dots, i_c, i')$ such that $0 \leq i_j \leq u_j$, $0 \leq i' \leq n - \sum_{j=1}^{c} u_j$, and $i_1 + i_2 + \ldots + i_c + i' = i$. For $\delta \in \Delta$ set $\mu_\delta = \bigotimes_{j=1}^{c} \varphi^j_{i_j} \in \operatorname{Irr} S$. Put $\varphi = \varphi_i$. Proposition 2.36 implies that $\varphi|S$ is a direct sum of components isomorphic to μ_δ with $\delta \in \Delta$ and each μ_δ has a nonzero multiplicity. (Now we do not need these multiplicities.) Then $d_\varphi(x) = \max\{d_{\mu_\delta}(x) \mid \delta \in \Delta\}$. Let $d_\varphi(x) = d_{\mu_\gamma}(x)$ for some $\gamma \in \Delta$. Since $H \cong S_{\mathbb{C}}$, we can regard H as a subgroup of $G_{\mathbb{C}}$ and identify the elements h_f with z_f for $0 \leq f \leq s$. Set $\nu = \varphi(\omega(\mu_\gamma)) \in \operatorname{Irr} H$. Recall that by Lemma 11.12 one can assume that either $u_1 = p^s + b_0$ with $b_0 < p$ or $i < p$. The construction of θ implies that $\theta(\omega_i) = \omega(\mu_\kappa)$ for some $\kappa \in \Delta$. Now we can deduce from Rules 1–9, Algorithm 1.6, and Proposition 1.12 that one of the following holds:

1) $u_1 = p^s + b_0$ with $0 < b_0 < p$ and $d_\nu(h_s) < d_{\pi_i}(z_s)$;
2) $i < p$, $u_1 \geq p^s + p$, and $d_\nu(h_1) < d_{\pi_i}(z_1)$;
3) $d_\nu(h_0) \leq d_{\sigma_i}(h_0)$.

Here it is essential that Rule 6 enables one to get $d_{\sigma_i}(h_s) = d_{\pi_i}(z_s)$ if $u_1 = p^s + b_0$, Rule 7 permits us to obtain $d_{\sigma_i}(h_1) = d_{\pi_i}(z_1)$ if $i < p$ and $u_1 \geq p^s + p$, and Rule 9 yields the maximal value of $d_{\sigma_i}(h_0)$ that is possible under Rules 1–8. Since ν is a component of $\pi_i|H$ by Proposition 2.36, Results 2.40, 2.11, and 2.13 imply that $d_{\mu_\gamma}(x^{p^f}) \leq d_\nu(h_f) \leq d_{\pi_i}(z_f)$ for $0 \leq f \leq s$. So, using Proposition 2.5 a), one concludes that $d_{\mu_\gamma}(x^{p^s}) < d_{\pi_i}(z_s)$ and $d_\varphi(x) \leq p^s(d_{\pi_i}(z_s) - 1)$ in Case 1), $d_{\mu_\gamma}(x^p) < d_{\pi_i}(z_1)$ and $d_\varphi(x) \leq p(d_{\pi_i}(z_1) - 1)$ in Case 2), and $d_\varphi(x) \leq d_\nu(h_0) \leq d_{\sigma_i}(h_0)$ in Case 3). By Proposition 1.12, $d_{\pi_i}(z_s) = d_{\sigma_i}(h_s)$ in Case 1) and $d_{\pi_i}(z_1) = d_{\sigma_i}(h_1)$ in Case 2). Applying Formula (11.2) in Case 1) and Corollary 11.6 in Case 2), one concludes that $d_\varphi(x) \leq d_{\sigma_i}(h_0) - 1 < d_{\sigma_i}(h_0)$ in these cases. This completes the proof for $G = A_r(K)$.

Now assume that $G \neq A_r(K)$, $i < r$ for $G = B_r(K)$, and $i < r - 1$ for $G = D_r(K)$. Define G^+ as in Proposition 11.9. One can pass to a subgroup $G_0 \subset G^+$ that is the image of G under the canonical mapping to G^+ and regard x as a unipotent element of G^+. Let $\varphi^+_i \in \operatorname{Irr} G^+$ and $\omega(\varphi^+_i) = \omega_i$. Construct the integer d^+_0 for the representation φ^+_i as in Proposition 11.9. It is well known that φ_i is a composition factor of $\varphi^+_i|G_0$. Hence $d_{\varphi_i}(x) \leq d_{\varphi^+_i}(x)$. By Proposition 11.9, $d^+_0 = d_{\sigma_i}(h_0)$. Now our claim follows from the result for $A_r(K)$ proved above.

Finally, let $p^s < u_1 < p^s + 2p$ and either $G = B_r(K)$ and $i = r$, or $G = D_r(K)$ and $i = r - 1$ or r. Use the notation of Proposition 2.27 and Lemma 11.11, but set

$\varphi_{n_j}^{u_j} = \varphi_{n_j}^j$ to simplify it a bit. It follows from Lemma 11.11 that $d_{\varphi_i}(x) = d_\sigma(x)$ where $\sigma \cong \Lambda \bigotimes \bigotimes_{j\in J_2} \varphi_{n_j}^j \in \operatorname{Irr} S$. Set $S^1 = \prod_{j\in J_1} S_j$ and $S^2 = \prod_{j\in J_2} S_j$, $\gamma_j = \varphi_{n_j}^j$ and $\delta_j = d_{\gamma_j}(x_j) - 1$. Then $x = x^1x^2$ where $x^i \in S^i$ is a regular unipotent element for $i = 1, 2$ and $x^2 = \prod_{j\in J_2} x_j$ where x_j are regular unipotent elements of the groups S_j. By Corollary 2.13, $d_{\varphi_i}(x) \le d_\Lambda(x^1) + \sum_{j\in J_2} \delta_j$. Set $H^+ = \prod_{j\in J_1} H_j$. We have $h_0 = h^+ \prod_{j\in J_2} h_{0,j}$ where $h^+ \in H^+$ and $h_{0,j} \in H_j$ are regular unipotent elements, $\gamma_j \in \operatorname{Irr} S_j$, and $\omega(\gamma_j) = \omega_{n_j}$ with $0 \le n_j \le u_j$. Corollary 2.41 and Algorithm 1.6 imply that $d_\Lambda(x^1) \le d_{\Lambda_\mathbb{C}}(h^+)$ and $\delta_j \le n_j(u_j - n_j)$. Recall that u_j is even for all $j \in J_2$ since they are improper for G. Set $u_j = 2u_j'$ for these j. By (2.8), $n_j(u_j - n_j) \le (u_j')^2$ and $\le (u_j')^2 - 1$ if $n_j \ne u_j'$. For $j \in J_2$ let $\mu_j, \nu_j \in \operatorname{Irr} H_j$, $\omega(\mu_j) = \omega_{u_j'}$, and $\omega(\nu_j) = \omega_{u_j'+1}$. By Lemma 11.10, $\psi \cong \Lambda_\mathbb{C} \bigotimes \bigotimes_{j\in J_2} \psi_j$ with $\psi_j \in \operatorname{Irr} H_j$ and either all $\psi_j = \mu_j$, or $J(x)$ contains no blocks of odd size, one of the representations ψ_j is equal to ν_j and others are μ_j. Lemmas 2.37 and 11.11 imply that in the latter case all indices n_j cannot be equal to u_j'. Now Algorithm 1.6 completes the proof. □

COROLLARY 11.17. *One has $d_\varphi(x) \le d_0$.*

PROOF. For $1 \le i \le r$ set $\delta_i(x) = d_{\varphi_i}(x) - 1$. Lemma 2.51 and Corollary 2.13 imply that $d_\varphi(x) \le 1 + \sum_{i=1}^r a_i\delta_i(x)$. Now the assertion of the corollary follows from Lemma 11.16 and Proposition 1.5. □

COROLLARY 11.18. *Assume that one of the following holds:*

a) $p^s + p \le u_1 < p^{s+1}$, $a_i = 0$ for $p \le i \le n - p$ if $G = A_r(K)$, and $a_i = 0$ for $i \ge p$ otherwise;

b) $u_1 = p + b_0$ with $b_0 < p$;

c) $G = B_r(K)$ or $D_r(K)$, $2p \le u_1 < 3p$, $a_i = 0$ for $p \le i < r$ if $G = B_r(K)$ and $a_i = 0$ for $p \le i < r - 1$ for $G = D_r(K)$.

Then $d_\varphi(x) \le d(x, \varphi)$.

PROOF. First assume that $G = D_r(K)$, all u_j are even, $c > 1$, $u_c = 2p$, $2p \le u_j < 3p$ for all $j \le c$, and $a_{r-1} + a_r \ne 0$. Algorithm 1.6, Corollary 11.8, and Lemma 11.10 imply that in this case $d_0 > d_{0,1} \ge p^2 + 1$ and $d_1 > d_{1,1} \ge p + 1$. Hence $d(x, \varphi) > p^2 \ge d_\varphi(x)$. So we can exclude this case. In all other situations, by Proposition 1.12, under our assumptions $d_f = d_{\varphi_\mathbb{C}}(z_f)$ for $0 < f \le s$. Hence by Proposition 2.5 a) and Corollary 2.40, $d_\varphi(x) \le p^f d_\varphi(x^{p^f}) \le d_f$ for such f. Corollary 11.17 completes the proof. □

Recall that $u_j = b_{s,j}p^s + \ldots + b_{0,j}$ is the p-adic expansion of u_j for $1 \le j \le c$. Fix maximal e with $u_e > p^s$. In Lemmas 11.19–11.22 below $u_1 = p^s + b_{0,1}$ with $s > 1$ and $b_{0,1} < p$. If $j \le e$ and $\theta_j(\overline{\omega}) = \sum_{i=1}^{r_j} a_{ij}\omega_i$, set

$$\Delta_j = \sum_{i=1}^{b_{0,j}-1} a_{ij}\omega_i + \sum_{i=u_j+1-b_{0,j}}^{r_j} a_{ij}\omega_i, \; \Lambda_j = a_{b_{0,j},j}\omega_{b_{0,j}} + a_{u_j-b_{0,j},j}\omega_{u_j-b_{0,j}}$$

for $H_j = A_{r_j}(\mathbb{C})$ and

$$\Delta_j = \sum_{i=1}^{b_{0,j}-1} a_{ij}\omega_i, \; \Lambda_j = a_{b_{0,j},j}\omega_{b_{0,j}}$$

otherwise; in all cases put $\Sigma_j = \theta_j(\omega) - \Delta_j - \Lambda_j$.

LEMMA 11.19. *Assume that* $\Sigma_j = 0$ *for all* $j \leq e$ *and* $\theta_g(\omega) \neq 0$ *for some* $g > e$. *Then* $u_g > p^s - b_{0,e}$.

PROOF. We use the notation $\mathbf{j}(i)$ and $\nu(\varepsilon_k^j)$ introduced before stating Rules 1–9. The assumptions of the lemma and Rules 1–9 imply that there exists i with $i \leq (r+1)/2$ for $G = A_r(K)$ such that $g = \mathbf{j}(i)$ and $\varepsilon^e_{b_{0,e}+1} \neq \theta_e(\varepsilon_d)$ for $d < i$. Observe that $\nu(\varepsilon^e_{b_{0,e}+1}) = p^s - b_{0,e} - 1$. Now Rule 9 forces that $\nu(\varepsilon_1^g) > p^s - b_{0,e} - 1$. Hence $u_g > p^s - b_{0,e}$ as desired. □

Set $\omega^* = \overline{\omega(\varphi^*)}$.

LEMMA 11.20. *In the assumptions of Lemma* 11.19 $d(x,\varphi) = \min\{p^s d_s, p^{s+1}\}$ *unless* $u_{e+1} < p^s$, $\overline{\omega}$ *or* $\omega^* = c_b\omega_b + \omega_{b+1} + c_{n-b}\omega_{n-b}$ *for* $G = A_r(K)$ *and* $\overline{\omega} = \overline{a_b}\omega_b + \omega_{b+1}$ *otherwise. In the exceptional case* $d(x,\varphi) = \min\{d_0, p^{s+1}\}$.

PROOF. Recall that $b = \sum_{j=1}^e b'_{0,j}$ with $b'_{0,j} = b_{0,j}$ for proper and $2b_{0,j}$ for improper u_j. Our assumptions yield that $a_i \neq 0$ and $\theta_g(\omega_i) \neq 0$ for some i with $b < i < n - b$ for $G = A_r(K)$ and $i > b$ otherwise. Fix minimal such i. If $G = A_r(K)$, passing to φ^* if necessary, one can assume that $i \leq (r+1)/2$.

Corollary 11.15 implies the assertion of our lemma if $d_s > p$. Hence assume that $d_s \leq p$. For $0 \leq f \leq s$ set $D_f^1 = \sum_{j=1}^e (d_{f,j} - 1)$ and $D_f^2 = \sum_{j=e+1}^c (d_{f,j} - 1)$. By Lemma 11.5,

$$d_f = D_f^1 + D_f^2 + 1,\ D^2 f \geq d_{f,g} - 1. \tag{11.3}$$

Obviously, $D_s^2 = 0$. By Proposition 2.5 b), for $f > 0$ the element $h_{f,j}$ has $b_{0,j}$ blocks of size $p^{s-f} + 1$ and $p^f - b_{0,j}$ blocks of size p^{s-f}. Since $u_g > p^s - b_{0,e}$ by Lemma 11.19, Proposition 2.5 b) also forces that $h_{f,g}$ has blocks of size p^{s-f} for $f > 1$. So Proposition 1.5 and Algorithm 1.6 yield that $D_0^1 \geq D_s^1 p^s$, $D_f^1 = D_s^1 p^{s-f}$ for $1 \leq f < s$, and $d_{f,g} \geq p^{s-f}$ for $f > 0$. Then (11.3) implies that $p^f d_f \geq (D_s^1 + 1)p^s = p^s d_s$ for $0 < f < s$ and $d_0 \geq p^s d_s$ if $d_{0,g} \geq p^s$. In the latter case $d(x,\varphi) = p^s d_s$. Hence it suffices to find out when $d_{0,g} < p^s$.

Set $\omega^b = \overline{\omega}(1,b) + \overline{\omega}(r-b+1,r)$ for $G = A_r(K)$ and $\omega^b = \overline{\omega}(1,b)$ otherwise.Write $\overline{\omega} = \omega^b + \lambda$. Obviously, $\theta_g(\omega^b) = 0$. By our assumptions, $\lambda = \omega_i + \sigma$ with $\sigma \in \mathbf{X}^+$. Since $i \leq (r+1)/2$ for $G = A_r(K)$, the construction of θ implies that for $H_g = A_{r_g}(K)$ in the canonical decomposition $\theta_g(\lambda) = \sum_{k=1}^{r_g} l_k\omega_k$ (here ω_k are the fundamental weights of H_g) for some $k \leq (r_g+2)/2$ the coefficient $l_k \neq 0$. The form of $\theta_j(\overline{\omega})$ for $j \leq e$ and Lemma 11.10 yield that $a_r = 0$ for $G = B_r(K)$ and $a_{r-1} = a_r = 0$ for $G = D_r(K)$ (here it is essential that $s > 1$). Hence $\theta_g(\overline{\omega}) \neq \omega_{r_g}$ if $H_g = B_{r_g}(\mathbb{C})$. Now Lemmas 3.1 and 11.19 imply that $d_{0,g} \geq p^s$ if $\theta_g(\overline{\omega}) \neq \omega_1$ or $u_g = p^s$. (The arguments above exclude the case where $H_g = A_{r_g}(\mathbb{C})$ and $\theta_g(\overline{\omega}) = \omega_{r_g}$.) So assume that $\theta_g(\overline{\omega}) = \omega_1$ and $u_g < p^s$. Then $b_{0,e} > 1$ as $u_g > p^s - b_{0,e}$. Now we claim that either $D_0^2 \geq p^s - 1$ or $\theta_j(\overline{\omega}) = 0$ if $j > e$ and $j \neq g$. Assume that $\theta_k(\overline{\omega}) \neq 0$ for some $k > e$ and $k \neq g$. By Lemma 11.19, $u_k > p^s - b_{0,e}$. Since $s > 1$, Proposition 1.5 and Algorithm 1.6 imply that $S_0^2 \geq (d_{0,g} - 1) + (d_{0,k} - 1) > p^s$. Now we can and shall suppose that $\theta_j(\overline{\omega}) = 0$ if $j > e$ and $j \neq g$. This forces $\lambda = \omega_i$ as $\theta_g(\sigma) = 0$. Next, we have to show that $i = b + 1$. Assume that $i > b + 1$. According to Rule 5, $\theta(\varepsilon_{b+1}) \neq 0$. Our assumptions and the facts proved above yield that $\theta_j(\varepsilon_{b+1}) = \theta_j(\varepsilon_{b+2}) = 0$ for $j \neq g$. Now Rule 9 implies that $\theta_g(\varepsilon_{b+1}) = \theta_g(\varepsilon_{b+1}) = \varepsilon_1^g$ and $\theta_g(\varepsilon_{b+2}) \neq 0$

since $u_g > p^s - b_{0,e} > 2p$. But then $\theta_g(\lambda)$ cannot be equal to ω_1 which yields a contradiction.

Assume that $\omega^b \neq \overline{a_b}\omega_b + \overline{a_{n-b}}\omega_{n-b}$ for $G = A_r(K)$ and $\omega^b \neq \overline{a_b}\omega_b$ for other groups. We have $\nu(\varepsilon^j_{b_{0,j}}) = p^s - b_{0,j} + 1 < p^s < \nu(\varepsilon^k_1)$ for $j, k \leq e$. Now one can deduce from Rules 1–9 that $\Delta_q \neq 0$ for some $q \leq e$. Proposition 1.5 and Algorithm 1.6 imply that $d_{0,q} \geq p^s(d_{s,q} - 1) + b_{0,q}$ and $d_{0,j} \geq 1 + p^s(d_{s,j} - 1)$ for each $j \leq e$. One easily concludes that $D_0^2 + 1 = d_{0,g} = u_g$. Hence $D_0^2 + 1 > p^s - b_{0,e}$ by Lemma 11.19. Recall that $b_{0,q} \geq b_{0,e}$. Therefore $D_0^1 \geq p^s D_s^1 + b_{0,e} - 1$ and $d_0 \geq p^s d_s$ by (11.3). This forces that $d(x, \varphi) = p^s d_s$ as required.

In the exceptional case the construction of θ implies that $\theta_{e+1}(\overline{\omega}) = \omega_1$ and $\theta_j(\overline{\omega}) = 0$ for $j > e+1$. By Proposition 1.5, Algorithm 1.6, and Formula (11.1), in this case $d_0 = (d_s - 1)p^s + u_{e+1} < d_s p^s$. We have already seen that $d_f \geq d_s p^s$ for $0 < f < s$. This completes the proof. □

LEMMA 11.21. *Assume that $\theta_j(\overline{\omega})$ is such as in Lemma 11.19 for $j \leq e$ and $\theta_j(\overline{\omega}) = 0$ for $j > e$. Then $d(x, \varphi) = \min\{d_0, pd_1, p^{s+1}\}$.*

PROOF. Using Propositions 1.5 and 2.5, Algorithm 1.6, and Lemma 11.5 and arguing as for D_f^1 in the proof of Lemma 11.20, one can conclude that $d_0 \geq 1 + p^s(d_s - 1)$ and $d_f = 1 + p^{s-f}(d_s - 1)$ for $0 < f < s$. This implies that $d(x, \varphi) = p^{s+1}$ for $d_s > p$, otherwise $d(x, \varphi) = d_0$ if $d_0 - p^s(d_s - 1) \leq p$ and $d(x, \varphi) = pd_1$ if $d_0 - p^s(d_s - 1) > p$. This yields the lemma. □

LEMMA 11.22. *Assume that $\Sigma_j \neq 0$ for some $j \leq e$. Put $\Omega = \{0, \omega_1, \omega_{b_{0,1}-1}, \omega_{u_1+1-b_{0,1}}, \omega_{r_1}\}$ for $H_1 = A_{r_1}(\mathbb{C})$ and $\Omega = \{0, \omega_1, \omega_{b_{0,1}-1}\}$ otherwise. Then $d(x, \varphi) = \min\{p^s d_s, p^{s+1}\}$ or one of the following holds:*

1) $d_s \leq p$, $b_{0,j} = b_{0,e}$, $b_{0,j-1} > b_{0,j}$ if $j > 1$; $\Sigma_j = \omega_{b_{0,j}+1}$ or $H_j = A_{r_j}(\mathbb{C})$ and $\Sigma_j = \omega_{u_j - b_{0,j} - 1}$; $\Sigma_a = 0$ if $a \leq e$ and $a \neq j$; $\theta_g(\overline{\omega}) = 0$ for $g > e$; $\Delta_a = 0$ for $1 < a \leq e$, and either $\Delta_1 = 0$, or $j = 1$ and $\Delta_1 \in \Omega$;

2) $p = 3$, $s = 2$, $G = B_r(K)$ or $D_r(K)$, $\overline{\omega} = \omega_r$ for $G = B_r(K)$ and $\overline{\omega} \in \{\omega_{r-1}, \omega_r\}$ for $G = D_r(K)$, $b_{0,1} = 2$, $u_2 \leq 3$, $u_3 \leq 1$, $\Sigma_1 = \omega_5$; $\Delta_1 \in \Omega$ if $u_2 = 1$ and $\Delta_1 = 0$ otherwise; $\theta_2(\overline{\omega}) = \omega_1$ for $u_2 = 2$ or 3.

In the exceptional cases 1) and 2) $d(x, \varphi) = d_0$.

PROOF. By Corollary 11.15, the lemma holds if $d_s \geq p$, so one can suppose that $d_s \leq p$. Lemma 11.7 implies that

$$d_f \geq d_{f,j,\gamma} \tag{11.4}$$

if $\overline{\omega} = \gamma + \delta$ with $\gamma, \delta \in \mathbf{X}^+$. Since $s > 1$, we have

$$b_{0,j} + 1 < [(u_j + 1)/2] < u_j - b_{0,j} - 1. \tag{11.5}$$

First assume that $H_j = B_{r_j}(\mathbb{C})$ and $a_{r_j,j} = 1$. The construction of θ implies that in this case $G = B_r(K)$ with $\overline{a_r} = 1$ or $G = D_r(K)$ with $\overline{a_{r-1}} + \overline{a_r} = 1$. Set $\gamma = \omega_r$ for $G = B_r(K)$ and $\gamma = \overline{a_{r-1}}\omega_{r-1} + \overline{a_r}\omega_r$ for $G = D_r(K)$. We have $b_{0,j} = 2q$ with $q \in \mathbb{Z}^+$ and $q > 0$, and $r_j = (p^s - 1)/2 + q$. Let $0 < f < s$. By Proposition 2.5 b), the element $h_{f,j}$ has $2q$ Jordan blocks of size $p^{s-f} + 1$ and $p^f - 2q$ blocks of size p^{s-f} on the standard H_j-module. Hence both p^{s-f} and $p^{s-f} - 2$ occur in the sequence (1.2) in Algorithm 1.6 for the H_j-conjugacy class of this element $2q$ times and both $p^{s-f} - 1$ and $p^{s-f} - 3$ occur in this sequence $p^f - 2q$ times. Therefore it follows from Proposition 2.5 and Algorithm 1.6 that

$$d_{0,j,\gamma} = 1 + ((p^s - 1)/2 + q)((p^s + 1)/2 + q)/2,$$
$$d_{f,j,\gamma} \geq 1 + 2q(p^{s-f} - 1) + (p^f - 2q)(p^{s-f} - 2) = 1 + p^s - 2p^f + 2q$$

for $0 < f < s$, and $d_{s,j,\gamma} = q + 1 \leq (p+1)/2 \leq p - 1$. One can conclude that $d_{0,j,\gamma} > p^{s+1}$ if $s > 2$ or $p > 7$ and check directly that $d_{0,j,\gamma} > p^s d_{s,j,\gamma}$ for $p^s = 25$ or 49. Obviously, $2p^{2f} \leq (p-1)p^{s+f-1}$. Hence $p^f d_{f,j,\gamma} > p^{s+1} \geq p^s d_s$ for $1 < f < s$. If $s > 2$, we have $pd_{1,j,\gamma} > (p-1)p^s \geq p^s d_{s,j,\gamma}$ since $2p^2 < p^s$. For $s = 2$ one can see that $pd_{1,j,\gamma} > (p-2)p^s$ and therefore $pd_{1,j,\gamma} \geq p^s d_{s,j,\gamma}$ for $p > 3$. Now it follows from Corollary 11.14 and Formula (11.4) that $p^f d_f > p^s d_s$ for $0 \leq f < s$ unless $p = 3$ and $s = 2$. In the former case $d(x, \varphi) = p^s d_s$.

Assume that $p = 3$ and $s = 2$. Then $q = 1$, $d_{0,j,\gamma} = 16$, $d_{1,j,\gamma} = 6$, and $d_{2,j,\gamma} = 2$ by Proposition 2.5 and Algorithm 1.6. If $u_a > 9$ for some $a \neq j$, we get $d_{0,a} \geq 16$, $d_{1,a} > 5$, and $d_{2,a} > 1$ by (11.4). Since $d_2 \leq 3$, one gets $d_{2,a} = 2$ by Lemma 11.5. Hence $3^f(d_{f,j} + d_{f,a} - 1) > 9(d_{2,j} + d_{2,a} - 1)$ for $f = 0$ or 1. Now Lemma 11.13 and Corollary 11.8 imply that $3^f d_f > 9d_2$ for such f which yields that $d(x, \varphi) = p^s d_s$. Suppose that $e = j = 1$. Then $d_2 = d_{2,1}$. Since $d_2 \leq 3$ by our assumptions, Algorithm 1.6 forces that $\Lambda_1 = 0$, $\Sigma_1 = \omega_5$, and $\Delta_1 = 0$ or ω_1. Observe that $d_{0,g} > 1$ if $u_g > 1$. If $\Delta_1 = \omega_1$, one can compute that $d_2 = 3$, $d_{0,1} = 26$, and $d_{1,1} = 9$. Corollary 11.8 implies that in this case $d(x, \varphi) = d_0$ if $u_2 = 1$ and $d(x, \varphi) = 27 = p^{s+1}$ otherwise. Now assume that $\Delta_1 = 0$. The construction of θ yields that $\overline{\omega} = \gamma$. Then $d_2 = d_{2,1} = 2$ and Lemma 11.5 yields that $3d_1 \geq 9d_2$. Hence $d(x, \varphi) = p^s d_s$ for $d_0 \geq 18$ and $d(x, \varphi) = d_0$ otherwise. Set $d_{0,3} = 1$ if $c \leq 2$ and $d_{0,2} = 1$ for $c = 1$. One easily deduces from Algorithm 1.6, Lemma 11.5, and the comments above that $d_0 \geq d_{0,1} + d_{0,2} - 1 \geq 15 + d_{0,2} + d_{0,3} - 1$, $d_{0,g} > 3$ if $u_g > 3$, $d_{0,2} = 2$ if $u_2 = 2$ or 3, and $d_0 \geq 18$ if $u_3 > 1$. Hence $d(x, \varphi) = p^s d_s$ if $u_2 > 3$ or $u_3 > 1$ and $d(x, \varphi) = d_0$ if $u_2 \leq 3$ and $c \leq 2$ or $u_3 = 1$. This completes the analysis of the case where $H_j = B_{r_j}(\mathbb{C})$ and $a_{r_j,j} = 1$.

Now assume that $H_j \neq B_{r_j}(\mathbb{C})$ or $a_{r_j,j} \neq 1$. First suppose that $H_j = B_{r_j}(\mathbb{C})$ and $a_{r_j,j}$ is odd. Then $G = B_r(K)$ and $a_r \neq 0$ or $G = D_r(K)$ and $a_{r_1} + a_r \neq 0$. Set $\gamma = \omega_r$ for $G = B_r(K)$ and $\gamma = \omega_w$ with $w \geq r-1$ and $a_w \neq 0$ for $G = D_r(K)$. Since φ is not p-large, the construction of θ implies that $p > 3$. Actually we have proved before that $p^f d_{f,j,\gamma} > p^s d_s$ for $0 \leq f < s$. Hence Lemma 11.13 and Corollary 11.14 imply that $d(x, \varphi) = p^s d_s$.

Next, suppose that $H_j \neq B_{r_j}(\mathbb{C})$ or $a_{r_j,j}$ is even. Let both Σ_j and its dual $\Sigma_j^* \neq \omega_{b_{0,j}+1}$. Then one can deduce from Formula (2.8), Proposition 1.5, and Algorithm 1.6 that $d_{0,j} > p^s(d_{s,j} - 1) + 2(p^s - 2 - b_{0,j})$ and $d_{f,j} > p^{s-f} d_{s,j}$ for $0 < f < s$. Since $s > 1$ and $b_{0,j} \leq p - 1$, we get that $2(p^s - 2 - b_{0,j}) > p^s$. Hence $p^f d_{f,j} > p^s d_{s,j}$ for $0 \leq f < s$. So Lemma 11.13 and Corollary 11.14 imply that $p^f d_f > p^s d_s$ for such f and therefore $d(x, \varphi) = p^s d_s$.

Now assume that Σ_j or $\Sigma_j^* = \omega_{b_{0,j}+1}$. Since $s > 1$ and $b_{0,j} < p$, Lemma 11.10 forces that $a_r = 0$ if $G = B_r(K)$ and $a_{r-1} + a_r = 0$ if $G = D_r(K)$. The construction of θ implies that there exists i with $\overline{a_i} = 1$ such that $\theta_j(\omega_i) = \Sigma_j + \sigma$ where $\sigma = 0$ or $H_j = A_{r_j}(\mathbb{C})$ and $\sigma = \omega_{r_j - b_{0,j}}$. Set $D_j = d_{s,j,\omega_i} - 1$. Proposition 1.5 and Algorithm 1.6 yield that $d_{0,j,\omega_i} = D_j p^s + p^s - b_{0,j}$ and $d_{f,j,\omega_i} = (D_j + 1)p^{s-f}$ for $0 < f < s$. It follows from Lemma 11.13 and Corollary 11.8 that $d(x, \varphi) = d_0$ or $p^s d_s$. We claim that $d(x, \varphi) = p^s d_s$ if $\Sigma_g \neq 0$ for some $g < e$, $g \neq j$. Indeed, the arguments above enable one to reduce the question to the case where Σ_g or

$\Sigma_g^* = \omega_{b_{0,g}+1}$. Put $D = d_{s,j} + d_{s,g} - 2$. Using Proposition 1.5, Algorithm 1.6, and Lemma 11.7, we conclude that $d_{0,j} + d_{0,g} - 1 \geq Dp^s + 2p^s - b_{0,j} - b_{0,g} - 1 > (D+1)p^s$ since $b_{0,j} + b_{0,g} + 1 < 2p$ and $s > 1$. Here we find i' that plays for Σ_g the same role as i for Σ_j, write $\overline{\omega} = \lambda + \omega_i = \nu + \omega_{i'}$ with $\lambda, \nu \in \mathbf{X}^+$ and use the equalities $d_{f,j} = d_{f,j,\omega_i} + d_{f,j,\lambda} - 1$ and $d_{f,g} = d_{f,g,\omega_{i'}} + d_{f,j,\mu} - 1$ for $f = 0$ and s. Now Corollary 11.14 forces that $d_0 > p^s d_s$ and proves the claim.

Next, suppose that $\Sigma_g = 0$ if $g < e$ and $g \neq j$. Passing to φ^* if necessary, one can assume that $i \leq n/2$ for $G = A_r(K)$ (here (11.5) is essentially used). Then one can deduce from Rule 9 and the definition of the values $\nu(\varepsilon_k^j)$ used to construct θ that $b_{0,e} = b_{0,j}$ and $b_{0,j} < b_{0,j-1}$ if $j > 1$. Now our goal is to show that $d(x, \varphi) = p^s d_s$ if $\theta_g(\overline{\omega}) \neq 0$ for some $g > e$. First assume that $\theta_g(\omega_i) \neq 0$. Then Rules 1–9 yield that there exist a and $k \leq i$ such that $\theta_j(\varepsilon_a) = \varepsilon^j_{b_{0,j}+1}$ and $\theta_g(\varepsilon_k) = \varepsilon_1^g$ and that $\theta_j(\varepsilon_q) \neq \varepsilon^j_{b_{0,j}+2}$ for $q \leq i$. Analyzing the values of the function ν on the elements of $\mathcal{E}(H_j)$ and $\mathcal{E}(H_g)$ and the construction of θ, one can conclude that $u_g > p^s - b_{0,j} - 2$. If $\theta_g(\omega_i) = 0$, there exists $i_1 \neq i$ such that $\theta_g(\omega_{i_1}) \neq 0$ and $a_{i_1} \neq 0$. Then $\Delta_j + \Lambda_j - \theta_j(\omega_{i_1}) \in \mathbf{X}^+(H_j)$. Arguing as in the proof of Lemma 11.19, we show that $u_g > p^s - b_{0,j}$. Hence in all cases $u_g > p^s - b_{0,j} - 2$. Proposition 1.5 and Algorithm 1.6 imply that $d_{0,g} \geq p^s - b_{0,j} - 1$. So $d_{0,j} + d_{0,g} - 1 > (D_j + 1)p^s$ since $p^s > 2b_{0,j} + 2$. Then Corollaries 11.8 and 11.14 imply that $d_0 > p^s d_s$. Hence $d(x, \varphi) = p^s d_s$.

Now assume that $\theta_g(\overline{\omega}) = 0$ for $g > e$. If $\Delta_a = 0$ for all $a \leq e$, one can deduce from Corollary 11.8, Algorithm 1.6, and the facts proved above that $d(x, \varphi) = d_0$. Let $\Delta_a \neq 0$ for some $a \leq e$. Hence $b_{0,a} > 1$. Rules 1–9 and the analysis of the values of ν on the elements of $\mathcal{E}(H_a)$ enable us to suppose that $b_{0,a} = b_{0,1}$. Hence $a < j$ or $a = j = 1$. First let $a < j$. Put $D_a = d_{s,a} - 1$. Algorithm 1.6 and Propositions 1.5 and 2.5 a) yield that $d_{0,a} \geq D_a p^s + b_{0,1}$. Hence $d_{0,a} + d_{0,j} - 1 \geq (D_a + D_j + 1)p^s$ since $b_{0,1} > b_{0,j}$. Now Corollaries 11.8 and 11.14 imply that $d(x, \varphi) = d_0$.

Next, let $a = j = 1$ and $\Delta_g = 0$ for $1 < g \leq e$. Put $\mathcal{B} = 2b_{0,1} - 4$ for $b_{0,1} > 3$ and $\mathcal{B} = 2b_{0,1} - 2$ otherwise. Then Proposition 1.5, Algorithm 1.6, and Formula (2.8) imply that $d_{0,g} - 1 = p^s(d_{s,g} - 1)$ for $1 < g \leq e$ and

$$d_{0,1} \geq D_1 p^s + p^s - b_{0,1} + \mathcal{B} \geq (D_1 + 1)p^s$$

if $\Delta_1 \notin \Omega$. If $\Delta_1 \in \Omega$, we get $d_{0,1} = D_1 p^s + p^s - 1$. Using Lemma 11.5, we conclude that $d(x, \varphi) = p^s d_s$ in the former case and d_0 in the latter one. This completes the proof. □

COROLLARY 11.23. *Let $u_1 = p^s + b_{0,1}$ with $s > 1$ and $0 < b_{0,1} < p$. For $1 \leq j \leq c$ if $|x_j| = p^{s_j+1}$, set $N_j = \min\{p^{s_j+1},\ p^f d_{f,j} \mid 0 \leq f \leq s_j\}$. Then one of the following holds:*

a) $d_s > p$;

b)$N_j = p^{s+1}$ or $p^s d_s$ for some $j \leq e$;

c) $N_g = p^s$ for some g (naturally, here $g > e$);

d) there exist distinct j and g such that $j, g \leq e$ and $N_j + N_g > p^s(d_{s,j} + d_{s,g} - 1)$;

e) there exist $j \leq e$ and $g > e$ with $p^s d_s < N_j + N_g$;

f) $p^s - p < N_g, N_h \leq p^s$ for some distinct g and h with $g, h > e$;

g) $p = 3$, $s = 2$, $e = 1$, $1 \leq u_2, u_3 \leq 3$, $N_1 + N_2 + N_3 - 1 > 9d_{2,1}$;

h) $\overline{\omega}$ satisfies the assumptions of Lemma 11.21;

i) one of the exceptional cases of Lemmas 11.20 and 11.22 occurs.

PROOF. This can be deduced from the proofs of Lemmas 11.20 and 11.22. □

COROLLARY 11.24. *Let* $u_1 = p^s + b_0$ *with* $s > 1$ *and* $0 < b < p$. *Then* $d_\varphi(x) \leq d(x, \varphi)$.

PROOF. Since $d_\varphi(x) \leq |x|$, assume that $d(x, \varphi) < p^{s+1}$. Results 11.17 and 11.20–11.22 reduce the question to the case where either $d(x, \varphi) = p^s d_s$ or $d(x, \varphi) = pd_1$ and the assumptions of Lemma 11.21 hold. In the latter situation Proposition 2.5 and the construction of θ imply that $d_1 = d_{\varphi_C}(z_1)$. By Proposition 1.12, $d_s = d_{\varphi_C}(z_s)$. It remains to apply Proposition 2.5 a) and Corollary 2.40. □

DEFINITION 11.1. Let $G = B_r(K)$ or $D_r(K)$. Set $r' = r$ for $G = B_r(K)$ and $r' = r - 1$ for $G = D_r(K)$. We say that (x, φ) is a special pair if $p^s + p \leq u_1 < p^s + 2p$, $a_r \neq 0$ for $G = B_r(K)$, $a_{r-1} + a_r \neq 0$ for $G = D_r(K)$, and $a_i = 0$ for $p \leq i < r'$.

LEMMA 11.25. *Let* (x, φ) *be a special pair with* $s > 1$. *Assume that* $d_s \leq p$. *Then* $k_2 < p^s + p$ *and hence* u_1 *is proper for* G.

PROOF. Suppose that $k_2 \geq p^s + p$. Since $s > 1$, one can observe that z_s has at least $2p$ Jordan blocks of size 2 and some blocks of size 1 in this case. But then Propositions 1.5 and 1.12 and Algorithm 1.6 force $d_s > p$ yielding a contradiction. □

LEMMA 11.26. *In the assumptions of Lemma* 11.25 *either* $d(x, \varphi) = p^s d_s$, *or* $p = 3$, $\overline{\omega} = \omega_r$ *for* $G = B_r(K)$ *and* $\overline{\omega} \in \{\omega_{r-1}, \omega_r\}$ *for* $G = D_r(K)$, $u_1 = 13$, $u_2 \leq 5$, $u_3 \leq 3$, $c = 2$ *if* $u_2 = 4$, *and* $c \leq 3$ *if* $u_2 = 5$. *In the exceptional cases* $d(x, \varphi) = d_0$ *if* $c \leq 3$ *and* $d(x, \varphi) = pd_1$ *for* $c > 3$.

PROOF. Set $a = \overline{a_r}$, $\gamma = a\omega_r$ for $G = B_r(K)$ and $a = \overline{a_{r-1}} + \overline{a_r}$, $\gamma = \overline{a_{r-1}}\omega_{r-1} + \overline{a_r}\omega_r$ for $G = D_r(K)$. As $d_s \leq p$, Proposition 1.5, Algorithm 1.6, and Lemma 11.25 force $a = 1$. Corollary 11.8 and Lemma 11.13 imply that $d(x, \varphi) = p^s d_s$ if

$$p^f d_{f,1,\gamma} \geq p^s d_{s,1,\gamma} \tag{11.6}$$

for $0 \leq f < s$. So it remains to find out when (11.6) does not hold. By Lemma 11.25, u_1 is proper for G, i.e. u_1 is odd. Hence $u_1 = p^s + p + 2b' + 1$ with $b' \in \mathbb{Z}^+$, $r_1 = (p^s + p)/2 + b'$, and $d_{s,1,\gamma} = (p+3)/2 + b'$. Proposition 2.5 b) implies that $h_{1,1}$ has p Jordan blocks of size $> p^{s-1}$ and at least one of them has size $> p^{s-1} + 1$; for $1 < f < s$ both p^{s-f} and $p^{s-f} - 2$ occur at least $p + 1$ times in the sequence (1.2) of Algorithm 1.6 constructed for the element $h_{f,1}$. Now Algorithm 1.6 yields that $d_{0,1,\gamma} = 1 + ((p^s + p)/2 + b')((p^s + p + 2)/2 + b')/2$; if $p^{s-1} > 5$, one gets $d_{1,1,\gamma} > 1 + p(2p^{s-1} - 3) > p^s$, and $d_{f,1,\gamma} > (p + 1)(p^{s-f} - 1) \geq p^{s-f+1}$ for $1 < f < s$. Therefore $d_{0,1,\gamma} > p^{s+1} \geq p^s d_s$ if $p^{s-1} > 8$, $pd_{1,1,\gamma} > p^{s+1}$ if $p^{s-1} > 6$, and in all cases $p^f d_{f,1,\gamma} \geq p^{s+1}$ for $1 < f < s$. Hence (11.6) holds if $s > 2$ or $p > 7$. Observe that $b' = 0$ for $p = 3$, $b' < 2$ for $p = 5$, and $b' < 3$ for $p = 7$. Taking this into account, one can directly verify that $d_{0,1,\gamma} > p^2 d_{2,1,\gamma}$ for $s = 2$ and $p = 5$ or 7 and that $d_{1,1,\gamma} \geq 25$ if $s = 2$ and $p = 5$. So (11.6) holds unless $s = 2$ and $p = 3$.

Now assume that $s = 2$ and $p = 3$. Then $d_{0,1,\gamma} = 22$, $d_{1,1,\gamma} = 8$, and $d_{2,1,\gamma} = 3$ by Algorithm 1.6. Corollary 11.8 yields that $u_2 \leq 9$ and $\theta_1(\overline{\omega} - \gamma) = 0$ since otherwise we would get $d_{2,2,\gamma} > 1$ or $d_{2,1} > p$ (by Proposition 1.5) and in both cases $d_2 > p$ which contradicts the assumptions of the lemma. Now the construction of θ implies that $\theta_j(\overline{\omega} - \gamma) = 0$ for $j > 1$. Hence $\overline{\omega} = \gamma$. Using Algorithm 1.6 and Lemma 11.5, one can show the following: $d_{0,2} \geq 7$ and $d_{1,2} \geq 3$ if $u_2 \geq 6$; $d_{0,2} = 4$

and $d_{1,2} = 2$ for $u_2 = 5$; $d_{0,2} = 5$ and $d_{1,2} = 2$ for $u_2 = 4$; $d_1 = d_{1,1} = 8$ if $u_2 \leq 3$; $d_1 \geq 9$ if $u_2 > 3$; $d_0 \geq 27$ if $u_3 > 3$, or $u_2 \geq 6$, or $u_2 = 5$ and $u_4 > 1$, or $u_2 = 4$ and $u_3 > 1$; $d_0 \geq 24$ if $u_3 > 1$; $d_0 \leq 26$ if $u_2 = 5$, $u_3 \leq 3$, and $c \leq 3$ or $u_2 = 4$ and $c = 2$; and $d_0 < 24$ if $c \leq 2$ and $u_2 \leq 3$. This completes the proof. □

COROLLARY 11.27. *Let (x, φ) be a special pair for G. For $1 \leq j \leq c$ define N_j as in Corollary 11.23. Then one of the following holds:*
a) $d_s > p$;
b) $e = 1$ and $N_1 = p^s d_{s,1} = p^s d_s$;
c) $G = B_r(K)$ or $D_r(K)$, $s = 2$, $p = 3$, $\overline{\omega} = \omega_r$ for $G = B_r(K)$ and $\overline{\omega} \in \{\omega_{r-1}, \omega_r\}$ for $G = D_r(K)$, $u_2 \leq 9$, and either $\sum_{j=1}^{c} N_j - c + 1 \geq 27$, or the exceptional case of Lemma 11.26 occurs.

PROOF. This follows immediately from the proof of Lemma 11.26. □

COROLLARY 11.28. *If (x, φ) is a special pair, then $d_\varphi(x) \leq d(x, \varphi)$.*

PROOF. One easily observes that in those exceptional cases of Lemma 11.26 where $d(x, \varphi) = pd_1$, we get $d_1 = d_{\varphi_{\mathbb{C}}}(z_1)$. Now argue as in the proof of Corollary 11.24 applying Propositions 1.12 and 2.5 a), Lemmas 11.25 and 11.26, and Corollaries 2.40, 11.17, and 11.15. □

COROLLARY 11.29. *One has $d_\varphi(x) \leq d(x, \varphi)$.*

PROOF. This follows immediately from Lemma 11.12 and Corollaries 11.18, 11.24, and 11.28. □

Now we start another part of the proof of Theorem 1.10 concerned with showing that $d_\varphi(x) \geq d(x, \varphi)$ for $\varphi \in \mathrm{Irr}_p$ and using properties of tensor products.

COROLLARY 11.30. *If $d_\varphi(x) \geq p^f d_f$ for some f with $0 \leq f \leq s$ or $d_\varphi(x) = |x|$, then $d_\varphi(x) = d(x, \varphi)$.*

PROOF. This follows directly from Corollary 11.29 since $p^f d_f$ and $|x| \geq d(x, \varphi)$ by definition. □

LEMMA 11.31. *Let $\omega = \sum_{a=0}^{g} p^a \mu_a$ where $\mu_a \in \mathbf{X}^+$ and are p-restricted. Then $d_f = 1 + \sum_{a=0}^{g} (d_{f,\mu_a} - 1)$ for $0 \leq f \leq s$.*

PROOF. Recall that $\overline{\omega} = \sum_{a=0}^{g} \mu_a$ and apply Corollary 11.8. □

The following proposition yields Theorem 1.10 in the exceptional cases of Lemma 11.4.

PROPOSITION 11.32. *Let $G = C_r(K)$, $\varphi \in \mathrm{Irr}_p$, and $\sum_{i=1}^{r} a_i < p \leq a_1 + 2\sum_{i=2}^{r} a_i$. Then either $d_\varphi(x) = p^{s+1}$, or $d_\varphi(x) = p^s d_s = p^s d_{s,1}$, or χ_1 is p-restricted, χ_j is trivial for each $j > 1$, and $d_\varphi(x) = d_0 = d_{0,1}$. Hence Theorem 1.10 holds for φ.*

PROOF. We distinguish 4 subcases concerned with different possibilities for the integers k_i.

1. Assume that $k_1 = p^s + 1$ and $k_2 \leq p^s$. Hence u_1 is proper for G. Then χ_1 is p-restricted since φ is not p-large. Hence $\psi_1 = (\chi_1)_{\mathbb{C}}$. Obviously, $\sum_{i=2}^{r} a_i > 0$. Set $\omega^1 = \theta_1(\omega)$ and write $\omega^1 = \sum_{i=1}^{r_1} a_{i1}\omega_i$. The construction of θ yields that $\sum_{i=1}^{r_1} a_{i1} = \sum_{i=1}^{r} a_i$. We say that ω^1 is of type I if $\omega^1 = a_{11}\omega_1$, of type II if

$\omega^1 = a_{11}\omega_1 + \omega_2$, and of type III in all other cases. Proposition 2.5 b) implies that z_s is a transvection and $h_{f,1}$ has one block of size $p^{s-f}+1$ and $p^f - 1$ blocks of size p^{s-f} for $0 < f \leq s$. Using Algorithm 1.6 and Propositions 1.5 and 1.12, one can make the following conclusions: $d_s = d_{s,1} = d_{\varphi_C}(z_s) = 1 + \sum_{i=1}^{r} a_i$; if ω^1 is of type II or III, then $p^f d_{f,1} \geq p^s d_s$ for $0 < f < s$; if ω^1 is of type II, then $d_{0,1} = d_s p^s - 1$; and $d_{0,1} \geq p^s d_s$ for ω^1 of type III. Now our goal is to prove that in all situations $d_\chi(x) \geq p^f d_f$ for some f. Then Corollary 11.30 enables one to complete the proof in Case 1 as $d_\varphi(x) \geq d_\chi(x)$.

First suppose that ω^1 is of type I. Since $\omega \neq a_1\omega_1$, Rules 1–9 yield that $u_2 = p^s$ and χ_2 is nontrivial. One easily deduces from Proposition 1.5 and Algorithm 1.6 that $d_{f,1} = (d_s - 1)p^{s-f} + 1$ for $0 \leq f < s$. Hence Conjecture (r,s) implies that $d_{\chi_1}(x_1) = 1 + (d_s - 1)p^s$ and $d_{\chi_2}(x_2) = p^s$. Notice that we do not analyze whether χ_2 is p-restricted. By Corollary 2.14, $d_\chi(x) \geq p^s d_s$.

Next, assume that ω^1 is of type II. Let $\omega = a_1\omega_1 + \omega_2$ or $k_2 = 1$. Then $\theta_j(\omega) = 0$ for $j > 1$ if θ_j is determined. Hence $d_f = d_{f,1}$ for $0 \leq f \leq s$. Conjecture (r,s) and the inequalities for $p^f d_{f,1}$ above yield that $d_\chi(x) = d_{\chi_1}(x_1) = d_0$.

Now suppose that $\omega \neq a_1\omega_1 + \omega_2$ and $k_2 > 1$. Then $a_i \neq 0$ for some $i > 2$ and $\mathbf{j}(\varepsilon_3) > 1$. We can deduce from Rules 1–9 that χ_2 is nontrivial. Hence $d_{\chi_2}(x_2) \geq 2$. As above, $d_{\chi_1}(x_1) = p^s d_s - 1$ and hence Lemma 2.10 and Corollary 2.14 yield that $d_\chi(x) \geq p^s d_s$.

Finally, assume that ω^1 is of type III. It is clear that the integers $d_{f,1}$, $0 \leq f \leq s$, play the same role for χ_1 and x_1 as d_f for φ and x. Since now $p^f d_{f,1} \geq p^s d_s$ for $f < s$, using Conjecture (r,s), we conclude that $d_\chi(x) \geq d_{\chi_1}(x_1) = p^s d_s$. This completes the analysis of Case 1.

Now let $k_1 > p^s + 1$ or $k_2 > p^s$. Then x^{p^s} has at least 2 blocks of size > 1 by Proposition 2.5 b). Thus it follows from Proposition 1.5 and Algorithm 1.6 that $d_s - 1 \geq a_1 + 2\sum_{i=2}^{r} a_i \geq p$ and hence $d(x,\varphi) = p^{s+1}$ by Corollary 11.15. So we shall show that $d_\varphi(x) = p^{s+1}$.

2. Suppose that $k_1 = k_2 = p^s + 1$. Define ω^1 as in Case 1 and set $\omega^2 = \theta_2(\omega)$. Write $\omega^2 = \sum_{i=1}^{r_2} a_{i2}\omega_i$. Both k_1 and k_2 are proper for G. Set $a = \sum_{i=1}^{r} a_i$ and $a' = \sum_{i=2}^{r} a_i$. Rules 1–9 yield that χ_1 and χ_2 are p-restricted, $\sum_{i=1}^{r_1} a_{i1} = a$ and $\sum_{i=1}^{r_2} a_{i2} = a'$. Using Proposition 1.5, Algorithm 1.6 and Conjecture (r,s) and arguing as in Case 1, we deduce that $d_{\chi_1}(x_1) > ap^s$ and $d_{\chi_2}(x_2) > a'p^s$. Since $a + a' \geq p$, Theorem 2.9 forces that $d_\varphi(x) = d_\chi(x) = p^{s+1}$.

3. Assume that $k_1 > p^s + 1$ if k_1 is even and $k_1 < r$ if k_1 is odd. Corollary 2.28 implies that $\overline{\mathrm{cl}(x)}$ contains an element x' with $J(x') = (k_1, 1, \ldots, 1)$ if k_1 is even and $J(x') = (k_1, k_1, 1, \ldots, 1)$ if $k_1 = k_2$ is odd. Set $G_1 = G(2, \ldots, r)$. Our assumptions yield that we can choose $x' \in G_1$. Now we shall construct a composition factor μ of $\varphi|G_1$ with $d_\mu(x') = p^{s+1}$. Then $d_\varphi(x') = p^{s+1}$ and $d_\varphi(x) = p^{s+1}$ by Lemma 2.19. It is clear that $r > 2$. Set $u = v$ if $a_1 = a_2 = 0$, $u = X_{-1,a_1}v$ if $a_1 \neq 0$ and $a_2 = 0$, and $u = v(1,2,a_2)$ if $a_2 \neq 0$. By Lemma 2.46, $u \neq 0$ and is fixed by $\mathcal{X}_i$ for $i > 1$. Hence u generates an indecomposable KG_1-module M' with highest weight $\omega' = \omega_{G_1}(u)$. Observe that M' has a composition factor $\mu = \varphi(\omega')$. We have $\omega' = a_1\omega_1 + (a_2 + a_3)\omega_2 + \omega_+(3, r-1, 1)$. For $0 \leq f \leq s$ let d'_f be the integer constructed for x' and μ in the same manner as d_f has been defined for x and φ. Arguing as for d_s, one can deduce that $d'_s > p$. Then Lemma 11.12 forces that $d(x', \mu) = p^{s+1}$. It is clear that μ is not an exceptional representation mentioned

in Theorem 1.10. Hence $d_\mu(x') = p^{s+1}$ by Conjecture (r, s). This completes the analysis of Case 3.

4. Now let $k_1 = r$ be odd. Set $G_A = G(1, 2, \ldots, r-1)$. One can assume that $x \in G_A$. We shall indicate a composition factor λ of $\varphi|G_A$ such that $d_\lambda(x) = p^{s+1}$. Then $d_\varphi(x) = p^{s+1}$.

To find λ, we construct a nonzero vector $w \in M$ such that w is fixed by $U^+(G_A)$ and generates an indecomposable G_A-module with relevant highest weight. The construction of w depends upon the coefficients a_i. Observe that $r > 2$.

a) Assume that $a_{r-1}+2a_r \geq p$. Then there exists $d \leq a_r$ such that $a_{r-1}+2d = p$ or $p+1$. Set $w = X_{-(r-1)}X_{-r,d}v$ in the first case and $w = X_{-(r-1),2}X_{-r,d}v$ in the second one. Using Lemma 2.1, one can conclude that $X_{r-1}w = 0$ and $X_{r-1}X_r w \neq 0$. Hence $w \neq 0$ and $\mathcal{X}_{r-1}$ fixes w. It is clear that the groups $\mathcal{X}_i$ with $i < r-1$ fix w. Hence w is fixed by $U^+(G_A)$.

b) Now let $a_{r-1} + 2a_r \leq p-1$. If $(\sum_{i=1}^{r-1} a_i) + 2a_r \geq p$, put $w = X_{-r,a_r}v$. Otherwise there exist k and h such that $1 < k < r$, $0 < h \leq a_k$, and $h + (\sum_{i=1}^{k} a_i) + 2\sum_{i=k+1}^{r} a_i + h = p$. In this case set $w = v(r,k,h)$. By Lemmas 2.46 and 2.1, $w \neq 0$ and is fixed by $U^+(G_A)$.

Set $\delta = \omega_{G_A}(w)$ and $\lambda = \varphi(\delta)$. Observe that w generates an indecomposable KG_A-module with highest weight δ and hence $\varphi|G_A$ has a composition factor isomorphic to λ. Since φ is not p-large and $p > 2$, one can conclude that $a_{r-2} < p-2$ in Case a). Taking this into account and computing δ, we deduce that either $\delta = 2\omega_{r-2} + (p-3)\omega_{r-1}$ or $\omega_{r-2} + (p-2)\omega_{r-1}$, or λ is p-large. In the latter case Conjecture (r, s) and Theorem 1.1 imply that $d_\lambda(x) = |x|$. For the exceptional weights δ denote by $d_{f,A}$, $0 \leq f \leq s$, the integers determined for x and λ in the same way as d_f for x and φ. Using Propositions 1.5 and 2.5, Algorithm 1.6, and Corollary 11.15 and arguing as before Case 2, we conclude that $d_{f,A} > p^{s+1}$ for these δ and $0 \leq f \leq s$. Now Conjecture (r, s) yields that $d_\lambda(x) = p^{s+1}$ and completes the proof. □

LEMMA 11.33. *Let $G = D_{2p}(K)$ and $\omega = \omega_{r-1}$ or ω_r. Assume that $k_1 = k_2 = 2p$. Then $d_\varphi(x) = p^2$ if $(\mathrm{cl}(x), \omega) = (C_1, \omega_{r-1})$ or (C_2, ω_r) and $d_\varphi(x) = p^2 - p + 2$ if $(\mathrm{cl}(x), \omega) = (C_2, \omega_{r-1})$ or (C_1, ω_r).*

PROOF. Say that the pair (x, φ) is of type I if $(\mathrm{cl}(x), \omega) = (C_1, \omega_{r-1})$ or (C_2, ω_r) and of type II otherwise. We have $S \cong A_{2p-1}(K)$. As before, we assume that the weights ω_0 and $\omega_{2p} \in \mathbf{X}(S)$ are zero. Set $I_1 = \{p + a \mid -p \leq a \leq p,\ a \text{ is odd}\}$ and denote by I_2 the similar set with even a. Observe that I_1 and I_2 are equal to the sets of even and odd nonnegative integers i, respectively, with $0 \leq i \leq 2p$. For such i set $\rho_i = \varphi_i^{2p} \in \mathrm{Irr}\, S$. Now one can deduce from Lemma 11.11 that $\varphi|S \cong \bigoplus_{i \in I_1} \rho_i$ if (x, φ) is of type I and $\varphi|S \cong \bigoplus_{i \in I_2} \rho_i$ if (x, φ) is of type II. In the first case $\varphi|S$ has a composition factor $\sigma_1 \cong \rho_{p-1}$ and in the second one a factor $\sigma_2 \cong \rho_p$. Until the end of the proof $d_{0,i}(S)$ and $d_{1,i}(S)$ denote the integers determined for the element $x \in S$ and the representation $\rho_i \in \mathrm{Irr}\, S$ in the same way as d_0 and d_1 were determined for x and φ. Algorithm 1.6 and Proposition 2.5 imply that $d_{0,p-1}(S) = p^2$, $d_{1,p-1}(S) = p$, and $d_{1,i}(S) \leq p-1$ for $i \leq p-2$ or $i \geq p+2$. Then Conjecture (r, s) and Corollary 7.3 force that $d_{\sigma_1}(x) = p^2$, $d_{\sigma_2}(x) = p^2 - p + 2$, and $d_{\rho_i}(x) \leq p^2 - p$ for $i \leq p-2$ or $i \geq p+2$. This yields the lemma. □

LEMMA 11.34. *Let $u_1 = 2p$, $G = B_r(K)$ or $r > 2p$ and $G = D_r(K)$. Assume that $\omega = \omega_r$ for $G = B_r(K)$ and $\omega \in \{\omega_{r-1}, \omega_r\}$ if $G = D_r(K)$. Then $d_\varphi(x) = p^2$.*

PROOF. We have $k_2 = k_1$ and $S_1 = A_{2p-1}(K)$. Let $\xi_1 = \varphi(\omega_{p-1}) \in \operatorname{Irr} S_1$. By Lemma 11.11, $\varphi|S$ has an irreducible component of the form $\xi_1 \otimes \xi_2$ where $\xi_2 = \bigotimes_{j=2}^{c} \delta_j$ with $\delta_j \in \operatorname{Irr} S_j$. The arguments in the proof of Lemma 11.33 yield that $d_{\xi_1}(x_1) = p^2$. Hence $d_\varphi(x) = p^2$ as desired. □

COROLLARY 11.35. *Let $G = B_r(K)$ or $D_r(K)$ and $u_1 = 2p$. Assume that $\varphi = \bigotimes_{g=0}^{q} \eta_g \operatorname{Fr}^g$ with $\eta_g \in \operatorname{Irr}_p$ and for some j we have $\omega(\eta_j) = \omega_a$ with $a = r$ for $G = B_r(K)$ and $a = r-1$ or r for $G = D_r(K)$. Suppose also that $\omega(\eta_l) \neq 0$ for some $l \neq j$. Then $d_\varphi(x) = p^2$.*

PROOF. Lemmas 11.33 and 11.34 imply that $d_{\eta_j}(x) \geq p^2 - p + 2$. Set $\lambda = \bigotimes_{g \neq j} \eta_g \operatorname{Fr}^g$. Then $d_\lambda(y) > 1$ as λ is a nontrivial representation of G. Proposition 2.5 a) forces that $d_\lambda(x) > p$. Then $d_\varphi(x) = p^2$ by Theorem 2.9. □

Proof of Theorem 1.9. Let φ and x satisfy the assumptions of Theorem 1.9. By Lemma 11.2, it suffices to consider the case where $\varphi \in \operatorname{Irr}_p$. Corollary 7.3 and results of Section 8 settle the case where $k_1 = n$, so assume that $k_1 < n$ and hence x has at least 2 Jordan blocks. Thus in the framework of this proof we have to consider one exceptional case where $s > 1$, $k_1 = p^s + p$, $0 < k_2 \leq p^s - p$, and ω or $\omega(\varphi^*) = \omega_p$, and to show that $d_\varphi(x) = |x|$ in all other cases under consideration.

1. First assume that $k_1 = p^s + p$ is proper for G, and ω or $\omega(\varphi^*) = \omega_p$. Then $G = A_r(K)$ or $C_r(K)$.

Suppose that $G = A_r(K)$ and $\omega = \omega_p$. Set $G_1 = G(1, 2, \ldots, k_1 - 1)$, $G_2 = G(k_1 + 1, \ldots, r)$ if $k_1 < r$ and $G_2 = 1$ otherwise, and $G_x = G_1 G_2$. One can assume that $x \in G_x$. Then x can be uniquely represented in the form $x = t_1 t_2$ with $t_i \in G_i$, $i = 1, 2$, and t_1 is a regular unipotent element in G_1. For $0 \leq i \leq k_1$ and $0 \leq j \leq n - k_1$ put $\varphi_i^1 = \varphi(\omega_i) \in \operatorname{Irr} G_1$, $\varphi_j^2 = \varphi(\omega_j) \in \operatorname{Irr} G_2$, and $\varphi_{ij} = \varphi_i^1 \otimes \varphi_j^2 \in \operatorname{Irr} G_x$. (We assume that φ_0^2 and φ_1^2 are trivial representations for $G_2 = 1$.) By Proposition 2.36,

$$\varphi|G_x = \bigoplus_{i+j=p,\ j \leq n-k_1} \varphi_{ij}.$$

Set $\sigma_1 = \varphi_{p-1}^1$, $\sigma_2 = \varphi_p^1$, $\mu = \varphi_1^2$, $\delta_1 = \varphi_{p-1,1}$, and $\delta_2 = \varphi_{p0}$. Then δ_1 and δ_2 are composition factors of $\varphi|G_x$. Denote by d_f^* the analogs of the integers d_f $(0 \leq f \leq s)$ for the group G_1, the element t_1, and the representation σ_1. Algorithm 1.6 and Conjecture (r, s) imply that $d_0^* = (p-1)p^s + p$, $d_f^* = (p-1)p^{s-f} + 1$ for $0 < f \leq s$ and hence $d_{\sigma_1}(t_1) = (p-1)p^s + p$. In particular, $d_{\sigma_1}(t_1) = p^{s+1}$ if $s = 1$. By Conjecture (r, s), $d_{\sigma_2}(t_1) = p^{s+1} - p + 2$.

Let $k_2 \leq p^s - p$ (this is possible only for $s > 1$). By Proposition 2.5 b), y has p Jordan blocks of size 2 and $p^s - p$ blocks of size 1. Then $d_{\varphi_{ij}}(y) \leq p-1$ and $d_{\varphi_{ij}}(x) \leq (p-1)p^s$ for $i < p-1$ by Algorithm 1.6, Theorem 1.14, and Proposition 2.5 a). Observe that $y = t_1^{p^{p^s}}$. Proposition 2.5 a), Lemma 2.10, and Corollary 2.13 yield that $d_{\sigma_1}(t_1^p) = (p-1)p^{s-1} + 1$, $d_\mu(t_2^p) \leq p^{s-1} - 1$, $d_{\delta_1}(x^p) \leq p^s - 1$, and $d_{\delta_1}(x) \leq p^{s+1} - p$. Hence $d_\varphi(x) = d_{\delta_2}(x) = p^{s+1} - p + 2$.

If $k_2 > p^s - p$, Theorem 2.9 forces that $d_\varphi(x) = d_{\delta_1}(x) = p^{s+1}$.

If $\omega = \omega_{n-p}$, pass to the dual representation to find $d_\varphi(x)$. This completes the analysis of our case for $G = A_r(K)$.

Next, assume that $G = C_r(K)$. Then $r_1 = k_1/2$. Denote by R_1 the subset of R that consists of linear combinations of ε_i with $i \leq r_1$ and by R_2 the subset consisting of such combinations for $i > r_1$. Set $G_j = \langle \mathcal{X}_\alpha \mid \alpha \in R_j \rangle$ for $j = 1, 2$

and $G_x = G_1G_2$. As in Item 1, one can suppose that $x \in G_x$. Observe that $G_1 \cong C_{r_1}(K)$ and $G_2 \cong C_{r-r_1}(K)$. Define the elements $t_j \in G_j$ $(j = 1, 2)$ and the representations $\sigma_1, \sigma_2 \in \operatorname{Irr} G_1$ and $\delta_1, \delta_2 \in \operatorname{Irr} G_x$ as for $G = A_r(K)$. One easily observes that the G_x-module generated by v has a composition factor isomorphic to δ_2. Since $p \leq r_1 < r$, the weight $\kappa = \varepsilon_1 + \ldots + \varepsilon_{p-1} + \varepsilon_{r_1+1}$ lies in the same W-orbit with ω and hence is in $\mathbf{X}(\varphi)$. Analyzing the action of root elements on weight subspaces of M, we can conclude that a nonzero vector of weight κ is fixed by $U^+(G_x)$ and generates an indecomposable G_x-module with the head admitting δ_1. Hence $d_\varphi(x) \geq \max\{d_{\delta_1}(x), d_{\delta_2}(x)\}$. Using Conjecture (r, s) and arguing as for $G = A_r(K)$, we obtain that $d_{\sigma_1}(t_1) = (p-1)p^s + p$ and $d_{\sigma_2}(t_1) = d_{\delta_2}(x) = p^{s+1} - p + 2$. If $k_2 > p^s - p$, applying Theorem 2.9 as earlier in this proof, one gets $d_\varphi(x) = d_{\delta_1}(x) = p^{s+1}$.

Now suppose that $k_2 \leq p^s - p$. The group G can be naturally mapped into $G^+ = A_{n-1}(K)$. Let $\varphi^+ = \varphi(\omega_p) \in \operatorname{Irr} G^+$. It is well known that φ is a composition factor of $\varphi^+|G$. Applying Conjecture (r, s) to φ^+, one gets $d_{\varphi^+}(x) = p^{s+1} - p + 2 = d_{\delta_2}(x)$. This forces $d_\varphi(x) = p^{s+1} - p + 2$ and completes the analysis of Case 1.

2. Now let $G \neq A_r(K)$, $u_1 = r$ is improper for G, and $c = 1$. By the assumptions of the theorem, if $u_1 < p^s + 2p$, then $a_i \neq 0$ for some i with $p \leq i < r$ if $G = B_r(K)$ and $p \leq i < r - 1$ if $G = D_r(K)$. We have $\chi = \chi_1$. Proposition 11.32 enables one to assume that χ is p-restricted (recall that by Lemma 11.12, $d(x, \varphi) = p^{s+1}$ in the assumptions of Theorem 1.9). Write $\theta(\omega) = \sum_{i=1}^{r-1} l_i\omega_i$. According to Conjecture (r, s), Theorem 1.9 is valid for S. The construction of θ yields that one of the following holds:

a) $l_j + l_{r-j} > 1$ for some j with $(p + 1)/2 \leq j \leq (r - 2)/2$ if $G = B_r(K)$ or $D_r(K)$ and $(p + 1)/2 \leq j \leq (r - 1)/2$ if $G = C_r(K)$;

b) $l_{j+1} + l_{r-j} > 1$ for some j with $(p - 1)/2 \leq j \leq (r - 2)/2$ if $G = B_r(K)$, $(p-1)/2 \leq j \leq (r-1)/2$ if $G = C_r(K)$, and $(p-1)/2 \leq j \leq (r-4)/2$ if $G = D_r(K)$;

c) $G = B_r(K)$ or $D_r(K)$, $l_{r/2}$ or $l_{r/2+1} \neq 0$, and $r > p^s + 2p$.

Using Propositions 1.5 and 2.5, Algorithm 1.6, and Corollary 11.15, one can show that $p^f d_f > p^{s+1}$ for $0 \leq f \leq s$. (Actually, we have to check this only if χ does not satisfy the assumptions of Theorem 1.9). Our assumptions yield that χ is not an exceptional representation from Theorem 1.10 for S. Then Conjecture (r, s) forces that $d_\varphi(x) = d_\chi(x) = p^{s+1}$.

3. Next, let $G = D_r(K)$, $k_1 = 2r - 1$, and $k_2 = 1$. Then $S \cong B_{r-1}(K)$ and $\chi = \chi_1$. The arguments in the proof of Proposition 11.3 yield that χ satisfies the assumptions of Theorem 1.9. To complete the proof in this case, use Conjecture (r, s).

4. Finally, it remains to consider the general case under the following assumptions: a) $k_1 < n$ for $G = A_r(K)$ and $k_1 < n-1$ for other groups; b) if k_1 is improper for G, then $2k_1 < n - 1$; c) if $k_1 = p^s + p$, then $\omega \neq \omega_p$ and for $G = A_r(K)$ we have $\omega \neq \omega_{n-p}$ as well. Set $G_1 = G(2, 3, \ldots, r)$. Corollary 2.28 implies that $\overline{\operatorname{cl}(x)}$ contains an element $g \in G_1$ with $J(g) = (k_1, 1, \ldots, 1)$ if k_1 is proper for G and $J(g) = (k_1, k_1, 1, \ldots, 1)$ if k_1 is improper (there may be no blocks of size 1 in $J(g)$). We shall indicate a composition factor η of $\varphi|G_1$ such that $d_\eta(g) = p^{s+1}$. Then $d_\varphi(x) = p^{s+1}$ by Lemma 2.19. Passing to φ^* if necessary, one can assume that $a_i \neq 0$ for some i with $p \leq i \leq n/2$ if $G = A_r(K)$. Fix minimal i with $a_i \neq 0$ and $i \geq p$. Recall that by our assumptions, $i < r - 1$ if $G = D_r(K)$ and $k_1 < p^s + 2p$ and $i < r$ for $G = B_r(K)$ and such k_1. Construct a vector w as

follows. If $G = C_{p+1}(K)$, $k_1 = 2p$, and $\omega = \omega_{p+1}$, put $w = X_{-1} \dots X_{-p}X_{-(p+1)}v$. If $i > p$ with $(i, k_1) \neq (p+1, p^s + p)$ or $i = p$ and $G = B_{p+1}(K)$ or $D_{p+2}(K)$, set $w = X_{-1,a_1}v$. In all other situations put $w = v(1, i, a_i)$. Set $\mu = \omega_{G_1}(w)$. We claim that $w \neq 0$ and is fixed by $\mathcal{X}_t$ for $t > 1$. Indeed, observe that in the first case $\omega(w) + \alpha_t \notin \mathbf{X}(\varphi)$ for $t > 1$ and apply Lemmas 2.1 and 2.46. Hence w generates an indecomposable G_1-module with highest weight μ and therefore $\eta = \varphi(\mu)$ is a composition factor of $\varphi|G_1$. First assume that either $G = C_{p+1}(K)$, $k_1 = 2p$, and $\omega = \omega_{p+1}$, or $i = p$ and $G = B_{p+1}(K)$ or $D_{p+2}(K)$. In the second case $k_1 = 2p + 1$ since $k_1 \leq n$ and $J(x)$ cannot contain two blocks of size $\geq 2p$. Define the integers d'_0 and d'_1 for g and η in the same way as d_0 and d_1 were determined for x and φ. Proposition 1.5 and Algorithm 1.6 force that d'_0 and $pd'_1 \geq p^2$. In all other situations η satisfies the assumptions of Theorem 1.9 and is not an exceptional representation for that theorem. Here we take into account that φ is not p-large and that $k_1 \neq 2p$ if $G = B_{p+2}(K)$ or $D_{p+3}(K)$. Now Conjecture (r, s) implies that in all situations under consideration $d_\eta(x) = p^{s+1}$ and completes the proof. □

Proof of Theorem 1.10. Now we can complete the proof of Theorem 1.10. We assume that Conjecture (r, s) holds and that Theorem 1.9 is valid for G and x. As before, the symbol $d(z)$ denotes the degree of the minimal polynomial for a tensor product z of a fixed sequence of unipotent Jordan blocks.

Let $\varphi \in \mathrm{Irr}$. By Theorem 2.2, $\varphi \cong \bigotimes_{g=0}^{q} \varphi_g \,\mathrm{Fr}^g$ where $\varphi_g \in \mathrm{Irr}_p$. Corollaries 4.11 and 10.4 and Lemma 11.1 enable one to assume that φ is not p-large and that φ is tensor decomposable if $G \neq D_r(K)$ and x is a regular unipotent element. Using Theorem 1.9 and Lemma 11.33, we can exclude all representations satisfying the assumptions of that theorem and all exceptional cases in Theorem 1.10. Set

$$\mu_g = \omega(\varphi_g), \quad d(f, j, g) = d_{f,j,\mu_g}, \quad \delta(f, j, g) = d(f, j, g) - 1$$

for $0 \leq f \leq s$, $1 \leq j \leq c$, $0 \leq g \leq q$. Let $\chi^g \in \mathrm{Irr}\, S$ and $\omega(\chi^g) = \theta(\mu_g)$. We have $\chi^g \cong \bigotimes_{j=1}^{c} \chi_j^g$ where $\chi_j^g \in \mathrm{Irr}\, S_j$. Recall that $S = G$ and $\chi^g = \varphi_g$ if $G \neq D_r(K)$ and x is a regular unipotent element. By Corollary 2.30, χ^g is a composition factor of $\varphi_g|S$. Set $m_{jg} = d_{\chi_j^g}(x_j)$ and $z = \bigotimes_{g=0}^{k} \bigotimes_{j=1}^{c} J_{m_{jg}}$. Hence Lemma 2.10 and Corollary 11.29 yield that it suffices to show that $d(z) = d(x, \varphi)$. Actually we have to consider the tensor product of relevant blocks with $m_{jg} > 1$. Lemma 11.4, Proposition 11.32, and Theorem 2.9 yield that at least one of the following holds:

a) all $\chi_j^g \in \mathrm{Irr}_p\, S_j$;

b) $d_\varphi(x) = p^{s+1}$;

c) $k_1 = p^s + 1$, $k_2 \leq p^s$, $\chi_1^g \in \mathrm{Irr}_p\, S_1$ for all g, and $d_{\varphi_g}(x) = p^s d(s, 1, g)$ for some g.

Naturally, Conditions a), b), and c) do not exclude each other. Corollary 11.30 proves the theorem in the situation where $d_\varphi(x) = p^{s+1}$. So in what follows we assume that a) or c) holds. Hence $\chi_j^g \in \mathrm{Irr}_p\, S_j$ for all g if $|x_j| = p^{s+1}$.

Observe that Theorem 1.10 holds for x_j and χ_j^g. If x is a regular unipotent element and $G \neq D_r(K)$, this follows from Corollary 10.4. In other cases apply Conjecture (r, s). Furthermore, Lemmas 11.2, 11.33, and 11.34 and Corollary 11.35 permit to assume that for $k_1 = 2p$ all weights $\mu_g \neq \omega_r$ if $G = B_r(K)$ and all $\mu_g \notin \{\omega_{r-1}, \omega_r\}$ if $G = D_r(K)$. Now the construction of θ and Lemma 11.10 imply that χ_j^g cannot be one of the exceptional representations of Theorem 1.10 for S_j if

$|x_j| = p^{s+1}$. By Corollaries 11.8 and 11.14,

$$d_f = 1 + \sum_{g=0}^{q} \sum_{j=1}^{c} \delta(f,j,g), \quad p^f \delta(f,j,g) \geq p^s \delta(s,j,g) \tag{11.7}$$

for $0 \leq f \leq s$. Set $\delta(f,g) = \sum_{j=1}^{c} \delta(f,j,g)$ and $d(f,g) = 1 + \delta(f,g)$. Then

$$d_f = 1 + \sum_{g=0}^{q} \delta(f,g).$$

The arguments above yield that

$$m_{jg} = \min\{p^{s+1}, p^f d(f,j,g) \mid 0 \leq f \leq s\} \tag{11.8}$$

if $|x_j| = p^{s+1}$.

The following assertion appears to be helpful.

(*) Let I be a set of pairs (j,g) with $1 \leq j \leq c$ and $0 \leq g \leq q$ such that

$$d(\bigotimes_{(j,g)\in I} J_{m_{jg}}) = p^s(1 + \sum_{(j,g)\in I} p^s \delta(s,j,g)).$$

Then $d(z) = \min\{p^{s+1}, p^s d_s\}$.

Indeed, set $z' = \bigotimes_{(j,g)\in I} J_{m_{jg}}$. Naturally, there is nothing to prove if I coincides with the set of all relevant pairs (j,g). Otherwise fix an ordering on the set of remaining pairs. Denote by J^t the block $J_{m_{j,g}}$ that corresponds to the tth pair under this ordering. Set $\prod_t = z' \bigotimes \bigotimes_{a=1}^{t} J^a$, $n_0 = d(z')$, and $n_t = d(\prod_t)$ for $t > 0$. Obviously, $n_t = d(J_{n_{t-1}} \otimes J^t)$. Put $N_0 = 1 + \sum_{(j,g)\in I} \delta(s,j,g)$ and for $t > 0$ set $N_t = N_{t-1} + \delta(s,j,g)$ where (j,g) is the tth pair in our ordering. Recall that $\delta(s,j,g) = 0$ if $|x_j| < p^{s+1}$. If $|x_j| = p^{s+1}$, the representation $\chi_j^g \in \operatorname{Irr}_p S_j$ for all g, in this case $m_{jg} = p^{s+1}$ or $p^f d(f,j,g)$ for some $f \leq s$ by (11.8). In the first case it is clear that $d(z) = p^{s+1}$, so assume that the second possibility holds for all j with $|x_j| = |x|$. Then (11.7) implies that $m_{jg} > \delta(s,j,g)$. Now several applications of Lemma 2.17 yield that $n_t = \min\{p^{s+1}, p^s N_t\}$ for all admissible t. Obviously, some $N_t = d_s$. Corollary 11.30 completes the proof of claim (*).

It is clear that (*) holds in Case c). Thus from now on we can and shall assume that all $\chi_j^g \in \operatorname{Irr}_p S_j$.

The assumptions made earlier in this proof show that it suffices to consider the following three cases.

1. $u_1 = p^s + b_0$ with $0 < b_0 < p$.
2. $G = B_r(K)$ or $D_r(K)$, $u_1 = p^s + p + b_0$ with $0 \leq b_0 < p$; $a_i = 0$ for $p \leq i < r$ if $G = B_r(K)$ and $a_i = 0$ for $p \leq i < r-1$ if $G = D_r(K)$; $a_r \neq 0$ for $G = B_r(K)$ and $a_{r-1} + a_r \neq 0$ for $G = D_r(K)$.
3. $u_1 \geq p^s + p$; $a_i = 0$ for $p \leq i \leq n - p$ if $G = A_r(K)$ and $a_i = 0$ for $i \geq p$ otherwise.

Next, we concentrate on the situations where

$$p^u \delta(u,j,g) \geq p^f \delta(f,j,g) \tag{11.9}$$

for $0 \leq u < f \leq s$ and all j and g. For $s = 1$ the inequality (11.9) follows from Lemma 11.13. Observe that in Case 3 Rules 1–9 imply that χ_j^g is trivial if $|x_j| < p^{s+1}$. Hence Corollary 11.6 implies (11.9) for Case 3. One can directly verify that (11.9) holds if u_1 and $\overline{\omega}$ satisfy the assumptions of Lemma 11.21. In all these cases one can conclude that the integers m_{jg} satisfy the assumptions of Proposition 2.16. That proposition completes the proof when (11.9) holds.

Next, assume that $s > 1$, Case 1 or 2 occurs, and $\overline{\omega}$does not satisfy the assumptions of Lemma 11.21. Lemmas 11.19–11.22 imply that $d(x, \varphi) = \min\{d_0, pd_1, p^s d_s, p^{s+1}\}$ in Case 1. Corollary 11.15 and Lemmas 11.25 and 11.26 yield that the same holds for Case 2.

First let $d(x, \varphi) = d_0$. Then $\overline{\omega}$ is one of the exceptional weights indicated in Lemmas 11.20, 11.22, and 11.26 and u_i satisfy the relevant assumptions of these lemmas. One can conclude that $m_{j,g} = d(0, j, g)$ and observe that the assumptions of Corollary 2.14 hold for m_{jg} with $\sum_{g=0}^{q} \sum_{j=1}^{c} (m_{jg} - 1) = d_0 - 1$. Then Corollary 2.14 settles this case.

Now suppose that x and $\overline{\omega}$ yield an exceptional case of Lemma 11.26 with $d(x, \varphi) = pd_1$. Then $p = 3$ and using Lemma 11.1, one can assume that $q = 0$. Conjecture (r, s) and Theorem 1.14 imply that $m_{01} = 22$ and $d(\bigotimes_{j=2}^{c} J_{m_{0j}}) = 3$. Then $d(u) = 24 = d(x, \varphi)$ by Theorem 2.9.

In all other cases we need to show that $d(u) = \min\{p^{s+1}, p^s d_s\}$. Assume that $d_s > p$. Let I_1 be the set of all pairs (j, g) for which $u_j > p^s$. Formulae (11.7) and (11.8) imply that $m_{jg} > \delta_{s,j,g}$ if $(j, g) \in I_1$. Set $z' = \bigotimes_{(j,g) \in I_1} J_{m_{jg}}$. Apply Lemma 2.18 to z' and deduce that $d(z) = d(z') = p^{s+1}$ as required.

Finally, suppose that $d(x, \varphi) = p^s d_s$ with $d_s \leq p$. First consider Case 1. For $0 \leq g \leq q$ set $d'_g = \min\{p^f d(f, g) \mid 0 \leq f \leq s\}$. Observe that $p^s d(s, g) \leq p^{s+1}$ for each g as $d(s, g) \leq d_s \leq p$. If $d'_g = p^s d(s, g)$ for some g, apply Corollary 11.23 and Lemma 2.17 to φ_g and deduce that a collection of some pairs (j, g) with $1 \leq j \leq c$ yields a set I required for (*). Here (11.8) forces $m_{jg} = N_j$ for relevant j in the notation of Corollary 11.23. So suppose there are no such g. Then each μ_g is either one of the exceptional weights of Lemmas 11.20 and 11.22, or satisfies the assumptions of Lemma 11.21. In particular, either

$$\theta_j(\mu_g) = \sum_{i=1}^{r_j} a_{ijg} \omega_i \quad \text{with} \quad a_{ijg} = 0 \tag{11.10}$$
$$\text{if} \quad p \leq i \leq r_j + 1 - p \quad \text{for} \quad S_j = A_{r_j}(K) \quad \text{and} \quad i \geq p \quad \text{otherwise,}$$

or $p = 3$, $s = 2$, $j = 1$, $G = B_r(K)$ or $D_r(K)$, and φ_g is a representation in Item 2 of Lemma 11.22. We say that φ_g is of type II in the second case. First assume that (11.10) holds for all j and g. Then Lemma 3.2 forces that (11.9) is valid for all j and g. We have already shown earlier that $d(x, \varphi) = d(z)$ in such cases. Now let φ_g be of type II for some g. Assume that φ_a be of type II for some $a \neq g$ as well. Since $d'_a = d(0, a)$ and $d_g = d(0, g)$, the arguments above for the case $d(x, \varphi) = d_0$ imply that $m_{1g} = d(0, 1, g)$ and $m_{1a} = d(0, 1, a)$. Hence $m_{1a} = m_{1g} = 16$. By Theorem 2.9, $d(J_{m_{1a}} \otimes J_{m_{1g}}) = 27 = p^{s+1}$ and hence $d_\varphi(x) = p^{s+1}$.

Next, let (11.10) hold for all $h \neq g$. Set $\eta = \bigotimes_{h \neq g} \varphi_h \operatorname{Fr}^h$ and $\omega_- = \omega(\eta)$. Obviously, η is nontrivial and $\varphi \cong \varphi_g \otimes \eta$. Set $d_{f,\eta} = 1 + \sum_{j=1}^{c} \sum_{h \neq g} \delta(f, j, h)$ for $0 \leq f \leq s$ and $d_\eta = \min\{3^f d_{f,\eta} \mid 0 \leq f \leq 2\}$. Actually it is already proved that $d_\eta(x) = d_\eta$. Indeed, the arguments above yield that Theorem 1.10 holds for η. Since $d_{2,\eta} \leq d_2 \leq 3$, we get $d_\eta = 27 = 9d_{2,\eta}$ if $9d_{2,\eta} \geq 27$. Since η is nontrivial, $d_\eta > 9$. Naturally, $d(z) \geq d(J_{m_{1g}} \otimes J_{d_\eta})$. Now Theorem 2.9 implies that $d_\varphi(x) = d(z) = 27$ if $d_\eta \geq 12$. Hence we can assume that $d_\eta = d_{0,\eta}$. Since $d(x, \varphi) = 9d_2$ and $d(x, \varphi_g) = d(0, g)$, we have $d_{\varphi_g}(x) + d_\eta = d_0 + 1 > 9d_2$. It follows from (11.8) that $d_{\varphi_g}(x) > 9\delta_{2,g}$ and $d_\eta > 9(d_{2,\eta} - 1)$. Now Lemma 2.17 yields that $d_\varphi(x) = 9d_2$ and completes the proof for Case 1.

It remains to consider Case 2 with $s > 1$, $d(x, \varphi) = p^s d_s$, and $d_s \leq p$. Define d'_g as in Case 1. We claim that $d'_g = p^s d(s, g)$ for some g. Indeed, otherwise Lemma 11.26 implies that $p = 3$, $u_1 = 13$, and each nonzero μ_g is an exceptional weight of that lemma. Since $d(x, \varphi) = p^s d_s$, there exist distinct g and h with $\mu_g, \mu_h \neq 0$. Then $d(2, 1, g) = d(2, 1, h) = 3$ which yields a contradiction as $d_2 \leq 3$ by our assumptions. Hence our claim holds. Now argue as for Case 1 applying Corollary 11.27, Lemma 2.17, and Formula (11.8) to show that there exists a collection of pairs (j, g) with $1 \leq j \leq c$ that satisfies the assumptions of (*).

Finally, all the possibilities have been considered. The theorem is proved. □

REMARK 11.36. Observe that we have not proved that always $d_\varphi(x) = d_\chi(x)$ in the assumptions of Theorem 1.9, though in many cases this equality holds under such assumptions. Furthermore, in some cases covered by Results 11.33–11.35 $d_\varphi(x) \neq d_\chi(x)$ and we deal with another composition factor of $\varphi|S(x)$ to compute $d_\varphi(x)$.

Proof of Proposition 1.11. Assume that $d_{\varphi_C}(z_s) > p$. Naturally, it suffices to consider the case where the assumptions of Theorem 1.9 do not hold. Then $d_\psi(h_s) > p$ by Proposition 1.12. To complete the proof, apply Corollary 11.15 and Theorem 1.10. □

Proof of Proposition 1.13. The assertion of the proposition follows immediately from Theorem 1.10 and Lemmas 11.20–11.22. □

So we have proved all main results of the article.

12. Some examples

In this section examples mentioned in the Introduction are discussed in more detail. Explicit computations for groups of small ranks that appear as maximal simple subgroups in exceptional simple algebraic groups are presented. Notation of the Introduction and Section 11 is used.

Proof of Lemma 1.2. Recall that $k_1 = n$ for $G \neq D_r(K)$ and $k_1 = n-1$, $k_2 = 1$ for $G = D_r(K)$. One can easily verify that $\langle \omega, \alpha \rangle = p-1$ for the maximal positive root α. If $s > 1$, the representation φ satisfies the assumptions of Lemma 11.21. That lemma and Theorem 1.10 imply that $d(x, \varphi) = \min\{d_0, pd_1, p^{s+1}\}$. By Proposition 2.5 b), x^p has one block of size $p^{s-1}+1$ for $G = A_r(K)$ or $C_r(K)$ and two such blocks for $G = B_r(K)$ or $D_r(K)$, other blocks are of size p^{s-1}. So Proposition 1.5 and Algorithm 1.6 yield that $d_0 = (p-1)p^s + 1$ and $d_1 = (p-1)p^{s-1} + 1$. This implies the lemma. □

Proof of Lemma 1.8. Rule 6 implies that $\mathbf{j}(\varepsilon_i) = 1$ for $i \leq l$. Hence $\theta_2(\omega) = 0$ and $\theta_1(\omega) = \omega_l$, except the case where $G = B_r(K)$ or $D_r(K)$ and $l = p-1$, in this case $\theta_1(\omega) = 2\omega_l$. Now apply Algorithm 1.6. One easily deduces that in all the situations under consideration $d_0 = lp + 1$ and $d_1 = d_{\varphi_C}(z_1) = l + 1$. Hence $d_\varphi(x) = lp + 1$. To compute $d_{\varphi_C}(z_0)$, one have to analyze the sequence (1.2) for x. In Items a)-j) below $t \in \mathbb{Z}$ and $\Sigma(x)$ is the non-increasing sequence consisting of l largest members of (1.2). The following subcases have to be distinguished:

a) $l = 4t$, $J(x) = (p+l, p)$, $\Sigma(x) = (p+4t-1, p+4t-3, \ldots, p+1, p-1, p-1, p-3, p-3, \ldots, p-2t+1, p-2t+1)$, and $d_{\varphi_C}(z_0) = lp + 2t^2$;

b) $l = 4t+1$, $J(x) = (p+l,p)$, $\Sigma(x) = (p+4t, p+4t-2, \dots, p+2, p, p-1, p-2, \dots, p-2t+1, p-2t)$, and $d_{\varphi_C}(z_0) = lp + 2t^2 + t$;

c) $l = 4t+2$, $J(x) = (p+l,p)$, $\Sigma(x) = (p+4t+1, p+4t-1, \dots, p+1, p-1, p-1, p-3, p-3, \dots, p-2t+1, p-2t+1, p-2t-1)$, and $d_{\varphi_C}(z_0) = lp + 2t^2 + 2t$;

d) $l = 4t+3$, $J(x) = (p+l,p)$, $\Sigma(x) = (p+4t+2, p+4t, \dots, p+2, p, p-1, p-2, \dots, p-2t+1, p-2t, p-2t-1)$, and $d_{\varphi_C}(z_0) = lp + 2t^2 + 3t + 1$;

e) $l = 6t$, $J(x) = (p+l,p,p)$, $\Sigma(x) = (p+6t-1, p+6t-3, \dots, p+1, p-1, p-1, p-1, p-3, p-3, p-3, \dots, p-2t+1, p-2t+1, p-2t+1)$, and $d_{\varphi_C}(z_0) = lp + 6t^2$;

f) $l = 6t+1$, $J(x) = (p+l,p,p)$, $\Sigma(x) = (p+6t, p+6t-2, \dots, p, p-1, p-1, p-2, p-3, p-3, p-4, \dots, p-2t+1, p-2t+1, p-2t)$, and $d_{\varphi_C}(z_0) = lp + 6t^2 + 2t$;

g) $l = 6t+2$, $J(x) = (p+l,p,p)$, $\Sigma(x) = (p+6t+1, p+6t-1, \dots, p+1, p-1, p-1, p-1, p-3, p-3, p-3, \dots, p-2t+1, p-2t+1, p-2t+1, p-2t-1)$, and $d_{\varphi_C}(z_0) = lp + 6t^2 + 4t$;

h) $l = 6t+3$, $J(x) = (p+l,p,p)$, $\Sigma(x) = (p+6t+2, p+6t, \dots, p, p-1, p-1, p-2, p-3, p-3, p-4, \dots, p-2t+1, p-2t+1, p-2t, p-2t-1)$, and $d_{\varphi_C}(z_0) = lp + 6t^2 + 6t + 1$;

i) $l = 6t+4$, $J(x) = (p+l,p,p)$, $\Sigma(x) = (p+6t+3, p+6t+1, \dots, p+1, p-1, p-1, p-1, p-3, p-3, p-3, \dots, p-2t+1, p-2t+1, p-2t+1, p-2t-1, p-2t-1)$, and $d_{\varphi_C}(z_0) = lp + 6t^2 + 8t + 2$;

j) $l = 6t+5$, $J(x) = (p+l,p,p)$, $\Sigma(x) = (p+6t+4, p+6t+2, \dots, p, p-1, p-1, p-2, p-3, p-3, p-4, \dots, p-2t+1, p-2t+1, p-2t, p-2t-1, p-2t-1)$, and $d_{\varphi_C}(z_0) = lp + 6t^2 + 10t + 4$.

This implies that in all cases $d_\varphi(x)$ is less than the value given by Formula (1.4) and completes the proof. □

The following lemma shows that $d(x,\varphi)$ can be equal to $p^f d_f$ with different f even for fundamental representations.

LEMMA 12.1. *Let $x \in G$ be a regular unipotent element and $\varphi = \varphi(\omega_2)$. Assume that $n = \sum_{f=0}^{s} b_f p^f$ with $0 \leq b_f < p$ and $b_s > 0$. If $2 \leq b_0 \leq (p+3)/2$ and $b_f \leq (p-1)/2$ for $f > 0$, one has $d(x,\varphi) = d_0$. If $b_f \leq (p-1)/2$ for $f > 0$ and $b_0 > (p+3)/2$, we get $d(x,\varphi) = pd_1$. If $1 < j \leq s$, $b_f \leq (p-1)/2$ for $f \geq j$ and $b_{j-1} > (p-1)/2$, then $d(x,\varphi) = p^j d_j$. If $b_s > (p-1)/2$, one has $d(x,\varphi) = p^{s+1}$.*

PROOF. For $0 < f \leq s$ set $n_f^+ = \sum_{i=f}^{s} b_i p^i$, $n_f^- = \sum_{i=0}^{f-1} b_i p^i$, and $n'_f = n_f^+/p^f$. By Proposition 2.5 b), x^{p^f} has n_f^- Jordan blocks of size $n'_f + 1$ and $p^f - n_f^-$ blocks of size n'_f. By Algorithm 1.6, $d_0 = 2n - 3$ and $p^f d_f = 2n_f^+ p^f$ if $f > 0$ and $n_f^- \geq 2$. In all cases if $0 < f < s$ and $b_f > (p-1)/2$, we have $p^t d_t \geq 2n_{t+1}^+ + p^{t+1}$ for $t \leq f$; and if $b_s > (p-1)/2$, one gets $p^f d_f \geq p^{s+1}$ for all $f \leq s$. Now one can easily check that all assumptions of the lemma hold. □

On the other hand, Lemmas 11.20–11.22 imply that the parameter $d(x,\varphi)$ can be equal to d_0, pd_1, $p^s d_s$, and p^{s+1} only if $p^s < k_1 < p^s + p$. Furthermore, for such k_1 in some cases the form of $\theta_j(\overline{\omega})$ shows that $d(x,\varphi) = \min\{p^s d_s, p^{s+1}\}$.

Now consider some classical groups of small ranks. We compute explicitly the degrees of the minimal polynomials of all unipotent elements whose order is more than p, in irreducible representations of the groups $A_r(K)$ and $D_r(K)$ with $5 \leq r \leq 8$, $B_3(K)$, $B_4(K)$, $C_3(K)$, and $C_4(K)$. Theorems 1.1 and 1.10 show that for a unipotent element $x \in G$ and an irreducible representation $\varphi \in \mathrm{Irr}$ the degree $d_\varphi(x)$ is completely determined by $\mathrm{cl}(x)$ and $\overline{\omega(\varphi)}$.

PROPOSITION 12.2. *Let $G = A_r(K)$ or $D_r(K)$ with $5 \leq r \leq 8$, $B_3(K)$, $B_4(K)$, $C_3(K)$, or $C_4(K)$. Assume that $x \in G$ is unipotent, $|x| > p$, $\varphi \in \mathrm{Irr}$, and $\omega(\varphi) \neq 0$. Set $\overline{\omega} = \omega(\varphi)$ and $d = d_\varphi(x)$. Then one of the following holds:*

a) $d = |x|$;

b) $\overline{\omega} = \omega_1$ or ω_r for $G = A_r(K)$ and $\overline{\omega} = \omega_1$ otherwise;

c) the collection $(G, p, \mathrm{cl}(x), \overline{\omega}, d)$ appears in Tables $I - -XII$.

PROOF. Apply Lemma 1.4 to determine the unipotent conjugacy classes in G and Theorems 1.1, 1.3, 1.10, and 1.9 to compute the minimal polynomials of their representatives. □

Recall that $\mathrm{cl}(x)$ is determined by $J(x)$, except some unipotent conjugacy classes of the group $D_r(K)$. In Tables I–XII we write down $J(x)$ in the columns for $\mathrm{cl}(x)$ and indicate which of two classes with the same $J(x)$ is considered where necessary. For these classes the notation of Lemma 1.4 ii) is used. If there are two classes with the same $J(x)$ and such class is not specified at a position in Tables IX–XII where this $J(x)$ appears, this means that d is the same for both these classes. We write $J(x) = (k_1, \dots, k_t, E_l)$ if $J(x) = (k_1, \dots, k_t, 1, \dots, 1)$ with l blocks of size 1. In Tables I-XII the indices in formulas for weights can take all values from 1 to r unless otherwise stated. For instance, we write "$\omega_i + \omega_j$, $2 \leq j \leq r-1$" if $\overline{\omega}$ is a sum of two fundamental weights, one of them satisfies the assumption above, and another is arbitrary. If for some $\overline{\omega}$ the degree d is the same for all x with $J(x) = (k_1, \dots, k_t, c_1, \dots, c_l)$, the notation $J(k_1, \dots, k_t, *)$ is used in the column for $J(x)$. The notation $J(k_1, \dots, k_t, < a, *)$ ($J(k_1, \dots, k_t, > a, *)$) is used if d is the same for fixed $\overline{\omega}$ and all x with $J(x) = (k_1, \dots, k_t, m_1, \dots, m_s)$ and $m_1 < a$ ($m_1 > a$, respectively).

Tables

TABLE I. $G = A_5(K)$

p	$\mathrm{cl}(x)$ $(J(x))$	$\overline{\omega}$	d
3	(6)	ω_3	8
3	(5, 1)	$\omega_i, \ 2 \le i \le 4$	7
3	$(4, *)$	$\omega_2, \ \omega_4$	5
3	$(4, *)$	$a\omega_1 + b\omega_5, \ a + b = 2$	7
3	$(4, *)$	$\omega_i + \omega_j, \ i \in \{1, 5\}, \ j \in \{2, 4\}$	8
3	(4, 2)	ω_3	6
3	$(4, E_2)$	ω_3	5
3	$(4, E_2)$	$\omega_1 + \omega_3, \ \omega_3 + \omega_5$	8
5	(6)	$\omega_2, \ \omega_4$	9
5	(6)	ω_3	10
5	(6)	$a\omega_1 + b\omega_5, \ a + b = 2$	11
5	(6)	$\omega_i + \omega_j, \ i \in \{1, 5\}, \ j \in \{2, 4\}$	14
5	(6)	$\omega_i + \omega_3, \ i \in \{1, 5\}, \ \omega_k + \omega_l, \ 2 \le k \le l \le 4$	15
5	(6)	$a\omega_1 + b\omega_5, \ a + b = 3$	16
5	(6)	$a\omega_1 + \omega_i + b\omega_5, \ a + b = 2, \ i = 2$ or 4	19
5	(6)	$a\omega_1 + \omega_3 + b\omega_5, \ a + b = 2, \ \omega_i + \omega_j + \omega_k, \ 2 \le j \le k \le 4$	20
5	(6)	$a\omega_1 + b\omega_5, \ a + b = 4$	21
5	(6)	$a\omega_1 + \omega_i + b\omega_5, \ a + b = 3, \ i = 2$ or 4	24

Table II. $G = A_6(K)$

p	cl(x) $(J(x))$	$\overline{\omega}$	d
3	$(5, *)$	$\omega_2,\ \omega_5$	7
3	$(5, 2)$	ω_3, ω_4	8
3	$(5, E_2)$	$\omega_3,\ \omega_4$	7
3	$(4, *)$	$a\omega_1 + b\omega_6,\ a + b = 2$	7
3	$(4, > 1, *)$	$\omega_3,\ \omega_4$	6
3	$(4, 3)$	$\omega_2,\ \omega_5$	6
3	$(4, < 3, *)$	$\omega_2,\ \omega_5$	5
3	$(4, < 3, *)$	$\omega_i + \omega_j,\ i \in \{1, 6\},\ j \in \{2, 5\}$	8
3	$(4, E_3)$	$\omega_3,\ \omega_4$	5
3	$(4, E_3)$	$\omega_i + \omega_j,\ i \in \{1, 6\},\ 3 \leq j \leq 4$	8
5	(7)	$\omega_2,\ \omega_5$	11
5	(7)	$a\omega_1 + b\omega_6,\ a + b = 2,\ \omega_3,\ \omega_4$	13
5	(7)	$\omega_i + \omega_j,\ i \in \{1, 6\},\ j \in \{2, 5\}$	17
5	(7)	$a\omega_1 + b\omega_6,\ a + b = 3,\ \omega_i + \omega_j,\ i \in \{1, 6\},\ 3 \leq j \leq 4$	19
5	(7)	$a\omega_2 + b\omega_5,\ a + b = 2$	21
5	(7)	$a\omega_1 + \omega_i + b\omega_6,\ \omega_i + \omega_j,\ a + b = 2,\ i \in \{2, 5\},\ 3 \leq j \leq 4$	23
5	$(6, 1)$	$\omega_2,\ \omega_5$	9
5	$(6, 1)$	$\omega_3,\ \omega_4$	10
5	$(6, 1)$	$a\omega_1 + b\omega_6,\ a + b = 2$	11
5	$(6, 1)$	$\omega_i + \omega_j,\ i \in \{1, 6\},\ j \in \{2, 5\}$	14
5	$(6, 1)$	$\omega_i + \omega_j,\ 2 \leq i \leq j \leq 5$ or $3 \leq j \leq 4$	15
5	$(6, 1)$	$a\omega_1 + b\omega_6,\ a + b = 3$	16
5	$(6, 1)$	$a\omega_1 + \omega_i + b\omega_6,\ a + b = 2,\ i \in \{2, 5\}$	19
5	$(6, 1)$	$\omega_i + \omega_j + \omega_k,\ 2 \leq j \leq k \leq 5$ or $3 \leq k \leq 4$	20
5	$(6, 1)$	$a\omega_1 + b\omega_6,\ a + b = 4$	21
5	$(6, 1)$	$a\omega_1 + \omega_i + b\omega_6,\ a + b = 3,\ i \in \{2, 5\}$	24

TABLE III. $G = A_7(K)$

p	cl(x) $(J(x))$	$\overline{\omega}$	d
3	$(5,*)$	$\omega_2,\ \omega_6$	7
3	$(5,2,1)$	$\omega_i,\ 3\le i\le 5$	8
3	$(5,E_3)$	$\omega_i,\ 3\le i\le 5$	7
3	$(4,*)$	$a\omega_1+b\omega_7,\ a+b=2$	7
3	$(4,4)$	$\omega_2,\ \omega_6$	7
3	$(4,4)$	$\omega_3,\ \omega_5$	8
3	$(4,3,1),\ (4,2,*)$	$\omega_i,\ 3\le i\le 5$	6
3	$(4,3,1)$	$\omega_2,\ \omega_6$	6
3	$(4,<3,*)$	$\omega_2,\ \omega_6$	5
3	$(4,<3,*)$	$\omega_i+\omega_j,\ i\in\{1,7\},\ j\in\{2,6\}$	8
3	$(4,E_4)$	$\omega_i,\ 3\le i\le 5$	5
3	$(4,E_4)$	$\omega_i+\omega_j,\ i\in\{1,7\},\ 3\le j\le 5$	8
5	(8)	$\omega_2,\ \omega_6$	13
5	(8)	$a\omega_1+b\omega_7,\ a+b=2$	15
5	(8)	$\omega_3,\ \omega_5$	16
5	(8)	ω_4	17
5	(8)	$\omega_i+\omega_j,\ i\in\{1,7\},\ j\in\{2,6\},\ a\omega_1+b\omega_7,\ a+b=3$	20
5	(8)	$\omega_i+\omega_j,\ i\in\{1,7\},\ j\in\{3,5\}$	23
5	(8)	$\omega_1+\omega_4,\ \omega_4+\omega_7$	24
5	$(7,1)$	$\omega_2,\ \omega_6$	11
5	$(7,1)$	$a\omega_1+b\omega_7,\ a+b=2,\ \omega_i,\ 3\le i\le 5$	13
5	$(7,1)$	$\omega_i+\omega_j,\ i\in\{1,7\},\ j\in\{2,6\}$	17
5	$(7,1)$	$a\omega_1+b\omega_7,\ a+b=3,\ \omega_i+\omega_j,\ i\in\{1,7\},\ 3\le j\le 5$	19
5	$(7,1)$	$a\omega_2+b\omega_6,\ a+b=2$	21
5	$(7,1)$	$a\omega_1+\omega_i+b\omega_7,\ \omega_i+\omega_j, a+b=2,\ i\in\{2,6\},\ 3\le j\le 5$	23
5	$(6,*)$	$\omega_2,\ \omega_6$	9
5	$(6,*)$	$\omega_i,\ 3\le i\le 5,$	10
5	$(6,*)$	$a\omega_1+b\omega_7,\ a+b=2$	11
5	$(6,*)$	$\omega_i+\omega_j,\ i\in\{1,7\},\ j\in\{2,6\}$	14
5	$(6,*)$	$\omega_i+\omega_j,\ i\in\{1,7\},\ 3\le j\le 5,\ \omega_k+\omega_l,\ 2\le k\le l\le 5$	15
5	$(6,*)$	$a\omega_1+b\omega_7,\ a+b=3$	16
5	$(6,*)$	$a\omega_1+\omega_i+b\omega_7,\ a+b=2,\ i\in\{2,6\}$	19
5	$(6,*)$	$\omega_i+\omega_j+\omega_k,\ 3\le k\le 5$ or $2\le j\le k\le 6$	20
5	$(6,*)$	$a\omega_1+b\omega_7,\ a+b=4$	21
5	$(6,*)$	$a\omega_1+\omega_i+b\omega_7,\ a+b=3,\ i\in\{2,6\}$	24

TABLE III CONTINUED

p	cl(x) ($J(x)$)	$\overline{\omega}$	d
7	(8)	$\omega_2, \ \omega_6$	13
7	(8)	$\omega_i, \ 3 \le i \le 5$	14
7	(8)	$a\omega_1 + b\omega_7, \ a+b=2$	15
7	(8)	$\omega_i + \omega_j, \ i \in \{1,7\}, \ j \in \{2,6\}$	20
7	(8)	$\omega_i + \omega_j, \ 3 \le j \le 5$ or $2 \le i \le j \le 6$	21
7	(8)	$a\omega_1 + b\omega_7, \ a+b=3$	22
7	(8)	$a\omega_1 + \omega_i + b\omega_7, \ a+b=2, \ i \in \{2,6\}$	27
7	(8)	$\omega_i + \omega_j + \omega_k, \ 3 \le k \le 5$ or $2 \le j \le k \le 6$	28
7	(8)	$a\omega_1 + b\omega_7, \ a+b=4$	29
7	(8)	$a\omega_1 + \omega_i + b\omega_7, \ a+b=3, \ i \in \{2,6\}$	34
7	(8)	$\sum_{i=1}^7 a_i\omega_i, \ \sum_{i=1}^7 a_i = 4, \ a_1 + a_7 < 3$ or $a_3 + a_4 + a_5 \ne 0$	35
7	(8)	$a\omega_1 + b\omega_7, \ a+b=5$	36
7	(8)	$a\omega_1 + \omega_i + b\omega_7, \ a+b=4, \ i \in \{2,6\}$	41
7	(8)	$\sum_{i=1}^7 a_i\omega_i, \ \sum_{i=1}^7 a_i = 5, \ a_1 + a_7 < 4$ or $a_3 + a_4 + a_5 \ne 0$	42
7	(8)	$a\omega_1 + b\omega_7, \ a+b=6$	43
7	(8)	$a\omega_1 + \omega_i + b\omega_7, \ a+b=5, \ i \in \{2,6\}$	48

TABLE IV. $G = A_8(K)$

p	$\mathrm{cl}(x)$ $(J(x))$	$\overline{\omega}$	d
3	$(5, 4)$	$\omega_2, \ \omega_7$	8
3	$(5, < 4, *)$	$\omega_2, \ \omega_7$	7
3	$(5, 2, *)$	$\omega_3, \ \omega_6$	8
3	$(5, 2, E_2)$	$\omega_4, \ \omega_5$	8
3	$(5, E_4)$	$\omega_i, \ 2 \leq i \leq 7$	7
3	$(4, *)$	$a\omega_1 + b\omega_8, \ a + b = 2$	7
3	$(4, 4, 1)$	$\omega_2, \ \omega_7$	7
3	$(4, 4, 1)$	$\omega_3, \ \omega_6$	8
3	$(4, 3, *), \ (4, 2, *)$	$\omega_i, \ 3 \leq i \leq 6$	6
3	$(4, 3, *)$	$\omega_2, \ \omega_7$	6
3	$(4, < 3, *)$	$\omega_2, \ \omega_7$	5
3	$(4, < 3, *)$	$\omega_i + \omega_j, \ i \in \{1, 8\}, \ j \in \{2, 7\}$	8
3	$(4, E_5)$	$\omega_i, \ 3 \leq i \leq 6$	5
3	$(4, E_5)$	$\omega_1 + \omega_i, \ \omega_i + \omega_8, \ 3 \leq i \leq 6$	8
5	(9)	$a\omega_1 + b\omega_8, \ a + b = 2, \ \omega_2, \ \omega_7$	15
5	(9)	$\omega_3, \ \omega_6$	19
5	(9)	$\omega_i + \omega_j, \ i \in \{1, 8\}, \ j \in \{2, 7\}, \ a\omega_1 + b\omega_8, \ a + b = 3$	20
5	(9)	$\omega_4, \ \omega_5$	21
5	$(8, 1)$	$\omega_2, \ \omega_7$	13
5	$(8, 1)$	$a\omega_1 + b\omega_8, \ a + b = 2$	15
5	$(8, 1)$	$\omega_3, \ \omega_6$	16
5	$(8, 1)$	$\omega_4, \ \omega_5$	17
5	$(8, 1)$	$\omega_i + \omega_j, \ i \in \{1, 8\}, \ j \in \{2, 7\}, \ a\omega_1 + b\omega_8, \ a + b = 3$	20
5	$(8, 1)$	$\omega_i + \omega_j, \ i \in \{1, 8\}, \ j \in \{3, 6\}$	23
5	$(8, 1)$	$\omega_i + \omega_j, \ i \in \{1, 8\}, \ 4 \leq j \leq 5$	24
5	$(7, *)$	$\omega_2, \ \omega_7$	11
5	$(7, *)$	$a\omega_1 + b\omega_8, \ a + b = 2, \ \omega_3, \ \omega_6$	13
5	$(7, *)$	$\omega_i + \omega_j, \ i \in \{1, 8\}, \ j \in \{2, 7\}$	17
5	$(7, *)$	$a\omega_1 + b\omega_8, \ a + b = 3, \ \omega_i + \omega_j, \ i \in \{1, 8\}, \ j \in \{3, 6\}$	19
5	$(7, *)$	$a\omega_2 + b\omega_7, \ a + b = 2$	21
5	$(7, *)$	$a\omega_1 + \omega_i + b\omega_8, \ \omega_i + \omega_j, \ a + b = 2, \ i \in \{2, 7\}, \ j \in \{3, 6\}$	23
5	$(7, 2)$	$\omega_4, \ \omega_5$	14
5	$(7, 2)$	$\omega_i + \omega_j, \ i \in \{1, 8\}, \ 4 \leq j \leq 5$	20
5	$(7, E_2)$	$\omega_4, \ \omega_5$	13
5	$(7, E_2)$	$\omega_i + \omega_j, \ i \in \{1, 8\}, \ 4 \leq j \leq 5$	19
5	$(7, E_2)$	$\omega_i + \omega_j, \ i \in \{2, 7\}, \ 4 \leq j \leq 5$	23
5	$(6, *)$	$\omega_2, \ \omega_7$	9
5	$(6, *)$	$\omega_i, \ 3 \leq i \leq 6$	10
5	$(6, *)$	$a\omega_1 + b\omega_8, \ a + b = 2$	11
5	$(6, *)$	$\omega_i + \omega_j, \ i \in \{1, 8\}, \ j \in \{2, 7\}$	14
5	$(6, *)$	$\omega_i + \omega_j, \ 2 \leq i \leq j \leq 7$ or $3 \leq j \leq 6$	15
5	$(6, *)$	$a\omega_1 + b\omega_8, \ a + b = 3$	16
5	$(6, *)$	$a\omega_1 + \omega_i + b\omega_8, \ a + b = 2, \ i \in \{2, 7\}$	19
5	$(6, *)$	$\omega_i + \omega_j + \omega_k, \ 2 \leq j \leq k \leq 7$ or $3 \leq k \leq 6$	20
5	$(6, *)$	$a\omega_1 + b\omega_8, \ a + b = 4$	21
5	$(6, *)$	$a\omega_1 + \omega_i + b\omega_8, \ a + b = 3, \ i \in \{2, 7\}$	24
7	(9)	$\omega_2, \ \omega_7$	15
7	(9)	$a\omega_1 + b\omega_8, \ a + b = 2$	17

Table IV continued

p	cl(x) $(J(x))$	$\overline{\omega}$	d
7	(9)	$\omega_3,\ \omega_6$	19
7	(9)	$\omega_4,\ \omega_5$	21
7	(9)	$\omega_i+\omega_j,\ \ i\in\{1,8\},\ \ j\in\{2,7\}$	23
7	(9)	$a\omega_1+b\omega_8,\ \ a+b=3$	25
7	(9)	$\omega_i+\omega_j,\ \ i\in\{1,8\},\ \ j\in\{3,6\}$	27
7	(9)	$\omega_i+\omega_j,\ \ i\in\{1,8\},\ \ 4\leq j\leq 5$	28
7	(9)	$a\omega_2+b\omega_7,\ \ a+b=2$	29
7	(9)	$a\omega_1+\omega_i+b\omega_8,\ \ a+b=2,\ \ i\in\{2,7\}$	31
7	(9)	$a\omega_1+b\omega_8,\ \ a+b=4,\ \ \omega_i+\omega_j,\ \ i\in\{2,7\},\ \ j\in\{3,6\}$	33
7	(9)	$a\omega_1+\omega_i+b\omega_8,\ \ \omega_i+\omega_j,\ \ \omega_k+\omega_l,$ $a+b=2,\ \ 3\leq i,j\leq 6,\ \ k\in\{2,7\},\ \ 4\leq l\leq 5$	35
7	(9)	$\omega_i+\omega_j+\omega_k,\ \ i\in\{1,8\},\ \ j,k\in\{2,7\}$	37
7	(9)	$a\omega_1+\omega_i+b\omega_8,\ \ a+b=3,\ \ i\in\{2,7\}$	39
7	(9)	$a\omega_1+b\omega_8,\ \ a+b=5,\ \ \omega_i+\omega_j+\omega_k,$ $i\in\{1,8\},\ \ j\in\{2,7\},\ \ k\in\{3,6\}$	41
7	(9)	$a\omega_1+\omega_i+b\omega_8,\ \ a+b=3,\ \ 3\leq i\leq 6,\ \ \omega_j+\omega_k+\omega_l,$ $j\in\{1,8\},\ \ 3\leq k\leq l\leq 6,$ or $k\in\{2,7\},\ \ 4\leq l\leq 5$	42
7	(9)	$a\omega_2+b\omega_7,\ \ a+b=3$	43
7	(9)	$a\omega_1+\omega_i+\omega_j+b\omega_8,\ \ a+b=2,\ \ i,j\in\{2,7\}$	45
7	(9)	$a\omega_1+\omega_i+b\omega_8,\ \ a+b=4,\ \ i\in\{2,7\},$ $c\omega_2+\omega_j+e\omega_7,\ \ c+e=2,\ \ j\in\{3,6\}$	47
7	$(8,1)$	$\omega_2,\ \omega_7$	13
7	$(8,1)$	$\omega_i,\ 3\leq i\leq 6$	14
7	$(8,1)$	$a\omega_1+b\omega_8,\ \ a+b=2$	15
7	$(8,1)$	$\omega_i+\omega_j,\ \ i\in\{1,8\},\ \ j\in\{2,7\}$	20
7	$(8,1)$	$\omega_i+\omega_j,\ \ 3\leq j\leq 6$ or $2\leq i\leq j\leq 7$	21
7	$(8,1)$	$a\omega_1+b\omega_8,\ \ a+b=3$	22
7	$(8,1)$	$a\omega_1+\omega_i+b\omega_8,\ \ a+b=2,\ \ i\in\{2,7\}$	27
7	$(8,1)$	$\omega_i+\omega_j+\omega_k,\ \ 3\leq k\leq 6$ or $2\leq j\leq k\leq 7$	28
7	$(8,1)$	$a\omega_1+b\omega_8,\ \ a+b=4$	29
7	$(8,1)$	$a\omega_1+\omega_i+b\omega_8,\ \ a+b=3,\ \ i\in\{2,7\}$	34
7	$(8,1)$	$\omega_i+\omega_j+\omega_k+\omega_l,\ \ 3\leq l\leq 6$ or $2\leq k\leq l\leq 7$	35
7	$(8,1)$	$a\omega_1+b\omega_8,\ \ a+b=5$	36
7	$(8,1)$	$a\omega_1+\omega_i+b\omega_8,\ \ a+b=4,\ \ i\in\{2,7\}$	41
7	$(8,1)$	$\sum_{i=1}^{8}a_i\omega_i,\ \ \sum_{i=1}^{8}a_i=5,\ \ \sum_{i=2}^{7}a_i>1$ or $\sum_{i=3}^{6}a_i\neq 0$	42
7	$(8,1)$	$a\omega_1+b\omega_8,\ \ a+b=6$	43
7	$(8,1)$	$a\omega_1+\omega_i+b\omega_8,\ \ a+b=5,\ \ i\in\{2,7\}$	48

TABLE V. $G = B_3(K)$

p	$\mathrm{cl}(x)$ $(J(x))$	$\overline{\omega}$	d
3	(7)	ω_3	7
3	$(5, E_2)$	ω_3	4
3	$(5, E_2)$	$\omega_2,\ 2\omega_3$	7
3	$(5, E_2)$	$\omega_1 + \omega_3$	8
5	(7)	ω_3	7
5	(7)	ω_2	11
5	(7)	$a\omega_1 + b\omega_3,\ \ a + b = 2$	13
5	(7)	$\omega_1 + \omega_2,\ \ \omega_2 + \omega_3$	17
5	(7)	$a\omega_1 + b\omega_3,\ \ a + b = 3$	19
5	(7)	$2\omega_2$	21
5	(7)	$a\omega_1 + \omega_2 + b\omega_3,\ a + b = 2$	23

Table VI. $G = B_4(K)$

p	cl(x) ($J(x)$)	$\overline{\omega}$	d
3	$(7, E_2)$	ω_4	7
3	$(5, *)$	ω_2	7
3	$(5, >1, *)$	ω_4	5
3	$(5, 2, 2)$	ω_3	8
3	$(5, E_4)$	ω_4	4
3	$(5, E_4)$	$\omega_3,\ 2\omega_4$	7
3	$(5, E_4)$	$\omega_1 + \omega_4$	8
3	$(4, 4, 1)$	ω_4	5
3	$(4, 4, 1)$	$2\omega_1,\ \omega_2$	7
3	$(4, 4, 1)$	$\omega_3,\ \omega_1 + \omega_4$	8
5	(9)	ω_4	11
5	(9)	$2\omega_1,\ \omega_2$	15
5	(9)	$\omega_3,\ \omega_1 + \omega_4$	19
5	(9)	$3\omega_1,\ \omega_1 + \omega_2$	20
5	(9)	$2\omega_4$	21
5	$(7, E_2)$	ω_4	7
5	$(7, E_2)$	ω_2	11
5	$(7, E_2)$	$a\omega_1 + b\omega_4,\ a + b = 2,\ \omega_3$	13
5	$(7, E_2)$	$\omega_1 + \omega_2,\ \omega_2 + \omega_4$	17
5	$(7, E_2)$	$a\omega_1 + b\omega_4,\ a + b = 3,\ \omega_1 + \omega_3,\ \omega_3 + \omega_4$	19
5	$(7, E_2)$	$2\omega_2$	21
5	$(7, E_2)$	$a\omega_1 + \omega_2 + b\omega_4,\ a + b = 2,\ \omega_2 + \omega_3$	23
7	(9)	ω_4	11
7	(9)	ω_2	15
7	(9)	$2\omega_1$	17
7	(9)	$\omega_3,\ \omega_1 + \omega_4$	19
7	(9)	$2\omega_4$	21
7	(9)	$\omega_1 + \omega_2$	23
7	(9)	$3\omega_1,\ \omega_2 + \omega_4$	25
7	(9)	$2\omega_1 + \omega_4,\ \omega_1 + \omega_3$	27
7	(9)	$\omega_1 + 2\omega_4,\ \omega_3 + \omega_4,\ 3\omega_4$	28
7	(9)	$2\omega_2$	29
7	(9)	$2\omega_1 + \omega_2$	31
7	(9)	$4\omega_1,\ \omega_1 + \omega_2 + \omega_4,\ \omega_2 + \omega_3$	33
7	(9)	$a\omega_1 + b\omega_4,\ a + b = 4,\ b > 0,\ 2\omega_1 + \omega_3,\ \omega_1 + \omega_3 + \omega_4,\ 2\omega_3,$ $\omega_2 + 2\omega_4,\ \omega_3 + 2\omega_4$	35
7	(9)	$\omega_1 + 2\omega_2$	37
7	(9)	$3\omega_1 + \omega_2,\ 2\omega_2 + \omega_4$	39
7	(9)	$5\omega_1,\ \omega_1 + \omega_2 + \omega_3,\ 2\omega_1 + \omega_2 + \omega_4$	41
7	(9)	$a\omega_1 + b\omega_4,\ a + b = 5,\ b > 0,\ 3\omega_1 + \omega_3,\ \omega_1 + 2\omega_3,$ $2\omega_1 + \omega_3 + \omega_4,\ \omega_1 + \omega_2 + 2\omega_4,\ \omega_1 + \omega_3 + 2\omega_4,\ \omega_2 + \omega_3 + \omega_4,$ $2\omega_3 + \omega_4,\ \omega_2 + 3\omega_4,\ \omega_3 + 3\omega_4$	42
7	(9)	$3\omega_2$	43
7	(9)	$2\omega_1 + 2\omega_2$	45
7	(9)	$2\omega_2 + \omega_3,\ \omega_1 + 2\omega_2 + \omega_4$	47

TABLE VII. $G = C_3(K)$

p	cl(x) ($J(x)$)	$\overline{\omega}$	d
3	(6)	ω_3	8
3	$(4, *)$	ω_2	5
3	$(4, *)$	$2\omega_1$	7
3	$(4, *)$	$\omega_1 + \omega_2$	8
3	$(4, 2)$	ω_3	6
3	$(4, E_2)$	ω_3	5
3	$(4. E_2)$	$\omega_1 + \omega_3$	8
5	(6)	ω_2	9
5	(6)	ω_3	10
5	(6)	$2\omega_1$	11
5	(6)	$\omega_1 + \omega_2$	14
5	(6)	$\omega_1 + \omega_3,\ \ \omega_i + \omega_j,\ \ 2 \leq i \leq j \leq 3$	15
5	(6)	$3\omega_1$	16
5	(6)	$2\omega_1 + \omega_2$	19
5	(6)	$2\omega_1 + \omega_3,\ \ \omega_i + \omega_j + \omega_k,\ \ 2 \leq j \leq k \leq 3$	20
5	(6)	$4\omega_1$	21
5	(6)	$3\omega_1 + \omega_2$	24

TABLE VIII. $G = C_4(K)$

p	cl(x) $(J(x))$	$\overline{\omega}$	d
3	$(4, *)$	$2\omega_1$	7
3	$(4, 4)$	ω_2	7
3	$(4, 4)$	ω_3	8
3	$(4, < 4, *)$	ω_2	5
3	$(4, < 4, *)$	$\omega_1 + \omega_2$	8
3	$(4, 2, *)$	$\omega_3,\ \omega_4$	6
3	$(4, E_4)$	$\omega_3,\ \ \omega_4$	5
3	$(4, E_4)$	$\omega_1 + \omega_3,\ \ \omega_1 + \omega_4$	8
5	(8)	ω_2	13
5	(8)	$2\omega_1$	15
5	(8)	ω_3	16
5	(8)	ω_4	17
5	(8)	$3\omega_1,\ \ \omega_1 + \omega_2$	20
5	(8)	$\omega_1 + \omega_3$	23
5	(8)	$\omega_1 + \omega_4$	24
5	$(6, *)$	ω_2	9
5	$(6, *)$	$\omega_3,\ \ \omega_4$	10
5	$(6, *)$	$2\omega_1$	11
5	$(6, *)$	$\omega_1 + \omega_2$	14
5	$(6, *)$	$\omega_i + \omega_j,\ 3 \le j \le 4$ or $2 \le i \le j \le 4$	15
5	$(6, *)$	$3\omega_1$	16
5	$(6, *)$	$2\omega_1 + \omega_2$	19
5	$(6, *)$	$\omega_i + \omega_j + \omega_k,\ \ 3 \le k \le 4$ or $2 \le j \le k \le 4$	20
5	$(6, *)$	$4\omega_1$	21
5	$(6, *)$	$3\omega_1 + \omega_2$	24
7	(8)	ω_2	13
7	(8)	$\omega_3,\ \ \omega_4$	14
7	(8)	$2\omega_1$	15
7	(8)	$\omega_1 + \omega_2$	20
7	(8)	$\omega_i + \omega_j,\ \ 3 \le j \le 4$ or $2 \le i \le j \le 4$	21
7	(8)	$3\omega_1$	22
7	(8)	$2\omega_1 + \omega_2$	27
7	(8)	$\omega_i + \omega_j + \omega_k,\ \ 3 \le k \le 4$ or $2 \le j \le k \le 4$	28
7	(8)	$4\omega_1$	29
7	(8)	$3\omega_1 + \omega_2$	34
7	(8)	$\sum_{i=1}^4 a_i\omega_i,\ \ \sum_{i=1}^4 a_i = 4,\ \ a_1 < 3$ or $a_3 + a_4 \neq 0$	35
7	(8)	$5\omega_1$	36
7	(8)	$4\omega_1 + \omega_2$	41
7	(8)	$\sum_{i=1}^4 a_i\omega_i,\ \ \sum_{i=1}^4 a_i = 5,\ \ a_1 < 4$ or $a_3 + a_4 \neq 0$	42
7	(8)	$6\omega_1$	43
7	(8)	$5\omega_1 + \omega_2$	48

Table IX. $G = D_5(K)$

p	cl(x) $(J(x))$	$\overline{\omega}$	d
3	$(7,3)$	$\omega_4,\ \omega_5$	8
3	$(7,E_3),\ (5,5)$	$\omega_4,\ \omega_5$	7
3	$(5,<5,*)$	ω_2	7
3	$(5,3,E_2),\ (5,2,2,1)$	$\omega_4\ ,\ \omega_5$	5
3	$(5,2,2,1)$	ω_3	8
3	$(5,E_5)$	$\omega_4,\ \omega_5$	4
3	$(5,E_5)$	$\omega_3,\ a\omega_4+b\omega_5,\ a+b=2$	7
3	$(5,E_5)$	$\omega_1+\omega_4,\ \omega_1+\omega_5$	8
3	$(4,4,E_2)$	$\omega_4,\ \omega_5$	5
3	$(4,4,E_2)$	$2\omega_1,\ \omega_2$	7
3	$(4,4,E_2)$	$\omega_3,\ \omega_1+\omega_4,\ \omega_1+\omega_5$	8
5	$(9,1)$	$\omega_4,\ \omega_5$	11
5	$(9,1)$	$2\omega_1,\ \omega_2$	15
5	$(9,1)$	$\omega_3,\ \omega_1+\omega_4,\ \omega_1+\omega_5$	19
5	$(9,1)$	$3\omega_1,\ \omega_1+\omega_2$	20
5	$(9,1)$	$a\omega_4+b\omega_5,\ a+b=2$	21
5	$(7,*)$	ω_2	11
5	$(7,*)$	$2\omega_1,\ \omega_3$	13
5	$(7,*)$	$\omega_1+\omega_2$	17
5	$(7,*)$	$3\omega_1,\ \omega_1+\omega_3$	19
5	$(7,*)$	$2\omega_2$	21
5	$(7,*)$	$2\omega_1+\omega_2,\ \omega_2+\omega_3$	23
5	$(7,3)$	$\omega_4,\ \omega_5$	8
5	$(7,3)$	$\omega_1+\omega_4,\ \omega_1+\omega_5$	14
5	$(7,3)$	$a\omega_4+b\omega_5,\ a+b=2$	15
5	$(7,3)$	$\omega_2+\omega_4,\ \omega_2+\omega_5$	18
5	$(7,3)$	$a\omega_1+b\omega_4+c\omega_5,\ a+b+c=3,\ a<3,\ \omega_3+\omega_4,\ \omega_3+\omega_5$	20
5	$(7,3)$	$\omega_1+\omega_2+\omega_4,\ \omega_1+\omega_2+\omega_5$	24
5	$(7,E_3)$	$\omega_4,\ \omega_5$	7
5	$(7,E_3)$	$a\omega_1+b\omega_4+c\omega_5,\ a+b+c=2$	13
5	$(7,E_3)$	$\omega_2+\omega_4,\ \omega_2+\omega_5$	17
5	$(7,E_3)$	$a\omega_1+b\omega_4+c\omega_5,\ a+b+c=3,\ \omega_3+\omega_4,\ \omega_3+\omega_5$	19
5	$(7,E_3)$	$a\omega_1+\omega_2+b\omega_4+c\omega_5,\ a+b+c=2$	23

Table IX continued

p	cl(x) ($J(x)$)	$\overline{\omega}$	d
7	$(9,1)$	$\omega_4,\ \omega_5$	11
7	$(9,1)$	ω_2	15
7	$(9,1)$	$2\omega_1$	17
7	$(9,1)$	$\omega_3,\ \omega_1+\omega_4,\ \omega_1+\omega_5$	19
7	$(9,1)$	$a\omega_4+b\omega_5,\ a+b=2$	21
7	$(9,1)$	$\omega_1+\omega_2$	23
7	$(9,1)$	$3\omega_1,\ \omega_2+\omega_4,\ \omega_2+\omega_5$	25
7	$(9,1)$	$2\omega_1+\omega_4,\ 2\omega_1+\omega_5, \omega_1+\omega_3$	27
7	$(9,1)$	$a\omega_1+b\omega_4+c\omega_5,\ a+b+c=3,\ a<2,\ \omega_3+\omega_4,$ $\omega_3+\omega_5$	28
7	$(9,1)$	$2\omega_2$	29
7	$(9,1)$	$2\omega_1+\omega_2$	31
7	$(9,1)$	$4\omega_1,\ \omega_1+\omega_2+\omega_4,\ \omega_1+\omega_2+\omega_5,\ \omega_2+\omega_3$	33
7	$(9,1)$	$a\omega_1+b\omega_4+c\omega_5,\ a+b+c=4,\ a\neq 4,\ 2\omega_1+\omega_3,$ $\omega_1+\omega_3+\omega_4, \omega_1+\omega_3+\omega_5,\ 2\omega_3,\ \omega_i+e\omega_4+f\omega_5,$ $2\leq i\leq 3,\ e+f=2$	35
7	$(9,1)$	$\omega_1+2\omega_2$	37
7	$(9,1)$	$3\omega_1+\omega_2,\ 2\omega_2+\omega_4,\ 2\omega_2+\omega_5$	39
7	$(9,1)$	$5\omega_1,\ \omega_1+\omega_2+\omega_3,\ 2\omega_1+\omega_2+\omega_4,\ 2\omega_1+\omega_2+\omega_5$	41
7	$(9,1)$	$a\omega_1+b\omega_4+c\omega_5,\ a+b+c=5,\ a\neq 5,\ 3\omega_1+\omega_3,$ $\omega_1+2\omega_3, 2\omega_1+\omega_3+\omega_4, 2\omega_1+\omega_3+\omega_5,$ $\omega_1+\omega_i+a\omega_4+b\omega_5,\ \omega_i+c\omega_4+d\omega_5,\ 2\leq i\leq 3,$ $a+b=2,\ c+d=3,\ \omega_2+\omega_3+\omega_4,\ \omega_2+\omega_3+\omega_5,$ $2\omega_3+\omega_4,\ 2\omega_3+\omega_5$	42
7	$(9,1)$	$3\omega_2$	43
7	$(9,1)$	$2\omega_1+2\omega_2$	45
7	$(9,1)$	$4\omega_1+\omega_2,\ 2\omega_2+\omega_3,\ \omega_1+2\omega_2+\omega_4,\ \omega_1+2\omega_2+\omega_5$	47

TABLE X. $G = D_6(K)$

p	cl(x) ($J(x)$)	$\overline{\omega}$	d
3	$(11,1)$	$\omega_5,\ \omega_6$	16
3	$(11,1)$	ω_2	19
3	$(11,1)$	$2\omega_1$	21
3	$(11,1)$	ω_3	25
3	$(11,1)$	$\omega_1+\omega_5,\ \omega_1+\omega_6$	26
3	$(7,3,E_2),\ (7,2,2,1)$	$\omega_5,\ \omega_6$	8
3	$(7,E_5)$	$\omega_5,\ \omega_6$	7
3	$(6,6),\ C_1$	ω_6	8
3	$(6,6), C_2$	ω_5	8
3	$(5,5,E_2)$	$\omega_5,\ \omega_6$	7
3	$(5,<5,*)$	ω_2	7
3	$(5,3,3,1),\ (5,3,2,2)$	$\omega_5,\ \omega_6$	6
3	$(5,3,E_4),\ (5,2,2,E_3)$	$\omega_5,\ \omega_6$	5
3	$(5,2,2,E_3)$	ω_3	8
3	$(5,E_7)$	$\omega_5,\ \omega_6$	4
3	$(5,E_7)$	$\omega_3, \omega_4,\ a\omega_5+b\omega_6,\ a+b=2$	7
3	$(5,E_7)$	$\omega_1+\omega_5,\ \omega_1+\omega_6$	8
3	$(4,4,*)$	$2\omega_1,\ \omega_2$	7
3	$(4,4,3,1)$	$\omega_5,\ \omega_6$	6
3	$(4,4,<3,*)$	ω_3	8
3	$(4,4,2,2), C_1$	ω_5	5
3	$(4,4,2,2), C_1$	ω_6	6
3	$(4,4,2,2), C_1$	$\omega_1+\omega_5$	8
3	$(4,4,2,2), C_2$	ω_6	5
3	$(4,4,2,2), C_2$	ω_5	6
3	$(4,4,2,2), C_2$	$\omega_1+\omega_6$	8
3	$(4,4,E_4)$	$\omega_5,\ \omega_6$	5
3	$(4,4,E_4)$	$\omega_1+\omega_5,\ \omega_1+\omega_6$	8
5	$(11,1)$	$\omega_5,\ \omega_6$	16
5	$(11,1)$	ω_2	19
5	$(11,1)$	$2\omega_1$	21
5	$(9,*)$	$2\omega_1,\ \omega_2$	15
5	$(9,*)$	ω_3	19
5	$(9,*)$	$3\omega_1,\ \omega_1+\omega_2$	20
5	$(9,*)$	ω_4	21
5	$(9,3)$	$\omega_5,\ \omega_6$	12
5	$(9,3)$	$\omega_1+\omega_5,\ \omega_1+\omega_6$	20
5	$(9,3)$	$a\omega_5+b\omega_6,\ a+b=2$	23
5	$(9,E_3)$	$\omega_5,\ \omega_6$	11
5	$(9,E_3)$	$\omega_1+\omega_5,\ \omega_1+\omega_6$	19
5	$(9,E_3)$	$a\omega_5+b\omega_6,\ a+b=2$	21
5	$(7,*)$	ω_2	11
5	$(7,*)$	$2\omega_1$	13
5	$(7,*)$	$\omega_1+\omega_2$	17
5	$(7,*)$	$3\omega_1$	19
5	$(7,*)$	$2\omega_2$	21
5	$(7,*)$	$2\omega_1+\omega_2$	23
5	$(7,5)$	$\omega_5,\ \omega_6$	10

TABLE X CONTINUED

p	cl(x) ($J(x)$)	$\overline{\omega}$	d
5	$(7,5)$	$\omega_3,\ \omega_4,\ a\omega_1+b\omega_5+c\omega_6,\ a+b+c=2,\ a\neq 2$	15
5	$(7,5)$	$\omega_1+\omega_3,\ \omega_1+\omega_4,\ a\omega_1+b\omega_5+c\omega_6,$ $a+b+c=3,\ a\neq 3,\ \omega_i+\omega_j,$ $2\leq i\leq 4,\ 5\leq j\leq 6$	20
5	$(7,<5,*)$	ω_3	13
5	$(7,<5,*)$	$\omega_1+\omega_3$	19
5	$(7,<5,*)$	$\omega_2+\omega_3$	23
5	$(7,3,E_2),\ (7,2,2,1)$	$\omega_5,\ \omega_6$	8
5	$(7,3,E_2),\ (7,2,2,1)$	$\omega_1+\omega_5,\ \omega_1+\omega_6$	14
5	$(7,3,E_2),\ (7,2,2,1)$	$\omega_2+\omega_5,\ \omega_2+\omega_6$	18
5	$(7,3,E_2),\ (7,2,2,1)$	$\omega_1+\omega_4,\ a\omega_1+b\omega_5+c\omega_6,\ a+b+c=3,\ a\neq 3,$ $\omega_i+\omega_j,\ 3\leq i\leq 4,\ 5\leq j\leq 6$	20
5	$(7,3,E_2),\ (7,2,2,1)$	$\omega_1+\omega_2+\omega_5,\ \omega_1+\omega_2+\omega_6$	24
5	$(7,3,E_2),\ (7,2,2,1)$	$a\omega_5+b\omega_6,\ a+b=2$	15
5	$(7,3,E_2)$	ω_4	15
5	$(7,2,2,1)$	ω_4	14
5	$(7,2,2,1)$	$\omega_2+\omega_4$	24
5	$(7,E_5)$	$\omega_5,\ \omega_6$	7
5	$(7,E_5)$	$\omega_4,\ a\omega_1+b\omega_5+c\omega_6,\ a+b+c=2$	13
5	$(7,E_5)$	$\omega_2+\omega_5,\ \omega_2+\omega_6$	17
5	$(7,E_5)$	$a\omega_1+b\omega_5+c\omega_6,\ a+b+c=3,\ \omega_i+\omega_j,$ $3\leq i\leq 4,\ j\in\{1,5,6\}$	19
5	$(7,E_5)$	$\omega_2+\omega_i+\omega_j,\ i,j\in\{1,5,6\},\ \omega_2+\omega_4$	23
5	$(6,6)$	$2\omega_1,\ \omega_2$	11
5	$(6,6)$	ω_3	14
5	$(6,6)$	$\omega_4,\ a\omega_5+b\omega_6,\ a+b=2$	15
5	$(6,6)$	$3\omega_1,\ \omega_1+\omega_2$	16
5	$(6,6)$	$\omega_1+\omega_3$	19
5	$(6,6)$	$a\omega_1+b\omega_5+c\omega_6,\ a+b+c=3,\ a<3,$ $\omega_1+\omega_4,\ \omega_i+\omega_j,\ 3\leq i\leq 4, 5\leq j\leq 6$	20
5	$(6,6)$	$4\omega_1,\ 2\omega_1+\omega_2,\ 2\omega_2$	21
5	$(6,6)$	$2\omega_1+\omega_3$	24
5	$(6,6),\ C_1$	ω_5	9
5	$(6,6),\ C_1$	ω_6	10
5	$(6,6),\ C_1$	$\omega_1+\omega_5$	14
5	$(6,6),\ C_1$	$\omega_1+\omega_6$	15
5	$(6,6),\ C_1$	$\omega_2+\omega_5$	19
5	$(6,6),\ C_1$	$\omega_2+\omega_6$	20
5	$(6,6),\ C_1$	$\omega_1+\omega_2+\omega_5$	24
5	$(6,6),\ C_2$	ω_6	9
5	$(6,6),\ C_2$	ω_5	10
5	$(6,6),\ C_2$	$\omega_1+\omega_6$	14
5	$(6,6),\ C_2$	$\omega_1+\omega_5$	15
5	$(6,6),\ C_2$	$\omega_2+\omega_6$	19
5	$(6,6),\ C_2$	$\omega_2+\omega_5$	20
5	$(6,6), C_2$	$\omega_1+\omega_2+\omega_6$	24
7	$(11,1)$	$\omega_5,\ \omega_6$	16
7	$(11,1)$	ω_2	19

Table X continued

p	cl(x) $(J(x))$	$\overline{\omega}$	d
7	$(11,1)$	$2\omega_1$	21
7	$(11,1)$	ω_3	25
7	$(11,1)$	$\omega_1+\omega_5,\ \omega_1+\omega_6$	26
7	$(11,1)$	$3\omega_1,\ \omega_1+\omega_2$	28
7	$(11,1)$	ω_4	29
7	$(11,1)$	$a\omega_5+b\omega_6,\ a+b=2$	31
7	$(11,1)$	$\omega_2+\omega_5,\ \omega_2+\omega_6$	34
7	$(11,1)$	$4\omega_1,\ 2\omega_1+\omega_i,\ i\in\{2,5,6\},\ \omega_1+\omega_3,\ 2\omega_2$	35
7	$(11,1)$	$\omega_1+\omega_4$	39
7	$(11,1)$	$\omega_3+\omega_5,\ \omega_3+\omega_6$	40
7	$(11,1)$	$\omega_1+a\omega_5+b\omega_6,\ a+b=2$	41
7	$(11,1)$	$5\omega_1,\ 3\omega_1+\omega_i,\ i\in\{2,5,6\},\ 2\omega_1+\omega_3,$ $\omega_1+2\omega_2,\ \omega_1+\omega_2+\omega_5,\ \omega_1+\omega_2+\omega_6,\ \omega_2+\omega_3$	42
7	$(11,1)$	$\omega_4+\omega_5,\ \omega_4+\omega_6$	44
7	$(11,1)$	$a\omega_5+b\omega_6,\ a+b=3$	46
7	$(11,1)$	$\omega_2+\omega_4$	47
7	$(9,*)$	ω_2	15
7	$(9,*)$	$2\omega_1$	17
7	$(9,*)$	ω_3	19
7	$(9,*)$	$\omega_4,\ a\omega_5+b\omega_6,\ a+b=2$	21
7	$(9,*)$	$\omega_1+\omega_2$	23
7	$(9,*)$	$3\omega_1$	25
7	$(9,*)$	$\omega_1+\omega_3$	27
7	$(9,*)$	$\omega_1+\omega_4,\ \omega_1+a\omega_5+b\omega_6,\ a+b=2,\ c\omega_5+d\omega_6,$ $c+d=3,\ \omega_i+\omega_j,\ 3\le i\le 4,\ 5\le j\le 6$	28
7	$(9,*)$	$2\omega_2$	29
7	$(9,*)$	$2\omega_1+\omega_2$	31
7	$(9,*)$	$4\omega_1,\ \omega_2+\omega_3$	33
7	$(9,*)$	$a\omega_1+b\omega_5+c\omega_6,\ a+b+c=4,\ a<4,$ $e\omega_1+\omega_i+f\omega_5+g\omega_6,\ \omega_i+\omega_j,\ e+f+g=2,$ $i,j\in\{3,4\},\ \omega_2+k\omega_5+l\omega_6,\ k+l=2,\ \omega_2+\omega_4$	35
7	$(9,*)$	$\omega_1+2\omega_2$	37
7	$(9,*)$	$3\omega_1+\omega_2$	39
7	$(9,*)$	$5\omega_1,\ \omega_1+\omega_2+\omega_3$	41
7	$(9,*)$	$a\omega_1+b\omega_5+c\omega_6,\ a+b+c=5,\ a<5,$ $e\omega_1+\omega_i+f\omega_5+g\omega_6,\ e+f+g=3,$ $2\le i\le 4,\ e<2$ or $i>2,$ $\omega_1+\omega_j+\omega_4,\ 2\le j\le 4,$ $\omega_1+2\omega_3,\ \omega_k+\omega_l+\omega_m,\ 2\le k\le l\le 4,$ $l>2,\ 5\le m\le 6$	42
7	$(9,*)$	$3\omega_2$	43
7	$(9,*)$	$2\omega_1+2\omega_2$	45
7	$(9,*)$	$4\omega_1+\omega_2,\ 2\omega_2+\omega_3$	47
7	$(9,3)$	$\omega_5,\ \omega_6$	12
7	$(9,3)$	$\omega_1+\omega_5,\ \omega_1+\omega_6$	20
7	$(9,3)$	$\omega_2+\omega_5,\ \omega_2+\omega_6$	26
7	$(9,3)$	$2\omega_1+\omega_5,\ 2\omega_1+\omega_6$	28
7	$(9,3)$	$\omega_1+\omega_2+\omega_5,\ \omega_1+\omega_2+\omega_6$	34

TABLE X CONTINUED

p	$\mathrm{cl}(x)$ $(J(x))$	$\overline{\omega}$	d
7	$(9,3)$	$2\omega_2+\omega_5,\ \ 2\omega_2+\omega_6$	40
7	$(9,3)$	$2\omega_1+\omega_2+\omega_5,\ \ 2\omega_1+\omega_2+\omega_6$	42
7	$(9,3)$	$\omega_1+2\omega_2+\omega_5,\ \ \omega_1+2\omega_2+\omega_6$	48
7	$(9,E_3)$	$\omega_5,\ \ \omega_6$	11
7	$(9,E_3)$	$\omega_1+\omega_5,\ \ \omega_1+\omega_6$	19
7	$(9,E_3)$	$\omega_2+\omega_5,\ \ \omega_2+\omega_6$	25
7	$(9,E_3)$	$2\omega_1+\omega_5,\ \ 2\omega_1+\omega_6$	27
7	$(9,E_3)$	$\omega_1+\omega_2+\omega_5,\ \ \omega_1+\omega_2+\omega_6$	33
7	$(9,E_3)$	$2\omega_2+\omega_5,\ \ 2\omega_2+\omega_6$	39
7	$(9,E_3)$	$2\omega_1+\omega_2+\omega_5,\ \ 2\omega_1+\omega_2+\omega_6$	41
7	$(9,E_3)$	$\omega_1+2\omega_2+\omega_5,\ \ \omega_1+2\omega_2+\omega_6$	47

TABLE XI. $G = D_7(K)$

p	cl(x) $(J(x))$	$\overline{\omega}$	d
3, 5	$(13,1)$	$\omega_6,\ \omega_7$	22
3, 5	$(13,1)$	ω_2	23
3, 5	$(13,1)$	$2\omega_1$	25
3, 5	$(11,*)$	ω_2	19
3, 5	$(11,*)$	$2\omega_1$	21
3	$(11,*)$	ω_3	25
3, 5	$(11,3)$	$\omega_6,\ \omega_7$	17
3, 5	$(11,E_3)$	$\omega_6,\ \omega_7$	16
3	$(11,E_3)$	$\omega_1+\omega_6,\ \omega_1+\omega_7$	26
3	$(7,3,E_4),\ (7,2,2,E_3)$	$\omega_6,\ \omega_7$	8
3	$(7,E_7)$	$\omega_6,\ \omega_7$	7
3	$(5,5,3,1),\ (5,5,2,2)$	$\omega_6,\ \omega_7$	8
3	$(5,5,E_4)$	$\omega_6,\ \omega_7$	7
3	$(5,4,4,1)$	$\omega_2,\ \omega_6,\ \omega_7$	8
3	$(5,<4,*)$	ω_2	7
3	$(5,3,>1,*)$	$\omega_6,\ \omega_7$	6
3	$(5,3,E_6),\ (5,2,2,E_5)$	$\omega_6,\ \omega_7$	5
3	$(5,2,*)$	ω_3	8
3	$(5,E_9)$	$\omega_6,\ \omega_7$	4
3	$(5,E_9)$	$\omega_i,\ 2\le i\le 5,\ a\omega_6+b\omega_7,\ a+b=2$	7
3	$(5,E_9)$	$\omega_1+\omega_6,\ \omega_1+\omega_7$	8
3	$(4,4,*)$	$\omega_2,\ 2\omega_1$	7
3	$(4,4,>1,*)$	$\omega_6,\ \omega_7$	6
3	$(4,4,<3)$	ω_3	8
3	$(4,4,E_6)$	$\omega_6,\ \omega_7$	5
3	$(4,4,E_6)$	$\omega_1+\omega_6,\ \omega_1+\omega_7$	8
5	$(9,*)$	$2\omega_1,\ \omega_2$	15
5	$(9,*)$	ω_3	19
5	$(9,*)$	$3\omega_1,\ \omega_1+\omega_2$	20
5	$(9,*)$	ω_4	21
5	$(9,>1,*)$	$\omega_1+\omega_6,\ \omega_1+\omega_7$	20
5	$(9,5)$	$\omega_6,\ \omega_7$	14
5	$(9,3,E_2),\ (9,2,2,1)$	$\omega_6,\ \omega_7$	12
5	$(9,3,E_2),\ (9,2,2,1)$	$a\omega_6+b\omega_7,\ a+b=2$	23
5	$(9,3,E_2)$	ω_5	23
5	$(9,2,2,1)$	ω_5	22
5	$(9,E_5)$	$\omega_6,\ \omega_7$	11
5	$(9,E_5)$	$\omega_1+\omega_6,\ \omega_1+\omega_7$	19
5	$(9,E_5)$	$\omega_5,\ a\omega_6+b\omega_7,\ a+b=2$	21
5	$(7,*)$	$2\omega_1$	13
5	$(7,*)$	$3\omega_1$	19
5	$(7,7)$	$\omega_2,\ \omega_6,\ \omega_7$	13
5	$(7,7)$	ω_3	17
5	$(7,7)$	$\omega_1+\omega_2,\ \omega_1+\omega_6,\ \omega_1+\omega_7$	19
5	$(7,7)$	ω_4	21
5	$(7,7)$	$\omega_1+\omega_3,\ \omega_5$	23
5	$(7,<7,*)$	ω_2	11
5	$(7,<7,*)$	$\omega_1+\omega_2$	17

TABLE XI CONTINUED

p	$\mathrm{cl}(x)$ $(J(x))$	$\overline{\omega}$	d
5	$(7,<7,*)$	$2\omega_2$	21
5	$(7,<7,*)$	$2\omega_1+\omega_2$	23
5	$(7,5,E_2)$, $(7,3,*)$, $(7,2,2,E_3)$	ω_5, $a\omega_6+b\omega_7$, $a+b=2$	15
5	$(7,5,E_2)$, $(7,3,*)$, $(7,2,2,E_3)$	$a\omega_1+b\omega_5+c\omega_6$, $a+b+c=3$, $a<3$, $\omega_1+\omega_4$, $\omega_1+\omega_5$, $\omega_i+\omega_j$, $3\le i\le 5$, $6\le j\le 7$	20
5	$(7,5,E_2)$, $(7,3,*)$	ω_4	15
5	$(7,5,E_2)$, $(7,3,>1,*)$	$\omega_1+\omega_6$, $\omega_1+\omega_7$	15
5	$(7,5,E_2)$	ω_6, ω_7	10
5	$(7,5,E_2)$	ω_3	15
5	$(7,5,E_2)$	$\omega_1+\omega_3$, $\omega_2+\omega_6$, $\omega_2+\omega_7$	20
5	$(7,<5,*)$	ω_3	13
5	$(7,<5,*)$	$\omega_1+\omega_3$	19
5	$(7,<5,*)$	$\omega_2+\omega_3$	23
5	$(7,3,>1,*)$	ω_6, ω_7	9
5	$(7,3,>1,*)$	$\omega_2+\omega_6$, $\omega_2+\omega_7$	19
5	$(7,3,E_4)$, $(7,2,2,E_3)$	ω_6, ω_7	8
5	$(7,3,E_4)$, $(7,2,2,E_3)$	$\omega_1+\omega_6$, $\omega_1+\omega_7$	14
5	$(7,3,E_4)$, $(7,2,2,E_3)$	$\omega_2+\omega_6$, $\omega_2+\omega_7$	18
5	$(7,3,E_4)$, $(7,2,2,E_3)$	$\omega_1+\omega_2+\omega_6$, $\omega_1+\omega_2+\omega_7$	24
5	$(7,2,2,E_3)$	ω_4	14
5	$(7,2,2,E_3)$	$\omega_2+\omega_4$	24
5	$(7,E_7)$	ω_6, ω_7	7
5	$(7,E_7)$	$\omega_1+\omega_6$, $\omega_1+\omega_7$, ω_4, ω_5	13
5	$(7,E_7)$	$\omega_2+\omega_6$, $\omega_2+\omega_7$	17
5	$(7,E_7)$	$a\omega_1+b\omega_5+c\omega_6$, $a+b+c=3$, $\omega_1+\omega_4$, $\omega_1+\omega_5$, $\omega_i+\omega_j$, $3\le i\le 5$, $6\le j\le 7$	19
5	$(7,E_7)$	$a\omega_1+\omega_2+b\omega_6+c\omega_7$, $a+b+c=2$, $\omega_2+\omega_4$, $\omega_2+\omega_5$	23
5	$(6,6,E_2)$	ω_6, ω_7	10
5	$(6,6,E_2)$	$2\omega_1$, ω_2	11
5	$(6,6,E_2)$	ω_3	14
5	$(6,6,E_2)$	$\omega_1+\omega_6$, $\omega_1+\omega_7$, ω_4, ω_5, $a\omega_6+b\omega_7$, $a+b=2$	15
5	$(6,6,E_2)$	$3\omega_1$, $\omega_1+\omega_2$	16
5	$(6,6,E_2)$	$\omega_1+\omega_3$	19
5	$(6,6,E_2)$	$a\omega_1+b\omega_6+c\omega_7$, $a+b+c=3$, $a<3$, $\omega_1+\omega_4$, $\omega_1+\omega_5$, $\omega_i+\omega_j$, $2\le i\le 5$, $6\le j\le 7$	20
5	$(6,6,E_2)$	$4\omega_1$, $2\omega_1+\omega_2$, $2\omega_2$	21
5	$(6,6,E_2)$	$2\omega_1+\omega_3$, $\omega_2+\omega_3$	24
7	$(13,1)$	$2\omega_1$, ω_2	21
7	$(13,1)$	ω_6, ω_7	22
7	$(13,1)$	$3\omega_1$, $\omega_1+\omega_2$, ω_3	28
7	$(13,1)$	$\omega_1+\omega_6$, $\omega_1+\omega_7$	34
7	$(13,1)$	$4\omega_1$, $2\omega_1+\omega_2$, $\omega_1+\omega_3$, $2\omega_2$, ω_4	35
7	$(13,1)$	ω_5	41

TABLE XI CONTINUED

p	cl(x) ($J(x)$)	$\overline{\omega}$	d
7	$(13,1)$	$5\omega_1,\ 3\omega_1+\omega_2,\ \omega_1+2\omega_2,\ 2\omega_1+\omega_3,$ $2\omega_1+\omega_6,\ 2\omega_1+\omega_7,$ $\omega_1+\omega_4,\ \omega_2+\omega_3,\ \omega_2+\omega_6,\ \omega_2+\omega_7$	42
7	$(13,1)$	$a\omega_6+b\omega_7,\ a+b=2$	43
7	$(11,*)$	ω_2	19
7	$(11,*)$	$2\omega_1$	21
7	$(11,*)$	ω_3	25
7	$(11,*)$	$3\omega_1,\ \omega_1+\omega_2$	28
7	$(11,*)$	ω_4	29
7	$(11,*)$	ω_5	31
7	$(11,*)$	$4\omega_1,\ 2\omega_1+\omega_2,\ 2\omega_1+\omega_6,\ 2\omega_1+\omega_7,$ $\omega_1+\omega_3,\ 2\omega_2$	35
7	$(11,*)$	$\omega_1+\omega_4$	39
7	$(11,*)$	$\omega_1+\omega_5$	41
7	$(11,*)$	$5\omega_1,\ 3\omega_1+\omega_2,\ 3\omega_1+\omega_6,\ 3\omega_1+\omega_7,$ $2\omega_1+\omega_3,\ \omega_1+\omega_2+\omega_6,\ \omega_1+\omega_2+\omega_7,$ $\omega_1+2\omega_2,\ \omega_2+\omega_3$	42
7	$(11,*)$	$\omega_2+\omega_4$	47
7	$(11,3)$	$\omega_6,\ \omega_7$	17
7	$(11,3)$	$\omega_1+\omega_6,\ \omega_1+\omega_7$	27
7	$(11,3)$	$a\omega_6+b\omega_7,\ a+b=2$	33
7	$(11,3)$	$\omega_2+\omega_6,\ \omega_2+\omega_7$	35
7	$(11,3)$	$\omega_3+\omega_6,\ \omega_3+\omega_7$	41
7	$(11,3)$	$\omega_1+a\omega_6+b\omega_7,\ a+b=2$	42
7	$(11,3)$	$\omega_4+\omega_6,\ \omega_4+\omega_7$	45
7	$(11,3)$	$\omega_5+\omega_6,\ \omega_5+\omega_7$	47
7	$(11,E_3)$	$\omega_6,\ \omega_7$	16
7	$(11,E_3)$	$\omega_1+\omega_6,\ \omega_1+\omega_7$	26
7	$(11,E_3)$	$a\omega_6+b\omega_7,\ a+b=2$	31
7	$(11,E_3)$	$\omega_2+\omega_6,\ \omega_2+\omega_7$	34
7	$(11,E_3)$	$\omega_3+\omega_6,\ \omega_3+\omega_7$	40
7	$(11,E_3)$	$\omega_1+a\omega_6+b\omega_7,\ a+b=2$	41
7	$(11,E_3)$	$\omega_4+\omega_6,\ \omega_4+\omega_7$	44
7	$(11,E_3)$	$\omega_5+\omega_6,\ \omega_5+\omega_7,\ a\omega_6+b\omega_7,\ a+b=3$	46
7	$(9,*)$	ω_2	15
7	$(9,*)$	$2\omega_1$	17
7	$(9,*)$	ω_3	19
7	$(9,*)$	$\omega_4,\ \omega_5,\ a\omega_6+b\omega_7,\ a+b=2$	21
7	$(9,*)$	$\omega_1+\omega_2$	23
7	$(9,*)$	$3\omega_1$	25
7	$(9,*)$	$\omega_1+\omega_3$	27
7	$(9,*)$	$a\omega_6+b\omega_7,\ a+b=3,\ \omega_1+c\omega_6+e\omega_7,$ $c+e=2,\ \omega_1+\omega_4,$ $\omega_1+\omega_5,\ \omega_i+\omega_j,\ 3\leq i\leq 5,\ 6\leq j\leq 7$	28
7	$(9,*)$	$2\omega_2$	29
7	$(9,*)$	$2\omega_1+\omega_2$	31
7	$(9,*)$	$4\omega_1,\ \omega_2+\omega_3$	33

TABLE XI CONTINUED

p	$\mathrm{cl}(x)$ $(J(x))$	$\overline{\omega}$	d
7	$(9,*)$	$a\omega_1+b\omega_6+c\omega_7,\ a+b+c=4,\ a<4,$ $2\omega_1+\omega_i,\ \omega_1+\omega_i+\omega_j,$ $3\le i\le 5,\ 6\le j\le 7, \omega_2+\omega_4,\ \omega_2+\omega_5,$ $\omega_k+e\omega_6+f\omega_7,\ 2\le k\le 5, e+f=2,$ $\omega_l+\omega_m,\ 3\le l\le m\le 5$	35
7	$(9,*)$	$\omega_1+2\omega_2$	37
7	$(9,*)$	$3\omega_1+\omega_2$	39
7	$(9,*)$	$5\omega_1,\ \omega_1+\omega_2+\omega_3$	41
7	$(9,*)$	$a\omega_1+b\omega_6+c\omega_7,\ a+b+c=5,\ a<5,$ $e\omega_1+\omega_i+f\omega_6+g\omega_7,\ e+f+g=3,$ $2\le i\le 5,\ i>2$ or $e<2,$ $\omega_j+\omega_k+\omega_l,\ j\in\{1,6,7\},\ 2\le k\le l\le 5,$ $l>2,\ j+k+l>6$	42
7	$(9,*)$	$3\omega_2$	43
7	$(9,*)$	$2\omega_1+2\omega_2$	45
7	$(9,*)$	$4\omega_1+\omega_2,\ 2\omega_2+\omega_3$	47
7	$(9,>1,*)$	$2\omega_1+\omega_6,\ 2\omega_1+\omega_7$	28
7	$(9,>1,*)$	$2\omega_1+\omega_2+\omega_6,\ 2\omega_1+\omega_2+\omega_7$	42
7	$(9,5)$	$\omega_6,\ \omega_7$	14
7	$(9,5)$	$\omega_1+\omega_6,\ \omega_1+\omega_7$	21
7	$(9,5)$	$\omega_2+\omega_6,\ \omega_2+\omega_7$	28
7	$(9,5)$	$\omega_1+\omega_2+\omega_6,\ \omega_1+\omega_2+\omega_7$	35
7	$(9,5)$	$2\omega_2+\omega_6,\ 2\omega_2+\omega_7$	42
7	$(9,3,E_2),\ (9,2,2,1)$	$\omega_6,\ \omega_7$	12
7	$(9,3,E_2),\ (9,2,2,1)$	$\omega_1+\omega_6,\ \omega_1+\omega_7$	20
7	$(9,3,E_2),\ (9,2,2,1)$	$\omega_2+\omega_6,\ \omega_2+\omega_7$	26
7	$(9,3,E_2),\ (9,2,2,1)$	$\omega_1+\omega_2+\omega_6,\ \omega_1+\omega_2+\omega_7$	34
7	$(9,3,E_2),\ (9,2,2,1)$	$2\omega_2+\omega_6,\ 2\omega_2+\omega_7$	40
7	$(9,3,E_2),\ (9,2,2,1)$	$\omega_1+2\omega_2+\omega_6,\ \omega_1+2\omega_2+\omega_7$	48
7	$(9,E_5)$	$\omega_6,\ \omega_7$	11
7	$(9,E_5)$	$\omega_1+\omega_6,\ \omega_1+\omega_7$	19
7	$(9,E_5)$	$\omega_2+\omega_6,\ \omega_2+\omega_7$	25
7	$(9,E_5)$	$2\omega_1+\omega_6,\ 2\omega_1+\omega_7$	27
7	$(9,E_5)$	$\omega_1+\omega_2+\omega_6,\ \omega_1+\omega_2+\omega_7$	33
7	$(9,E_5)$	$2\omega_2+\omega_6,\ 2\omega_2+\omega_7$	39
7	$(9,E_5)$	$2\omega_1+\omega_2+\omega_6,\ 2\omega_1+\omega_2+\omega_7$	41
7	$(9,E_5)$	$\omega_1+2\omega_2+\omega_6,\ \omega_1+2\omega_2+\omega_7$	47
11	$(13,1)$	$\omega_6,\ \omega_7$	22
11	$(13,1)$	ω_2	23
11	$(13,1)$	$2\omega_1$	25
11	$(13,1)$	ω_3	31
11	$(13,1)$	$\omega_1+\omega_6,\ \omega_1+\omega_7,\ \omega_4,\ \omega_5,\ a\omega_6+b\omega_7,$ $a+b=2$	33
11	$(13,1)$	$\omega_1+\omega_2$	35
11	$(13,1)$	$3\omega_1$	37
11	$(13,1)$	$\omega_1+\omega_3$	43

TABLE XI CONTINUED

p	cl(x) $(J(x))$	$\overline{\omega}$	d
11	$(13,1)$	$a\omega_1+b\omega_6+c\omega_7,\ a+b+c=3,\ a<3,$ $\omega_1+\omega_4,\ \omega_1+\omega_5,\ \omega_i+\omega_j,$ $2\le i\le 5,\ 6\le j\le 7$	44
11	$(13,1)$	$2\omega_2$	45
11	$(13,1)$	$2\omega_1+\omega_2$	47
11	$(13,1)$	$4\omega_1$	49
11	$(13,1)$	$\omega_2+\omega_3$	53
11	$(13,1)$	$\sum_{i=1}^7 a_i\omega_i,\ a_1+a_6+a_7+2\sum_{i=2}^5 a_i=4,$ and $a_1a_3+\sum_{i=4}^7 a_i\ne 0$ or $a_3>1$	55
11	$(13,1)$	$\omega_1+2\omega_2$	57
11	$(13,1)$	$3\omega_1+\omega_2$	59
11	$(13,1)$	$5\omega_1$	61
11	$(13,1)$	$\omega_1+\omega_2+\omega_3$	65
11	$(13,1)$	$\sum_{i=1}^7 a_i\omega_i,\ a_1+a_6+a_7+2\sum_{i=2}^5 a_i=5,$ and $\sum_{i=4}^7 a_i\ne 0,$ or $a_3>1,$ or $a_1>a_3=1$	66
11	$(13,1)$	$3\omega_2$	67
11	$(13,1)$	$2\omega_1+2\omega_2$	69
11	$(13,1)$	$4\omega_1+\omega_2$	71
11	$(13,1)$	$6\omega_1$	73
11	$(13,1)$	$2\omega_2+\omega_3$	75
11	$(13,1)$	$\sum_{i=1}^7 a_i\omega_i,\ a_1+a_6+a_7+2\sum_{i=2}^5 a_i=6,$ and $a_1a_3+\sum_{i=4}^7 a_i\ne 0$ or $a_3>1$	77
11	$(13,1)$	$\omega_1+3\omega_2$	79
11	$(13,1)$	$3\omega_1+2\omega_2$	81
11	$(13,1)$	$5\omega_1+\omega_2$	83
11	$(13,1)$	$7\omega_1$	85
11	$(13,1)$	$\omega_1+2\omega_2+\omega_3$	87
11	$(13,1)$	$\sum_{i=1}^7 a_i\omega_i,\ a_1+a_6+a_7+2\sum_{i=2}^5 a_i=7,$ and $\sum_{i=4}^7 a_i\ne 0,$ or $a_3>1,$ or $a_1>a_3=1$	88
11	$(13,1)$	$4\omega_2$	89
11	$(13,1)$	$2\omega_1+3\omega_2$	91
11	$(13,1)$	$4\omega_1+2\omega_2$	93
11	$(13,1)$	$6\omega_1+\omega_2$	95
11	$(13,1)$	$8\omega_1,\ 3\omega_2+\omega_3$	97
11	$(13,1)$	$\sum_{i=1}^7 a_i\omega_i,\ a_1+a_6+a_7+2\sum_{i=2}^5 a_i=8,$ and $a_1a_3+\sum_{i=4}^7 a_i\ne 0$ or $a_3>1$	99
11	$(13,1)$	$\omega_1+4\omega_2$	101
11	$(13,1)$	$3\omega_1+3\omega_2$	103
11	$(13,1)$	$5\omega_1+2\omega_2$	105
11	$(13,1)$	$7\omega_1+\omega_2$	107
11	$(13,1)$	$9\omega_1,\ \omega_1+3\omega_2+\omega_3$	109

TABLE XI CONTINUED

p	$\mathrm{cl}(x)$ $(J(x))$	$\overline{\omega}$	d
11	$(13,1)$	$\sum_{i=1}^{7} a_i\omega_i$, $a_1+a_6+a_7+2\sum_{i=2}^{5} a_i = 9$, and $\sum_{i=4}^{7} a_i \neq 0$, or $a_3 > 1$, or $a_1 > a_3 = 1$	110
11	$(13,1)$	$5\omega_2$	111
11	$(13,1)$	$2\omega_1 + 4\omega_2$	113
11	$(13,1)$	$4\omega_1 + 3\omega_2$	115
11	$(13,1)$	$6\omega_1 + 2\omega_2$	117
11	$(13,1)$	$8\omega_1 + \omega_2$, $4\omega_2 + \omega_3$	119

TABLE XII. $G = D_8(K)$

p	$\text{cl}(x)$ $(J(x))$	$\overline{\omega}$	d
3, 5	$(13,*)$	ω_2	23
3	$(13,*)$	$2\omega_1$	25
3, 5	$(13,3)$	$\omega_7,\ \omega_8$	23
3, 5	$(13,E_3)$	$\omega_7,\ \omega_8$	22
3	$(11,*)$	ω_2	19
3	$(11,*)$	$2\omega_1$	21
3	$(11,*)$	ω_3	25
3	$(11,5)$	$\omega_7,\ \omega_8$	18
3	$(11,3,E_2)$, $(11,2,2,1)$	$\omega_7,\ \omega_8$	17
3	$(11,E_5)$	$\omega_7,\ \omega_8$	16
3	$(11,E_5)$	$\omega_1+\omega_7,\ \omega_1+\omega_8$	26
3	$(7,3,E_6)$, $(7,2,2,E_5)$	$\omega_7,\ \omega_8$	8
3	$(7,E_9)$	$\omega_7,\ \omega_8$	7
3	$(5,5,3,E_3)$, $(5,5,2,2,E_2)$, $(5,4,4,E_3)$	$\omega_7,\ \omega_8$	8
3	$(5,5,E_6)$	$\omega_7,\ \omega_8$	7
3	$(5,4,4,*)$	ω_2	8
3	$(5,<4,*)$	ω_2	7
3	$(5,3,>1,*)$, $(5,2,2,2,2,E_3)$	$\omega_7,\ \omega_8$	6
3	$(5,3,E_8)$, $(5,2,2,E_7)$	$\omega_7,\ \omega_8$	5
3	$(5,2,2,*)$	ω_3	8
3	$(5,E_{11})$	$\omega_7,\ \omega_8$	4
3	$(5,E_{11})$	$\omega_i,\ 2\le i\le 6,\ a\omega_7+b\omega_8,\ a+b=2$	7
3	$(5,E_{11})$	$\omega_1+\omega_7,\ \omega_1+\omega_8$	8
3	$(4,4,*)$	$2\omega_1,\ \omega_2$	7
3	$(4,4,4,4)$, C_1	ω_7	8
3	$(4,4,4,4)$, C_2	ω_8	8
3	$(4,4,3,*)$, $(4,4,2,2,*)$	$\omega_7,\ \omega_8$	6
3	$(4,4,<3)$	ω_3	8
3	$(4,4,E_8)$	$\omega_7,\ \omega_8$	5
3	$(4,4,E_8)$	$\omega_1+\omega_7,\ \omega_1+\omega_8$	8
5	$(11,*)$	ω_2	19
5	$(11,*)$	$2\omega_1$	21
5	$(11,5)$	$\omega_7,\ \omega_8$	19
5	$(11,3,E_2)$, $(11,2,2,1)$	$\omega_7,\ \omega_8$	17
5	$(11,E_5)$	$\omega_7,\ \omega_8$	16
5	$(9,*)$	$2\omega_1,\ \omega_2$	15
5	$(9,*)$	$3\omega_1,\ \omega_1+\omega_2$	20
5	$(9,7)$	$\omega_7,\ \omega_8$	17
5	$(9,7)$	ω_3	20
5	$(9,<7)$	ω_3	19
5	$(9,<7)$	ω_4	21
5	$(9,5,E_2)$, $(9,3,*)$, $(9,2,2,E_3)$	$\omega_1+\omega_7,\ \omega_1+\omega_8$	20
5	$(9,5,E_2)$	$\omega_7,\ \omega_8$	14
5	$(9,3,*)$	ω_5	23

TABLE XII CONTINUED

p	cl(x) ($J(x)$)	$\overline{\omega}$	d
5	$(9,3,>1,*)$	$\omega_7,\ \omega_8$	13
5	$(9,3,2,2)$	ω_6	24
5	$(9,3,E_4),\ (9,2,2,E_3)$	$\omega_7,\ \omega_8$	12
5	$(9,3,E_4),\ (9,2,2,E_3)$	$\omega_6,\ a\omega_7+b\omega_8,\ a+b=2$	23
5	$(9,2,2,E_3)$	ω_5	22
5	$(9,E_7)$	$\omega_7,\ \omega_8$	11
5	$(9,E_7)$	$\omega_1+\omega_7,\ \omega_1+\omega_8$	19
5	$(9,E_7)$	$\omega_5,\ \omega_6,\ a\omega_7+b\omega_8,\ a+b=2$	21
5	$(8,8)$	$2\omega_1,\ \omega_2$	15
5	$(8,8)$	$3\omega_1,\ \omega_1+\omega_2,\ \omega_3$	20
5	$(8,8),C_1$	ω_7	16
5	$(8,8),C_1$	ω_8	17
5	$(8,8),C_1$	$\omega_1+\omega_7$	23
5	$(8,8),C_1$	$\omega_1+\omega_8$	24
5	$(8,8),\ C_2$	ω_8	16
5	$(8,8),\ C_2$	ω_7	17
5	$(8,8),C_2$	$\omega_1+\omega_8$	23
5	$(8,8),C_2$	$\omega_1+\omega_7$	24
5	$(7,*)$	$2\omega_1$	13
5	$(7,*)$	$3\omega_1$	19
5	$(7,7,E_2)$	$\omega_2,\ \omega_7,\ \omega_8$	13
5	$(7,7,E_2)$	ω_3	17
5	$(7,7,E_2)$	$\omega_1+\omega_2,\ \omega_1+\omega_7,\ \omega_1+\omega_8$	19
5	$(7,7,E_2)$	ω_4	21
5	$(7,7,E_2)$	$\omega_1+\omega_3,\ \omega_5$	23
5	$(7,<7,*)$	ω_2	11
5	$(7,<7,*)$	$\omega_1+\omega_2$	17
5	$(7,<7,*)$	$2\omega_2$	21
5	$(7,<7,*)$	$2\omega_1+\omega_2$	23
5	$(7,5,*),\ (7,4,4,1),$ $(7,3,*),\ (7,2,2,*)$	$a\omega_7+b\omega_8,\ a+b=2,\ \omega_5,\ \omega_6$	15
5	$(7,5,*),\ (7,4,4,1),$ $(7,3,*),\ (7,2,2,*)$	$a\omega_1+b\omega_7+c\omega_8,\ a+b+c=3,\ a<3,$ $\omega_1+\omega_i,\ 4\le i\le 6,\ \omega_j+\omega_7,\ \omega_j+\omega_8,\ 3\le j\le 6$	20
5	$(7,5,*),\ (7,4,4,1),$ $(7,3,*)$	ω_4	15
5	$(7,5,*),\ (7,4,4,1),$ $(7,3,>1,*),$ $(7,2,2,2,2,1)$	$\omega_1+\omega_7,\ \omega_1+\omega_8$	15
5	$(7,5,*),\ (7,4,4,1),$ $(7,3,3,3)$	$\omega_7,\ \omega_8$	10
5	$(7,5,*),\ (7,4,4,1),$ $(7,3,3,3)$	$\omega_2+\omega_7,\ \omega_2+\omega_8$	20
5	$(7,5,*),\ (7,4,4,1)$	$\omega_1+\omega_3$	20
5	$(7,5,*)$	ω_3	15
5	$(7,4,4,1)$	ω_3	14
5	$(7,4,4,1)$	$\omega_2+\omega_3$	24
5	$(7,<4,*)$	ω_3	13
5	$(7,<4,*)$	$\omega_1+\omega_3$	19

TABLE XII CONTINUED

p	$\mathrm{cl}(x)$ $(J(x))$	$\overline{\omega}$	d
5	$(7,<4,*)$	$\omega_2+\omega_3$	23
5	$(7,3,3,E_3)$, $(7,3,2,2,E_2)$, $(7,2,2,2,2,1)$	$\omega_7,\ \omega_8$	9
5	$(7,3,3,E_3)$, $(7,3,2,2,E_2)$, $(7,2,2,2,2,1)$	$\omega_2+\omega_7,\ \omega_2+\omega_8$	19
5	$(7,3,E_6)$, $(7,2,2,E_5)$	$\omega_7,\ \omega_8$	8
5	$(7,3,E_6)$, $(7,2,2,E_5)$	$\omega_1+\omega_7,\ \omega_1+\omega_8$	14
5	$(7,3,E_6)$, $(7,2,2,E_5)$	$\omega_2+\omega_7,\ \omega_2+\omega_8$	18
5	$(7,3,E_6)$, $(7,2,2,E_5)$	$\omega_1+\omega_2+\omega_7,\ \omega_1+\omega_2+\omega_8$	24
5	$(7,2,2,*)$	ω_4	14
5	$(7,2,2,*)$	$\omega_2+\omega_4$	24
5	$(7,E_9)$	$\omega_7,\ \omega_8$	7
5	$(7,E_9)$	$\omega_i+\omega_j,\ i,j\in\{1,7,8\},\ \omega_k,\ 3\le k\le 6$	13
5	$(7,E_9)$	$\omega_2+\omega_i,\ i\in\{1,7,8\}$	17
5	$(7,E_9)$	$\omega_i+\omega_j+\omega_k,\ \omega_i+\omega_l,\ i,j,k\in\{1,7,8\},\ 3\le l\le 6$	19
5	$(7,E_9)$	$\omega_2+\omega_i+\omega_j,\ i,j\in\{1,7,8\}$	23
5	$(6,6,*)$	$\omega_7,\ \omega_8$	10
5	$(6,6,*)$	$2\omega_1,\ \omega_2$	11
5	$(6,6,*)$	ω_3	14
5	$(6,6,*)$	$\omega_1+\omega_7,\ \omega_1+\omega_8,\ a\omega_7+b\omega_8,$ $a+b=2,\ \omega_i,\ 4\le i\le 6$	15
5	$(6,6,*)$	$3\omega_1,\ \omega_1+\omega_2$	16
5	$(6,6,*)$	$\omega_1+\omega_3$	19
5	$(6,6,*)$	$a\omega_1+b\omega_7+c\omega_8,\ a+b+c=3,\ a<3,$ $\omega_1+\omega_i,\ 4\le i\le 6,\ \omega_j+\omega_7,\ \omega_j+\omega_8,\ 2\le j\le 6$	20
5	$(6,6,*)$	$4\omega_1,\ 2\omega_1+\omega_2,\ 2\omega_2$	21
5	$(6,6,*)$	$2\omega_1+\omega_3,\ \omega_2+\omega_3$	24
7	$(15,1)$	ω_2	27
7	$(15,1)$	$2\omega_1,\ \omega_7,\ \omega_8$	29
7	$(15,1)$	ω_3	35
7	$(15,1)$	$\omega_1+\omega_2$	41
7	$(15,1)$	ω_4	42
7	$(15,1)$	$3\omega_1,\ \omega_1+\omega_7,\ \omega_1+\omega_8$	43
7	$(13,*)$	$2\omega_1,\ \omega_2$	21
7	$(13,*)$	$3\omega_1,\ \omega_1+\omega_2,\ \omega_3$	28
7	$(13,*)$	$4\omega_1,\ 2\omega_1+\omega_2,\ \omega_1+\omega_3,\ 2\omega_2,\ \omega_4$	35
7	$(13,*)$	ω_5	41
7	$(13,*)$	$5\omega_1,\ 3\omega_1+\omega_2,\ 2\omega_1+\omega_3,\ 2\omega_1+\omega_7,\ 2\omega_1+\omega_8,$ $\omega_1+2\omega_2,\ \omega_1+\omega_4,\ \omega_2+\omega_3,\ \omega_2+\omega_7,\ \omega_2+\omega_8$	42
7	$(13,*)$	ω_6	43
7	$(13,3)$	$\omega_7,\ \omega_8$	23
7	$(13,3)$	$\omega_1+\omega_7,\ \omega_1+\omega_8$	35
7	$(13,3)$	$a\omega_7+b\omega_8,\ a+b=2$	45
7	$(13,E_3)$	$\omega_7,\ \omega_8$	22
7	$(13,E_3)$	$\omega_1+\omega_7,\ \omega_1+\omega_8$	34
7	$(13,E_3)$	$a\omega_7+b\omega_8,\ a+b=2$	43

TABLE XII CONTINUED

p	cl(x) $(J(x))$	$\overline{\omega}$	d
7	$(11,*)$	ω_2	19
7	$(11,*)$	$2\omega_1$	21
7	$(11,*)$	ω_3	25
7	$(11,*)$	$3\omega_1,\ \omega_1+\omega_2$	28
7	$(11,*)$	ω_4	29
7	$(11,*)$	$4\omega_1,\ 2\omega_1+\omega_2,\ 2\omega_1+\omega_7,$ $2\omega_1+\omega_8,\ \omega_1+\omega_3,\ 2\omega_2$	35
7	$(11,*)$	$\omega_1+\omega_4$	39
7	$(11,*)$	$5\omega_1, 3\omega_1+\omega_2,\ 3\omega_1+\omega_7,\ 3\omega_1+\omega_8,\ 2\omega_1+\omega_3,$ $\omega_1+2\omega_2,\ \omega_1+\omega_2+\omega_7,\ \omega_1+\omega_2+\omega_8,\ \omega_2+\omega_3$	42
7	$(11,*)$	$\omega_2+\omega_4$	47
7	$(11,>1,*)$	$\omega_2+\omega_7,\ \omega_2+\omega_8$	35
7	$(11,>1,*)$	$\omega_1+\omega_6,\ \omega_1+a\omega_7+b\omega_8, a+b=2$	42
7	$(11,5)$	$\omega_7,\ \omega_8$	19
7	$(11,5)$	$\omega_1+\omega_7,\ \omega_1+\omega_8$	28
7	$(11,5)$	ω_5	33
7	$(11,5)$	$\omega_6,\ a\omega_7+b\omega_8, a+b=2$	35
7	$(11,5)$	$\omega_1+\omega_5,\ \omega_3+\omega_7,\ \omega_3+\omega_8$	42
7	$(11,5)$	$\omega_4+\omega_7,\ \omega_4+\omega_8$	47
7	$(11,<5,*)$	ω_5	31
7	$(11,<5,*)$	$\omega_1+\omega_5$	41
7	$(11,3,E_2),\ (11,2,2,1)$	$\omega_7,\ \omega_8$	17
7	$(11,3,E_2),\ (11,2,2,1)$	$\omega_1+\omega_7,\ \omega_1+\omega_8$	27
7	$(11,3,E_2),\ (11,2,2,1)$	$a\omega_7+b\omega_8, a+b=2$	33
7	$(11,3,E_2),\ (11,2,2,1)$	$\omega_3+\omega_7,\ \omega_3+\omega_8$	41
7	$(11,3,E_2),\ (11,2,2,1)$	$\omega_4+\omega_7,\ \omega_4+\omega_8$	45
7	$(11,3,E_2),\ (11,2,2,1)$	$\omega_5+\omega_7,\ \omega_5+\omega_8$	47
7	$(11,3,E_2)$	ω_6	33
7	$(11,2,2,1)$	ω_6	32
7	$(11,2,2,1)$	$\omega_6+\omega_7,\ \omega_6+\omega_8$	48
7	$(11,E_5)$	$\omega_7,\ \omega_8$	16
7	$(11,E_5)$	$\omega_1+\omega_7,\ \omega_1+\omega_8$	26
7	$(11,E_5)$	$\omega_6,\ a\omega_7+b\omega_8, a+b=2$	31
7	$(11,E_5)$	$\omega_2+\omega_7,\ \omega_2+\omega_8$	34
7	$(11,E_5)$	$\omega_3+\omega_7,\ \omega_3+\omega_8$	40
7	$(11,E_5)$	$\omega_1+\omega_6,\ \omega_1+a\omega_7+b\omega_8, a+b=2$	41
7	$(11,E_5)$	$\omega_4+\omega_7,\ \omega_4+\omega_8$	44
7	$(11,E_5)$	$\omega_5+\omega_7,\ \omega_5+\omega_8,\ \omega_6+\omega_7,\ \omega_6+\omega_8$ $a\omega_7+b\omega_8, a+b=3$	46
7	$(9,*)$	ω_2	15
7	$(9,*)$	$2\omega_1$	17
7	$(9,*)$	$\omega_i,\ 4\leq i\leq 6,\ a\omega_7+b\omega_8, a+b=2$	21
7	$(9,*)$	$\omega_1+\omega_2$	23
7	$(9,*)$	$3\omega_1$	25
7	$(9,*)$	$\omega_1+\omega_i,\ 4\leq i\leq 6,\ a\omega_1+b\omega_7+c\omega_8,$ $a+b+c=3,\ a<2,$ $\omega_j+\omega_7,\ \omega_j+\omega_8,\ 3\leq j\leq 6$	28
7	$(9,*)$	$2\omega_2$	29

TABLE XII CONTINUED

p	$\mathrm{cl}(x)$ $(J(x))$	$\overline{\omega}$	d
7	$(9,*)$	$2\omega_1+\omega_2$	31
7	$(9,*)$	$4\omega_1$	33
7	$(9,*)$	$2\omega_1+\omega_i,\ \omega_1+\omega_i+\omega_7,\ \omega_1+\omega_i+\omega_8,\ 3\leq i\leq 6,$ $a\omega_1+b\omega_7+c\omega_8,\ a+b+c=4,\ a<4,$ $\omega_j+\omega_k,\ 2\leq j\leq k\leq 6,\ j+k\geq 6,$ $\omega_l+e\omega_7+f\omega_8,\ 2\leq l\leq 6,\ e+f=2$	35
7	$(9,*)$	$\omega_1+2\omega_2$	37
7	$(9,*)$	$3\omega_1+\omega_2$	39
7	$(9,*)$	$5\omega_1$	41
7	$(9,*)$	$a\omega_1+b\omega_7+c\omega_8,\ a+b+c=5,\ a<5,$ $e\omega_1+\omega_i+f\omega_7+g\omega_8,\ e+f+g=3,$ $2\leq i\leq 6,\ i>2$ or $e<2,$ $\omega_j+\omega_k+\omega_l,\ j\in\{1,7,8\},\ 2\leq k\leq l\leq 6,$ $l>2,\ j+k+l>6$	42
7	$(9,*)$	$3\omega_2$	43
7	$(9,*)$	$2\omega_1+2\omega_2$	45
7	$(9,*)$	$4\omega_1+\omega_2$	47
7	$(9,>1,*)$	$2\omega_1++\omega_7,\ 2\omega_1+\omega_8$	28
7	$(9,>1,*)$	$2\omega_1+\omega_2+\omega_7,\ 2\omega_1+\omega_2+\omega_8$	42
7	$(9,>3,*),\ (9,3,>1,*)$	$\omega_1+\omega_7,\ \omega_1+\omega_8$	21
7	$(9,>3,*),\ (9,3,>1,*)$	$\omega_1+\omega_2+\omega_7,\ \omega_1+\omega_2+\omega_8$	35
7	$(9,>3,*)$	$\omega_7,\ \omega_8$	14
7	$(9,>3,*)$	$\omega_2+\omega_7,\ \omega_2+\omega_8$	28
7	$(9,>3,*)$	$2\omega_2+\omega_7,\ 2\omega_2+\omega_8$	42
7	$(9,7)$	ω_3	21
7	$(9,7)$	$\omega_1+\omega_3$	28
7	$(9,7)$	$\omega_2+\omega_3$	35
7	$(9,7)$	$\omega_1+\omega_2+\omega_3$	42
7	$(9,<7,*)$	ω_3	19
7	$(9,<7,*)$	$\omega_1+\omega_3$	27
7	$(9,<7,*)$	$\omega_2+\omega_3$	33
7	$(9,<7,*)$	$\omega_1+\omega_2+\omega_3$	41
7	$(9,<7,*)$	$2\omega_2+\omega_3$	47
7	$(9,3,>1,*)$	$\omega_7,\ \omega_8$	13
7	$(9,3,>1,*)$	$\omega_2+\omega_7,\ \omega_2+\omega_8$	27
7	$(9,3,>1,*)$	$2\omega_2+\omega_7,\ 2\omega_2+\omega_8$	41
7	$(9,3,E_4),\ (9,2,2,E_3)$	$\omega_7,\ \omega_8$	12
7	$(9,3,E_4),\ (9,2,2,E_3)$	$\omega_1+\omega_7,\ \omega_1+\omega_8$	20
7	$(9,3,E_4),\ (9,2,2,E_3)$	$\omega_2+\omega_7,\ \omega_2+\omega_8$	26
7	$(9,3,E_4),\ (9,2,2,E_3)$	$\omega_1+\omega_2+\omega_7,\ \omega_1+\omega_2+\omega_8$	34
7	$(9,3,E_4),\ (9,2,2,E_3)$	$2\omega_2+\omega_7,\ 2\omega_2+\omega_8$	40
7	$(9,3,E_4),\ (9,2,2,E_3)$	$\omega_1+2\omega_2+\omega_7,\ \omega_1+2\omega_2+\omega_8$	48
7	$(9,E_7)$	$\omega_7,\ \omega_8$	11
7	$(9,E_7)$	$\omega_1+\omega_7,\ \omega_1+\omega_8$	19
7	$(9,E_7)$	$\omega_2+\omega_7,\ \omega_2+\omega_8$	25
7	$(9,E_7)$	$2\omega_1+\omega_7,\ 2\omega_1+\omega_8$	27
7	$(9,E_7)$	$\omega_1+\omega_2+\omega_7,\ \omega_1+\omega_2+\omega_8$	33
7	$(9,E_7)$	$2\omega_2+\omega_7,\ 2\omega_2+\omega_8$	39

TABLE XII CONTINUED

p	$\mathrm{cl}(x)$ $(J(x))$	$\overline{\omega}$	d
7	$(9, E_7)$	$2\omega_1+\omega_2+\omega_7,\ 2\omega_1+\omega_2+\omega_8$	41
7	$(9, E_7)$	$\omega_1+2\omega_2+\omega_7,\ \omega_1+2\omega_2+\omega_8$	47
7	$(8,8)$	$\omega_7,\ \omega_8$	14
7	$(8,8)$	$2\omega_1,\ \omega_2$	15
7	$(8,8)$	ω_3	20
7	$(8,8)$	$\omega_1+\omega_7,\ \omega_1+\omega_8,\ \omega_i,\ 4\le i\le 6,$ $a\omega_7+b\omega_8,\ a+b=2$	21
7	$(8,8)$	$3\omega_1,\ \omega_1+\omega_2$	22
7	$(8,8)$	$\omega_1+\omega_3$	27
7	$(8,8)$	$\omega_1+\omega_i,\ 4\le i\le 6,\ a\omega_1+b\omega_7+c\omega_8,$ $a+b+c=3,\ a<3,$ $\omega_j+\omega_7,\ \omega_j+\omega_8,\ 2\le j\le 6$	28
7	$(8,8)$	$4\omega_1,\ 2\omega_1+\omega_2,\ 2\omega_2$	29
7	$(8,8)$	$2\omega_1+\omega_3,\ \omega_2+\omega_3$	34
7	$(8,8)$	$a\omega_1+b\omega_7+c\omega_8,\ a+b+c=4,\ a<4,$ $2\omega_1+\omega_i,\ 4\le i\le 6,$ $\omega_1+\omega_j+\omega_7,\ \omega_1+\omega_j+\omega_8,$ $\omega_j+e\omega_7+f\omega_8,\ 2\le j\le 6,\ e+f=2,$ $\omega_k+\omega_l,\ 2\le k\le l\le 6,\ k+l\ge 6,$	35
7	$(8,8)$	$5\omega_1,\ 3\omega_1+\omega_2,\ \omega_1+2\omega_2$	36
7	$(8,8)$	$3\omega_1+\omega_3,\ \omega_1+\omega_2+\omega_3$	41
7	$(8,8)$	$a\omega_1+b\omega_7+c\omega_8,\ a+b+c=5,\ a<5,$ $e\omega_1+\omega_i+f\omega_7+g\omega_8,\ e+f+g=3,\ 2\le i\le 6,\ i>3$ or $e<3,$ $\omega_j+\omega_k+\omega_l,\ j\in\{1,7,8\},\ 2\le k\le l\le 6,\ j+k+l>6$	42
7	$(8,8)$	$6\omega_1,\ 4\omega_1+\omega_2,\ 2\omega_1+2\omega_2,\ 3\omega_2$	43
7	$(8,8)$	$4\omega_1+\omega_3,\ 2\omega_1+\omega_2+\omega_3,\ 2\omega_2+\omega_3$	48
11	$(15,1)$	ω_2	27
11	$(15,1)$	$2\omega_1,\ \omega_7,\ \omega_8$	29
11	$(15,1)$	ω_3	37
11	$(15,1)$	$\omega_1+\omega_2$	41
11	$(15,1)$	$3\omega_1,\ \omega_1+\omega_7,\ \omega_1+\omega_8$	43
11	$(15,1)$	ω_4	45
11	$(15,1)$	$\omega_1+\omega_3,\ \omega_5$	51
11	$(15,1)$	$2\omega_2$	53
11	$(15,1)$	$4\omega_1,\ 2\omega_1+\omega_i,\ \omega_i+\omega_j,\ i,j\in\{2,7,8\},\ i+j>4,\ \omega_6$	55
11	$(15,1)$	$\omega_1+\omega_4$	59
11	$(15,1)$	$\omega_2+\omega_3$	63
11	$(15,1)$	$2\omega_1+\omega_3,\ \omega_1+\omega_5,\ \omega_3+\omega_7,\ \omega_3+\omega_8$	65
11	$(15,1)$	$5\omega_1,\ 3\omega_1+\omega_i,\ \omega_1+\omega_i+\omega_j,\ i,j\in\{2,7,8\},\ \omega_1+\omega_6$	66
11	$(15,1)$	$\omega_2+\omega_4$	71
11	$(15,1)$	$2\omega_3,\ 2\omega_1+\omega_4,\ \omega_4+\omega_7,\ \omega_4+\omega_8$	73
11	$(15,1)$	$6\omega_1,\ 4\omega_1+\omega_i,\ 3\omega_1+\omega_3,\ 2\omega_1+\omega_i+\omega_j,\ 2\omega_1+\omega_l,\ \omega_1+\omega_i+\omega_3,$ $\omega_i+\omega_j+\omega_k,\ \omega_i+\omega_l,\ i,j,k\in\{2,7,8\},\ 5\le l\le 6$	77
11	$(15,1)$	$\omega_3+\omega_4$	81
11	$(15,1)$	$\omega_1+\omega_2+\omega_4$	85
11	$(15,1)$	$3\omega_1+\omega_4,\ \omega_1+2\omega_3,\ \omega_1+\omega_4+\omega_7,\ \omega_1+\omega_4+\omega_8,\ \omega_3+\omega_5$	87

TABLE XII CONTINUED

p	$\mathrm{cl}(x)$ $(J(x))$	$\overline{\omega}$	d
11	$(15,1)$	$7\omega_1,\ 5\omega_1+\omega_i,\ 4\omega_1+\omega_3,\ 3\omega_1+\omega_i+\omega_j,\ 3\omega_1+\omega_l,\ 2\omega_1+\omega_i+\omega_3,$ $\omega_1+\omega_i+\omega_l,\ \omega_1+\omega_i+\omega_j+\omega_k,\ \omega_i+\omega_j+\omega_3,$ $i,j,k\in\{2,7,8\},\ 5\le l\le 6,\ \omega_3+\omega_6$	88
11	$(15,1)$	$2\omega_4$	89
11	$(15,1)$	$\omega_1+\omega_3+\omega_4,\ \omega_4+\omega_5$	95
11	$(15,1)$	$2\omega_2+\omega_4$	97
11	$(15,1)$	$8\omega_1,\ 6\omega_1+\omega_i,\ 5\omega_1+\omega_3,\ 4\omega_1+\omega_i+\omega_j,\ 4\omega_1+\omega_l,\ 3\omega_1+\omega_i+\omega_3,$ $2\omega_1+\omega_i+\omega_l,\ 2\omega_1+\omega_i+\omega_j+\omega_k,\ 2\omega_1+2\omega_3,\ \omega_1+\omega_i+\omega_j+\omega_3,$ $\omega_1+\omega_3+\omega_m,\ \omega_i+2\omega_3,\ i,j,k\in\{2,7,8\},\ 4\le l\le 6,\ 5\le m\le 6,$ $\omega_i+\omega_j+\omega_l,\ i+j+l>8,\ a\omega_2+b\omega_7+c\omega_8,\ a+b+c=4,$ $\omega_4+\omega_6,\ e\omega_5+f\omega_6,\ e+f=2$	99
11	$(15,1)$	$\omega_1+2\omega_4$	103
11	$(15,1)$	$\omega_2+\omega_3+\omega_4$	107
11	$(15,1)$	$2\omega_1+\omega_3+\omega_4,\ \omega_1+\omega_4+\omega_5,\ 3\omega_3,\ \omega_3+\omega_4+\omega_7,\ \omega_3+\omega_4+\omega_8$	109
11	$(15,1)$	$9\omega_1,\ 7\omega_1+\omega_i,\ 6\omega_1+\omega_3,\ 5\omega_1+\omega_i+\omega_j,\ 5\omega_1+\omega_l,\ 4\omega_1+\omega_i+\omega_3,$ $3\omega_1+\omega_i+\omega_l,\ 3\omega_1+\omega_i+\omega_j+\omega_k,\ 3\omega_1+2\omega_3,\ 2\omega_1+\omega_i+\omega_j+\omega_3,$ $2\omega_1+\omega_3+\omega_m,\ \omega_1+\omega_i+2\omega_3,\ \omega_1+\omega_i+\omega_j+\omega_l,\ \omega_3+\omega_i+\omega_m,$ $i,j,k\in\{2,7,8\},\ 4\le l\le 6,\ 5\le m\le 6,$ $\omega_1+a\omega_2+b\omega_7+c\omega_8,\ a+b+c=4,\ \omega_1+\omega_4+\omega_6,$ $\omega_1+e\omega_5+f\omega_6,\ e+f=2,\ \omega_3+q\omega_2+s\omega_7+t\omega_8,\ q+s+t=3$	110
11	$(15,1)$	$\omega_2+2\omega_4$	115
11	$(15,1)$	$2\omega_1+2\omega_4,\ 2\omega_3+\omega_4,\ 2\omega_4+\omega_7,\ 2\omega_4+\omega_8$	117
11	$(13,*)$	$\omega_7,\ \omega_8$	22
11	$(13,*)$	ω_2	23
11	$(13,*)$	$2\omega_1$	25
11	$(13,*)$	ω_3	31
11	$(13,*)$	$\omega_1+\omega_7,\ \omega_1+\omega_8,\ a\omega_7+b\omega_8,\ a+b=2,\ \omega_i,\ 4\le i\le 6$	33
11	$(13,*)$	$\omega_1+\omega_2$	35
11	$(13,*)$	$3\omega_1$	37
11	$(13,*)$	$\omega_1+\omega_3$	43
11	$(13,*)$	$\omega_1+\omega_i,\ 4\le i\le 6,\ a\omega_1+b\omega_7+c\omega_8,\ a+b+c=3,\ a<3,$ $\omega_j+\omega_7,\ \omega_j+\omega_8,\ 2\le j\le 6$	44
11	$(13,*)$	$2\omega_2$	45
11	$(13,*)$	$2\omega_1+\omega_2$	47
11	$(13,*)$	$4\omega_1$	49
11	$(13,*)$	$\omega_2+\omega_3$	53
11	$(13,*)$	$\sum_{i=1}^{8}a_i\omega_i,\ a_1+a_7+a_8+2\sum_{i=2}^{6}a_i=4,$ and $a_1a_3+\sum_{i=4}^{8}a_i\ne 0$ or $a_3>1$	55
11	$(13,*)$	$\omega_1+2\omega_2$	57
11	$(13,*)$	$3\omega_1+\omega_2$	59
11	$(13,*)$	$5\omega_1$	61
11	$(13,*)$	$\omega_1+\omega_2+\omega_3$	65
11	$(13,*)$	$\sum_{i=1}^{8}a_i\omega_i,\ a_1+a_7+a_8+2\sum_{i=2}^{6}a_i=5,$ and $\sum_{i=4}^{8}a_i\ne 0,$ or $a_3>1,$ or $a_1>a_3=1$	66
11	$(13,*)$	$3\omega_2$	67
11	$(13,*)$	$2\omega_1+2\omega_2$	69
11	$(13,*)$	$4\omega_1+\omega_2$	71
11	$(13,*)$	$6\omega_1$	73

TABLE XII CONTINUED

p	cl(x) ($J(x)$)	$\overline{\omega}$	d
11	$(13,*)$	$2\omega_2+\omega_3$	75
11	$(13,*)$	$\sum_{i=1}^{8} a_i\omega_i,\ a_1+a_7+a_8+2\sum_{i=2}^{6} a_i=6$, and $a_1a_3+\sum_{i=4}^{8} a_i \neq 0$ or $a_3>1$	77
11	$(13,*)$	$\omega_1+3\omega_2$	79
11	$(13,*)$	$3\omega_1+2\omega_2$	81
11	$(13,*)$	$5\omega_1+\omega_2$	83
11	$(13,*)$	$7\omega_1$	85
11	$(13,*)$	$\omega_1+2\omega_2+\omega_3$	87
11	$(13,*)$	$\sum_{i=1}^{8} a_i\omega_i,\ a_1+a_7+a_8+2\sum_{i=2}^{6} a_i=7$, and $\sum_{i=4}^{8} a_i\neq 0$, or $a_3>1$, or $a_1>a_3=1$	88
11	$(13,*)$	$4\omega_2$	89
11	$(13,*)$	$2\omega_1+3\omega_2$	91
11	$(13,*)$	$4\omega_1+2\omega_2$	93
11	$(13,*)$	$6\omega_1+\omega_2$	95
11	$(13,*)$	$8\omega_1,\ 3\omega_2+\omega_3$	97
11	$(13,*)$	$\sum_{i=1}^{8} a_i\omega_i,\ a_1+a_7+a_8+2\sum_{i=2}^{6} a_i=8$, and $a_1a_3+\sum_{i=4}^{8} a_i\neq 0$, or $a_3>1$	99
11	$(13,*)$	$\omega_1+4\omega_2$	101
11	$(13,*)$	$3\omega_1+3\omega_2$	103
11	$(13,*)$	$5\omega_1+2\omega_2$	105
11	$(13,*)$	$7\omega_1+\omega_2$	107
11	$(13,*)$	$9\omega_1,\ \omega_1+3\omega_2+\omega_3$	109
11	$(13,*)$	$\sum_{i=1}^{8} a_i\omega_i,\ a_1+a_7+a_8+2\sum_{i=2}^{6} a_i=9$, and $\sum_{i=4}^{8} a_i\neq 0$, or $a_3>1$, or $a_1>a_3=1$	110
11	$(13,*)$	$5\omega_2$	111
11	$(13,*)$	$2\omega_1+4\omega_2$	113
11	$(13,*)$	$4\omega_1+3\omega_2$	115
11	$(13,*)$	$6\omega_1+2\omega_2$	117
11	$(13,*)$	$8\omega_1+\omega_2,\ 4\omega_2+\omega_3$	119
13	$(15,1)$	$\omega_7,\ \omega_8$	26
13	$(15,1)$	ω_2	27
13	$(15,1)$	$2\omega_1$	29
13	$(15,1)$	ω_3	37
13	$(15,1)$	$\omega_1+\omega_7,\ \omega_1+\omega_8,\ a\omega_7+b\omega_8,\ a+b=2,\ \omega_i,\ 4\leq i\leq 6$	39
13	$(15,1)$	$\omega_1+\omega_2$	41
13	$(15,1)$	$3\omega_1$	43
13	$(15,1)$	$\omega_1+\omega_3$	51
13	$(15,1)$	$a\omega_1+b\omega_7+c\omega_8,\ a+b+c=3,\ a<3,\ \omega_1+\omega_i,\ 4\leq i\leq 6,$ $\omega_j+\omega_7,\ \omega_j+\omega_8,\ 2\leq j\leq 6$	52
13	$(15,1)$	$2\omega_2$	53
13	$(15,1)$	$2\omega_1+\omega_2$	55
13	$(15,1)$	$4\omega_1$	57
13	$(15,1)$	$\omega_2+\omega_3$	63
13	$(15,1)$	$\sum_{i=1}^{8} a_i\omega_i,\ a_1+a_7+a_8+2\sum_{i=2}^{6} a_i=4$, and $a_1a_3+\sum_{i=4}^{8} a_i\neq 0$ or $a_3>1$	65
13	$(15,1)$	$\omega_1+2\omega_2$	67
13	$(15,1)$	$3\omega_1+\omega_2$	69
13	$(15,1)$	$5\omega_1$	71

TABLE XII CONTINUED

p	cl(x) ($J(x)$)	$\overline{\omega}$	d
13	$(15,1)$	$\omega_1+\omega_2+\omega_3$	77
13	$(15,1)$	$\sum_{i=1}^{8} a_i\omega_i$, $a_1+a_7+a_8+2\sum_{i=2}^{6} a_i=5$, and $\sum_{i=4}^{8} a_i\neq 0$, or $a_3>1$, or $a_1>a_3=1$	78
13	$(15,1)$	$3\omega_2$	79
13	$(15,1)$	$2\omega_1+2\omega_2$	81
13	$(15,1)$	$4\omega_1+\omega_2$	83
13	$(15,1)$	$6\omega_1$	85
13	$(15,1)$	$2\omega_2+\omega_3$	89
13	$(15,1)$	$\sum_{i=1}^{8} a_i\omega_i$, $a_1+a_7+a_8+2\sum_{i=2}^{6} a_i=6$, and $a_1a_3+\sum_{i=4}^{8} a_i\neq 0$ or $a_3>1$	91
13	$(15,1)$	$\omega_1+3\omega_2$	93
13	$(15,1)$	$3\omega_1+2\omega_2$	95
13	$(15,1)$	$5\omega_1+\omega_2$	97
13	$(15,1)$	$7\omega_1$	99
13	$(15,1)$	$\omega_1+2\omega_2+\omega_3$	103
13	$(15,1)$	$\sum_{i=1}^{8} a_i\omega_i$, $a_1+a_7+a_8+2\sum_{i=2}^{6} a_i=7$, and $\sum_{i=4}^{8} a_i\neq 0$, or $a_3>1$, or $a_1>a_3=1$	104
13	$(15,1)$	$4\omega_2$	105
13	$(15,1)$	$2\omega_1+3\omega_2$	107
13	$(15,1)$	$4\omega_1+2\omega_2$	109
13	$(15,1)$	$6\omega_1+\omega_2$	111
13	$(15,1)$	$8\omega_1$	113
13	$(15,1)$	$3\omega_2+\omega_3$	115
13	$(15,1)$	$\sum_{i=1}^{8} a_i\omega_i$, $a_1+a_7+a_8+2\sum_{i=2}^{6} a_i=8$, and $a_1a_3+\sum_{i=4}^{8} a_i\neq 0$ or $a_3>1$, or $a_1>a_3=1$	117
13	$(15,1)$	$\omega_1+4\omega_2$	119
13	$(15,1)$	$3\omega_1+3\omega_2$	121
13	$(15,1)$	$5\omega_1+2\omega_2$	123
13	$(15,1)$	$7\omega_1+\omega_2$	125
13	$(15,1)$	$9\omega_1$	127
13	$(15,1)$	$\omega_1+3\omega_2+\omega_3$	129
13	$(15,1)$	$\sum_{i=1}^{8} a_i\omega_i$, $a_1+a_7+a_8+2\sum_{i=2}^{6} a_i=9$, and $\sum_{i=4}^{8} a_i\neq 0$, or $a_3>1$, or $a_1>a_3=1$	130
13	$(15,1)$	$5\omega_2$	131
13	$(15,1)$	$2\omega_1+4\omega_2$	133
13	$(15,1)$	$4\omega_1+3\omega_2$	135
13	$(15,1)$	$6\omega_1+2\omega_2$	137
13	$(15,1)$	$8\omega_1+\omega_2$	139
13	$(15,1)$	$10\omega_1$, $4\omega_2+\omega_3$	141
13	$(15,1)$	$\sum_{i=1}^{8} a_i\omega_i$, $a_1+a_7+a_8+2\sum_{i=2}^{6} a_i=10$, and $a_1a_3+\sum_{i=4}^{8} a_i\neq 0$ or $a_3>1$	143
13	$(15,1)$	$\omega_1+5\omega_2$	145
13	$(15,1)$	$3\omega_1+4\omega_2$	147
13	$(15,1)$	$5\omega_1+3\omega_2$	149
13	$(15,1)$	$7\omega_1+2\omega_2$	151
13	$(15,1)$	$9\omega_1+\omega_2$	153
13	$(15,1)$	$11\omega_1$, $\omega_1+4\omega_2+\omega_3$	155

TABLE XII CONTINUED

p	cl(x) ($J(x)$)	$\overline{\omega}$	d
13	$(15,1)$	$\sum_{i=1}^{8} a_i\omega_i$, $a_1+a_7+a_8+2\sum_{i=2}^{6} a_i = 11$, and $\sum_{i=4}^{8} a_i \neq 0$, or $a_3 > 1$, or $a_1 > a_3 = 1$	156
13	$(15,1)$	$6\omega_2$	157
13	$(15,1)$	$2\omega_1+5\omega_2$	159
13	$(15,1)$	$4\omega_1+4\omega_2$	161
13	$(15,1)$	$6\omega_1+3\omega_2$	163
13	$(15,1)$	$8\omega_1+2\omega_2$	165
13	$(15,1)$	$10\omega_1+\omega_2$, $5\omega_2+\omega_3$	167

Bibliography

[BC76] P Bala and R. W. Carter, *Classes of unipotent elements in simple algebraic groups, I, II*, Proc. Cambridge Phil. Soc. **79, 80** (1976), 401–425, 1–17.

[Bor70] A. Borel, *Properties and linear representations of Chevalley groups*, Seminar on Algebraic Groups and Related Finite Groups 1968/69 (Berlin) (A. Borel, ed.), Lecture Notes in Mathematics, vol. 131, Springer, 1970, pp. A1–A55.

[Bou68] B. Bourbaki, *Groupes et algebres de Lie, Chaps. IV–VI*, Hermann, Paris, 1968.

[Bou75] ———, *Groupes et algebres de Lie, Chaps. VII–VIII*, Hermann, Paris, 1975.

[Car85] R. W. Carter, *Finite groups of Lie type: conjugacy classes and complex characters*, Wiley, Chichester, 1985.

[CC76] R. W. Carter and E. Cline, *The submodule structure of Weyl modules for groups of type* A_1, Proceedings of the Conference on Finite Groups (Park City, Utah 1975) (W. R. Scott and F. Gross, eds.), Academic Press, New York/London, 1976, pp. 303–311.

[Che04] A. Chermak, *Quadratic pairs*, J. Algebra **277** (2004), 36–72.

[DMZ01] L. Di Martino and A.E. Zalesski, *Minimum polynomials and lower bounds for eigenvalue multiplicities in representations of classical groups*, J. Algebra **243** (2001), 228–263, Corrigendum: J. Algebra **296** (2006), 249–252.

[DMZ08] L. Di Martino and A. E. Zalesski, *Eigenvalues of unipotent elements in cross-characteristic representations of finite classical groups*, J. Algebra **319** (2008), 2668–2722.

[Fei82] W. Feit, *The representation theory of finite groups*, North-Holland, Amsterdam, 1982.

[GMST02] R. M. Guralnick, K. Magaard, J. Saxl, and P. H. Tiep, *Cross characteristic representations of symplectic groups and unitary groups*, J. Algebra **257** (2002), 291–347.

[GR85] P. M. Gudivok and V. P. Rudko, *Tensor products of representations of finite groups*, Uzhgorod Univ., Uzhgorod, 1985 (in Russian).

[Har86] B. Hartley, *Relative higher relations modules, and a property of irreducible K-representations of* $GL_n(K)$, J. Algebra **104** (1986), 113–125.

[Hes76] W. Hesselink, *Singularities in the nilpotent scheme of a classical group*, Trans. Amer. Math. Soc. **222** (1976), 1–32.

[HH56] P. Hall and G. Higman, *On the p-length of p-solvable groups and reduction theorem for Burnside's problem*, Proc. London Math. Soc. **7** (1956), no. 21, 1–42.

[Ho76] Chat-Yin Ho, *On the quadratic pairs*, J. Algebra **43** (1976), 338–358.

[Jam78] G. D. James, *The representation theory of the symmetric groups*, Lecture Notes in Mathematics, vol. 682, Springer, Berlin, 1978.

[Jan73] J. C. Jantzen, *Darstellungen halbeinfacher algebraicher gruppen und zugeordnete kontravariante formen*, Bonner math. Schr. **67** (1973).

[Jan03] ———, *Representations of algebraic groups. Second edition*, Mathematical surveys and monographs, vol. 107, American Mathematical Society, Orlando, 2003.

[KZ04] A. S. Kleshchev and A. E. Zalesski, *Minimal polynomials of elements of order p in p-modular projective representations of alternating groups*, Proc. Amer. Math. Soc. **132** (2004), 1605–1612.

[Law95] R. Lawther, *Jordan block sizes of unipotent elements in exceptional algebraic groups*, Commun. in Algebra **25** (1995), 4125–4156.

[Pom80] K. Pommerening, *Uber die unipotenten Klassen reduktiver Gruppen, I and II*, J. Algebra **49, 65** (1977, 1980), 525–536,373–398.

[PS83] A. A. Premet and I. D. Suprunenko, *Quadratic modules for Chevalley groups over fields of odd characteristics*, Math. Nachrichten **110** (1983), 65–96.

[Rob83] G. R. Robinson, *Some remarks on reduction (mod p) of complex linear groups*, J. Algebra **83** (1983), 477–483.

[Rob95] ———, *On elements with restricted eigenvalues in linear groups*, J. Algebra **178** (1995), 635–642.

[Sei87] G. M. Seitz, *The maximal subgroups of classical algebraic groups*, Memoirs of the AMS, vol. 365, American Mathematical Society, Providence, 1987.

[Sei00] ———, *Unipotent elements, tilting modules, and saturation*, Invent. Math. **141** (2000), 467–502.

[Shu65] E. Shult, *On groups admitting fixed points free abelian operator groups*, Illinois J. Math. **9** (1965), 701–720.

[Smi82] S. Smith, *Irreducible modules and parabolic subgroups*, J. Algebra **75** (1982), 286–289.

[Spa82] N. Spaltenstein, *Classes unipotentes et sous-groupes de Borel*, Lecture Notes in Mathematics, vol. 946, Springer, Berlin, Heidelberg, 1982.

[SS70] T. A. Springer and R. Steinberg, *Conjugacy classes*, Seminar on algebraic groups and related finite groups (Berlin), Lecture Notes in Mathematics, vol. 131, Springer, 1970, pp. 167–266.

[SS97] J. Saxl and G. M. Seitz, *Subgroups of algebraic groups containing regular unipotent elements*, J. London Math. Soc. **55** (1997), no. 2, 370–386.

[Ste63] R. Steinberg, *Representations of algebraic groups*, Nagoya Math. J. **22** (1963), 33–56.

[Ste68] ———, *Lectures on Chevalley groups*, Yale Univ. Math. Dept., New Haven, Conn., 1968.

[Sup96] I. D. Suprunenko, *Minimal polynomials of elements of order p in irreducible representations of Chevalley groups over fields of characteristic p*, Sib. Adv. Math. **6** (1996), no. 4, 97–150.

[Sup97] ———, *On Jordan blocks of elements of order p in irreducible representations of classical groups with p-large highest weights*, J. Algebra **191** (1997), 589–627.

[Sup98] ———, *Restrictions of large modular irreducible representations of the special linear group to naturally embedded small subgroups cannot be semisimple*, Dokl. NAN Belarusi **42** (1998), no. 3, 27–31 (in Russian).

[Sup01] ———, *The minimal polynomials of unipotent elements in irreducible representations of the classical groups over fields of odd characteristic*, Dokl. NAN Belarusi **45** (2001), no. 3, 41–44 (in Russian).

[Tho71] J. G. Thompson, *Quadratic pairs*, Actes du Congres Int. Math. (Nice 1970) (Paris), vol. 1, Gauthier-Villars, 1971, pp. 375–376.

[TZ] P. H. Tiep and A. E. Zalesski, *Hall-Higman type theorems for semisimple elements of finite classical groups*, To appear in Proc. London Math. Soc.

[TZ00] ———, *Some aspects of finite linear groups: A survey*, J. Math. Sciences **100** (2000), 1893–1914.

[TZ02] ———, *Mod p reducibility of unramified representations of finite groups of Lie type*, Proc. London Math. Soc. **84** (2002), 439–472.

[Zal] A. E. Zalesski, *Minimum polynomials of the elements of prime order in representations of quasi-simple groups*, To appear in J. Algebra.

[Zal88] ———, *Eigenvalues of matrices of complex representations of finite groups of Lie type*, Lecture Notes Mathematics, vol. 1352, pp. 206–218, Springer, Berlin, 1988.

[Zal96] ———, *The eigenvalues of matrices of projective complex representations of alternating groups*, Vestsi Akad. Navuk Belarusi, Ser. Fiz.-Mat. Navuk (1996), no. 3, 41–43 (in Russian).

[Zal99] A.E. Zalesski, *Minimal polynomials and eigenvalues of p-elements in representations of groups with a cyclic Sylow p-subgroup*, J. London Math. Soc. **59** (1999), 845–866.

[Zal06] A. E. Zalesski, *The number of distinct eigenvalues of elements in finite linear groups*, J. London Math. Soc. **74** (2006), 361–378.

Index

Editorial Information

To be published in the *Memoirs*, a paper must be correct, new, nontrivial, and significant. Further, it must be well written and of interest to a substantial number of mathematicians. Piecemeal results, such as an inconclusive step toward an unproved major theorem or a minor variation on a known result, are in general not acceptable for publication.

Papers appearing in *Memoirs* are generally at least 80 and not more than 200 published pages in length. Papers less than 80 or more than 200 published pages require the approval of the Managing Editor of the Transactions/Memoirs Editorial Board.

As of March 31, 2009, the backlog for this journal was approximately 12 volumes. This estimate is the result of dividing the number of manuscripts for this journal in the Providence office that have not yet gone to the printer on the above date by the average number of monographs per volume over the previous twelve months, reduced by the number of volumes published in four months (the time necessary for preparing a volume for the printer). (There are 6 volumes per year, each usually containing at least 4 numbers.)

A Consent to Publish and Copyright Agreement is required before a paper will be published in the *Memoirs*. After a paper is accepted for publication, the Providence office will send a Consent to Publish and Copyright Agreement to all authors of the paper. By submitting a paper to the *Memoirs*, authors certify that the results have not been submitted to nor are they under consideration for publication by another journal, conference proceedings, or similar publication.

Information for Authors

Memoirs are printed from camera copy fully prepared by the author. This means that the finished book will look exactly like the copy submitted.

Initial submission. The AMS uses Centralized Manuscript Processing for initial submissions. Authors should submit a PDF file using the Initial Manuscript Submission form found at `www.ams.org/peer-review-submission`, or send one copy of the manuscript to the following address: Centralized Manuscript Processing, MEMOIRS OF THE AMS, 201 Charles Street, Providence, RI 02904-2294 USA. If a paper copy is being forwarded to the AMS, indicate that it is for it Memoirs and include the name of the corresponding author, contact information such as email address or mailing address, and the name of an appropriate Editor to review the paper (see the list of Editors below).

The paper must contain a *descriptive title* and an *abstract* that summarizes the article in language suitable for workers in the general field (algebra, analysis, etc.). The *descriptive title* should be short, but informative; useless or vague phrases such as "some remarks about" or "concerning" should be avoided. The *abstract* should be at least one complete sentence, and at most 300 words. Included with the footnotes to the paper should be the 2000 *Mathematics Subject Classification* representing the primary and secondary subjects of the article. The classifications are accessible from `www.ams.org/msc/`. The list of classifications is also available in print starting with the 1999 annual index of *Mathematical Reviews*. The Mathematics Subject Classification footnote may be followed by a list of *key words and phrases* describing the subject matter of the article and taken from it. Journal abbreviations used in bibliographies are listed in the latest *Mathematical Reviews* annual index. The series abbreviations are also accessible from `www.ams.org/msnhtml/serials.pdf`. To help in preparing and verifying references, the AMS offers MR Lookup, a Reference Tool for Linking, at `www.ams.org/mrlookup/`.

Electronically prepared manuscripts. The AMS encourages electronically prepared manuscripts, with a strong preference for $\mathcal{A}_{\mathcal{M}}\mathcal{S}$-LaTeX. To this end, the Society has prepared $\mathcal{A}_{\mathcal{M}}\mathcal{S}$-LaTeX author packages for each AMS publication. Author packages include instructions for preparing electronic manuscripts, samples, and a style file that generates the particular design specifications of that publication series. Though $\mathcal{A}\mathcal{M}\mathcal{S}$-LaTeX is the highly preferred format of TeX, author packages are also available in $\mathcal{A}\mathcal{M}\mathcal{S}$-TeX.

Authors may retrieve an author package for *Memoirs of the AMS* from `www.ams.org/journals/memo/memoauthorpac.html` or via FTP to `ftp.ams.org` (login as `anonymous`, enter username as password, and type `cd pub/author-info`). The *AMS Author Handbook* and the *Instruction Manual* are available in PDF format from the author package link. The author package can also be obtained free of charge by sending email to `tech-support@ams.org` (Internet) or from the Publication Division, American Mathematical Society, 201 Charles St., Providence, RI 02904-2294, USA. When requesting an author package, please specify $\mathcal{A}\mathcal{M}\mathcal{S}$-LaTeX or $\mathcal{A}\mathcal{M}\mathcal{S}$-TeX and the publication in which your paper will appear. Please be sure to include your complete mailing address.

After acceptance. The final version of the electronic file should be sent to the Providence office (this includes any TeX source file, any graphics files, and the DVI or PostScript file) immediately after the paper has been accepted for publication.

Before sending the source file, be sure you have proofread your paper carefully. The files you send must be the EXACT files used to generate the proof copy that was accepted for publication. For all publications, authors are required to send a printed copy of their paper, which exactly matches the copy approved for publication, along with any graphics that will appear in the paper.

Accepted electronically prepared files can be submitted via the web at `www.ams.org/submit-book-journal/`, sent via FTP, or sent on CD-Rom or diskette to the Electronic Prepress Department, American Mathematical Society, 201 Charles Street, Providence, RI 02904-2294 USA. TeX source files, DVI files, and PostScript files can be transferred over the Internet by FTP to the Internet node `ftp.ams.org` (130.44.1.100). When sending a manuscript electronically via CD-Rom or diskette, please be sure to include a message identifying the paper as a Memoir.

Electronically prepared manuscripts can also be sent via email to `pub-submit@ams.org` (Internet). In order to send files via email, they must be encoded properly. (DVI files are binary and PostScript files tend to be very large.)

Electronic graphics. Comprehensive instructions on preparing graphics are available at `www.ams.org/authors/journals.html`. A few of the major requirements are given here.

Submit files for graphics as EPS (Encapsulated PostScript) files. This includes graphics originated via a graphics application as well as scanned photographs or other computer-generated images. If this is not possible, TIFF files are acceptable as long as they can be opened in Adobe Photoshop or Illustrator. No matter what method was used to produce the graphic, it is necessary to provide a paper copy to the AMS.

Authors using graphics packages for the creation of electronic art should also avoid the use of any lines thinner than 0.5 points in width. Many graphics packages allow the user to specify a "hairline" for a very thin line. Hairlines often look acceptable when proofed on a typical laser printer. However, when produced on a high-resolution laser imagesetter, hairlines become nearly invisible and will be lost entirely in the final printing process.

Screens should be set to values between 15% and 85%. Screens which fall outside of this range are too light or too dark to print correctly. Variations of screens within a graphic should be no less than 10%.

Inquiries. Any inquiries concerning a paper that has been accepted for publication should be sent to `memo-query@ams.org` or directly to the Electronic Prepress Department, American Mathematical Society, 201 Charles St., Providence, RI 02904-2294 USA.

Editors

This journal is designed particularly for long research papers, normally at least 80 pages in length, and groups of cognate papers in pure and applied mathematics. Papers intended for publication in the *Memoirs* should be addressed to one of the following editors. The AMS uses Centralized Manuscript Processing for initial submissions to AMS journals. Authors should follow instructions listed on the Initial Submission page found at www.ams.org/memo/memosubmit.html.

Algebra to ALEXANDER KLESHCHEV, Department of Mathematics, University of Oregon, Eugene, OR 97403-1222; email: ams@noether.uoregon.edu

Algebraic geometry to DAN ABRAMOVICH, Department of Mathematics, Brown University, Box 1917, Providence, RI 02912; email: amsedit@math.brown.edu

Algebraic geometry and its applications to MINA TEICHER, Emmy Noether Research Institute for Mathematics, Bar-Ilan University, Ramat-Gan 52900, Israel; email: teicher@macs.biu.ac.il

Algebraic topology to ALEJANDRO ADEM, Department of Mathematics, University of British Columbia, Room 121, 1984 Mathematics Road, Vancouver, British Columbia, Canada V6T 1Z2; email: adem@math.ubc.ca

Combinatorics to JOHN R. STEMBRIDGE, Department of Mathematics, University of Michigan, Ann Arbor, Michigan 48109-1109; email: JRS@umich.edu

Commutative and homological algebra to LUCHEZAR L. AVRAMOV, Department of Mathematics, University of Nebraska, Lincoln, NE 68588-0130; email: avramov@math.unl.edu

Complex analysis and harmonic analysis to ALEXANDER NAGEL, Department of Mathematics, University of Wisconsin, 480 Lincoln Drive, Madison, WI 53706-1313; email: nagel@math.wisc.edu

Differential geometry and global analysis to CHRIS WOODWARD, Department of Mathematics, Rutgers University, 110 Frelinghuysen Road, Piscataway, NJ 08854; email: ctw@math.rutgers.edu

Dynamical systems and ergodic theory and complex analysis to YUNPING JIANG, Department of Mathematics, CUNY Queens College and Graduate Center, 65-30 Kissena Blvd., Flushing, NY 11367; email: Yunping.Jiang@qc.cuny.edu

Functional analysis and operator algebras to DIMITRI SHLYAKHTENKO, Department of Mathematics, University of California, Los Angeles, CA 90095; email: shlyakht@math.ucla.edu

Geometric analysis to WILLIAM P. MINICOZZI II, Department of Mathematics, Johns Hopkins University, 3400 N. Charles St., Baltimore, MD 21218; email: trans@math.jhu.edu

Geometric topology to MARK FEIGHN, Math Department, Rutgers University, Newark, NJ 07102; email: feighn@andromeda.rutgers.edu

Harmonic analysis, representation theory, and Lie theory to ROBERT J. STANTON, Department of Mathematics, The Ohio State University, 231 West 18th Avenue, Columbus, OH 43210-1174; email: stanton@math.ohio-state.edu

Logic to STEFFEN LEMPP, Department of Mathematics, University of Wisconsin, 480 Lincoln Drive, Madison, Wisconsin 53706-1388; email: lempp@math.wisc.edu

Number theory to JONATHAN ROGAWSKI, Department of Mathematics, University of California, Los Angeles, CA 90095; email: jonr@math.ucla.edu

Number theory to SHANKAR SEN, Department of Mathematics, 505 Malott Hall, Cornell University, Ithaca, NY 14853; email: ss70@cornell.edu

Partial differential equations to GUSTAVO PONCE, Department of Mathematics, South Hall, Room 6607, University of California, Santa Barbara, CA 93106; email: ponce@math.ucsb.edu

Partial differential equations and dynamical systems to PETER POLACIK, School of Mathematics, University of Minnesota, Minneapolis, MN 55455; email: polacik@math.umn.edu

Probability and statistics to RICHARD BASS, Department of Mathematics, University of Connecticut, Storrs, CT 06269-3009; email: bass@math.uconn.edu

Real analysis and partial differential equations to DANIEL TATARU, Department of Mathematics, University of California, Berkeley, Berkeley, CA 94720; email: tataru@math.berkeley.edu

All other communications to the editors should be addressed to the Managing Editor, ROBERT GURALNICK, Department of Mathematics, University of Southern California, Los Angeles, CA 90089-1113; email: guralnic@math.usc.edu.

Titles in This Series

941 **Gelu Popescu,** Unitary invariants in multivariable operator theory, 2009

940 **Gérard Iooss and Pavel I. Plotnikov,** Small divisor problem in the theory of three-dimensional water gravity waves, 2009

939 **I. D. Suprunenko,** The minimal polynomials of unipotent elements in irreducible representations of the classical groups in odd characteristic, 2009

938 **Antonino Morassi and Edi Rosset,** Uniqueness and stability in determining a rigid inclusion in an elastic body, 2009

937 **Skip Garibaldi,** Cohomological invariants: Exceptional groups and spin groups, 2009

936 **André Martinez and Vania Sordoni,** Twisted pseudodifferential calculus and application to the quantum evolution of molecules, 2009

935 **Mihai Ciucu,** The scaling limit of the correlation of holes on the triangular lattice with periodic boundary conditions, 2009

934 **Arjen Doelman, Björn Sandstede, Arnd Scheel, and Guido Schneider,** The dynamics of modulated wave trains, 2009

933 **Luchezar Stoyanov,** Scattering resonances for several small convex bodies and the Lax-Phillips conjuecture, 2009

932 **Jun Kigami,** Volume doubling measures and heat kernel estimates of self-similar sets, 2009

931 **Robert C. Dalang and Marta Sanz-Solé,** Hölder-Sobolv regularity of the solution to the stochastic wave equation in dimension three, 2009

930 **Volkmar Liebscher,** Random sets and invariants for (type II) continuous tensor product systems of Hilbert spaces, 2009

929 **Richard F. Bass, Xia Chen, and Jay Rosen,** Moderate deviations for the range of planar random walks, 2009

928 **Ulrich Bunke,** Index theory, eta forms, and Deligne cohomology, 2009

927 **N. Chernov and D. Dolgopyat,** Brownian Brownian motion-I, 2009

926 **Riccardo Benedetti and Francesco Bonsante,** Canonical wick rotations in 3-dimensional gravity, 2009

925 **Sergey Zelik and Alexander Mielke,** Multi-pulse evolution and space-time chaos in dissipative systems, 2009

924 **Pierre-Emmanuel Caprace,** "Abstract" homomorphisms of split Kac-Moody groups, 2009

923 **Michael Jöllenbeck and Volkmar Welker,** Minimal resolutions via algebraic discrete Morse theory, 2009

922 **Ph. Barbe and W. P. McCormick,** Asymptotic expansions for infinite weighted convolutions of heavy tail distributions and applications, 2009

921 **Thomas Lehmkuhl,** Compactification of the Drinfeld modular surfaces, 2009

920 **Georgia Benkart, Thomas Gregory, and Alexander Premet,** The recognition theorem for graded Lie algebras in prime characteristic, 2009

919 **Roelof W. Bruggeman and Roberto J. Miatello,** Sum formula for SL_2 over a totally real number field, 2009

918 **Jonathan Brundan and Alexander Kleshchev,** Representations of shifted Yangians and finite W-algebras, 2008

917 **Salah-Eldin A. Mohammed, Tusheng Zhang, and Huaizhong Zhao,** The stable manifold theorem for semilinear stochastic evolution equations and stochastic partial differential equations, 2008

916 **Yoshikata Kida,** The mapping class group from the viewpoint of measure equivalence theory, 2008

915 **Sergiu Aizicovici, Nikolaos S. Papageorgiou, and Vasile Staicu,** Degree theory for operators of monotone type and nonlinear elliptic equations with inequality constraints, 2008

914 **E. Shargorodsky and J. F. Toland,** Bernoulli free-boundary problems, 2008

913 **Ethan Akin, Joseph Auslander, and Eli Glasner,** The topological dynamics of Ellis actions, 2008

912 **Igor Chueshov and Irena Lasiecka,** Long-time behavior of second order evolution equations with nonlinear damping, 2008

911 **John Locker,** Eigenvalues and completeness for regular and simply irregular two-point differential operators, 2008

910 **Joel Friedman,** A proof of Alon's second eigenvalue conjecture and related problems, 2008

909 **Cameron McA. Gordon and Ying-Qing Wu,** Toroidal Dehn fillings on hyperbolic 3-manifolds, 2008

908 **J.-L. Waldspurger,** L'endoscopie tordue n'est pas si tordue, 2008

907 **Yuanhua Wang and Fei Xu,** Spinor genera in characteristic 2, 2008

906 **Raphaël S. Ponge,** Heisenberg calculus and spectral theory of hypoelliptic operators on Heisenberg manifolds, 2008

905 **Dominic Verity,** Complicial sets characterising the simplicial nerves of strict ω-categories, 2008

904 **William M. Goldman and Eugene Z. Xia,** Rank one Higgs bundles and representations of fundamental groups of Riemann surfaces, 2008

903 **Gail Letzter,** Invariant differential operators for quantum symmetric spaces, 2008

902 **Bertrand Toën and Gabriele Vezzosi,** Homotopical algebraic geometry II: Geometric stacks and applications, 2008

901 **Ron Donagi and Tony Pantev (with an appendix by Dmitry Arinkin),** Torus fibrations, gerbes, and duality, 2008

900 **Wolfgang Bertram,** Differential geometry, Lie groups and symmetric spaces over general base fields and rings, 2008

899 **Piotr Hajłasz, Tadeusz Iwaniec, Jan Malý, and Jani Onninen,** Weakly differentiable mappings between manifolds, 2008

898 **John Rognes,** Galois extensions of structured ring spectra/Stably dualizable groups, 2008

897 **Michael I. Ganzburg,** Limit theorems of polynomial approximation with exponential weights, 2008

896 **Michael Kapovich, Bernhard Leeb, and John J. Millson,** The generalized triangle inequalities in symmetric spaces and buildings with applications to algebra, 2008

895 **Steffen Roch,** Finite sections of band-dominated operators, 2008

894 **Martin Dindoš,** Hardy spaces and potential theory on C^1 domains in Riemannian manifolds, 2008

893 **Tadeusz Iwaniec and Gaven Martin,** The Beltrami Equation, 2008

892 **Jim Agler, John Harland, and Benjamin J. Raphael,** Classical function theory, operator dilation theory, and machine computation on multiply-connected domains, 2008

891 **John H. Hubbard and Peter Papadopol,** Newton's method applied to two quadratic equations in $\mathbb{C}^2$ viewed as a global dynamical system, 2008

890 **Steven Dale Cutkosky,** Toroidalization of dominant morphisms of 3-folds, 2007

889 **Michael Sever,** Distribution solutions of nonlinear systems of conservation laws, 2007

For a complete list of titles in this series, visit the
AMS Bookstore at **www.ams.org/bookstore/**.